Tourism in the Caribbean

The Caribbean is one of the world's premier tourism destinations. Changes in travel patterns, markets and traveller motivations have brought both considerable growth and dramatic change to the region's tourism sector. In addition, persistent turbulence in other economic sectors in the region has served to enhance the relative importance of tourism as an economic development strategy. Tourism is therefore becoming increasingly crucial to the economic survival of local economies in most, if not all, island states and dependencies in the region.

The book is divided into three parts. The first gives an overview of existing tourism trends in the region. Part two addresses tourism development issues, including sustainability, ecotourism, heritage tourism, cruise tourism, community participation, management implications for tourism businesses, resorts, regional organisational structures and linkages with agriculture. Part three considers future trends, including an assessment of recent world events and their impact on tourism in the region, and considers prospective shifts in access and airlift, economic sustainability and markets.

Tourism in the Caribbean brings together a high-calibre team of international researchers to provide an up-to-date assessment of the scope and nature of tourism development in the Caribbean. It will be a valuable resource for students of Tourism and Caribbean studies, as well as governments, and national and regional tourism offices.

David Timothy Duval is Lecturer in the Department of Tourism at the University of Otago, New Zealand.

Contemporary Geographies of Leisure, Tourism and Mobility

Series Editor: Michael Hall is Professor at the Department of Tourism, University of Otago, New Zealand.

The aim of this series is to explore and communicate the intersections and relationships between leisure, tourism and human mobility within the social sciences.

It will incorporate both traditional and new perspectives on leisure and tourism from contemporary geography, e.g. notions of identity, representation and culture, while also providing for perspectives from cognate areas such as anthropology, cultural studies, gastronomy and food studies, marketing, policy studies and political economy, regional and urban planning, and sociology, within the development of an integrated field of leisure and tourism studies.

Also, increasingly, tourism and leisure are regarded as steps in a continuum of human mobility. Inclusion of mobility in the series offers the prospect to examine the relationship between tourism and migration, the sojourner, educational travel, and second home and retirement travel phenomena.

The series comprises two strands:

Contemporary Geographies of Leisure, Tourism and Mobility aims to address the needs of students and academics, and the titles will be published in hardback and paperback. Titles include:

The Moralisation of Tourism
Sun, sand . . . and saving the world?
Jim Butcher

The Ethics of Tourism Development
Mick Smith and Rosaleen Duffy

Tourism in the Caribbean
Trends, development, prospects
Edited by David Timothy Duval

Qualitative Research in Tourism
Ontologies, epistemologies and methodologies
Edited by Jenny Phillimore and Lisa Goodson

Routledge Studies in Contemporary Geographies of Leisure, Tourism and Mobility is a forum for innovative new research intended for research students and academics, and the titles will be available in hardback only. Titles include:

1 Living with Tourism
Negotiating identities in a Turkish village
Hazel Tucker

Tourism in the Caribbean

Trends, development, prospects

Edited by David Timothy Duval

Routledge
Taylor & Francis Group

LONDON AND NEW YORK

First published 2004
by Routledge
11 New Fetter Lane, London EC4P 4EE

Simultaneously published in the USA and Canada
by Routledge
29 West 35th Street, New York, NY 10001

Routledge is an imprint of the Taylor & Francis Group

© 2004 Editorial matter and selection, David Timothy Duval;
individual chapters, the contributors

Typeset in Times by
Florence Production Ltd, Stoodleigh, Devon
Printed and bound in Great Britain by
The Cromwell Press, Trowbridge, Wiltshire

British Library Cataloguing in Publication Data
A catalogue record for this book is available
from the British Library

Library of Congress Cataloging in Publication Data
Duval, David Timothy, 1970–
    Tourism in the caribbean: trends, development, prospects/
    David Timothy Duval.
        p. cm. – (Contemporary geographies of leisure, tourism
        and mobility; 3)
        Includes bibliographical references.
        1. Tourism – Caribbean Area   I. Title.   II. Series:Routledge/
    contemporary geographies of leisure, tourism and mobility; 3
    G155.C35D88 2004
    338.4′7917290453–dc21   2003014695

ISBN 0–415–30361–3 (hbk)
ISBN 0–415–30362–1 (pbk)

In honour of his many contributions, both scholarly and personal, this book is dedicated to Klaus de Albuquerque, formerly Professor of Sociology of the College of Charleston, Charleston, South Carolina, USA.

Klaus de Albuquerque
(1946–1999)

His name was like an archipelago,
his skin a blend of German India,
his blood laced with Mombasa brine,
a multi-layered fabric fashioned
from the whole earth catalogue.

Once the data were assembled
he could draft a paper in a day
despite the crowded classes
and a parlour clutch of colleagues
and the throw of his heart
    always heeling towards his son.

Hidden underneath the arc of his pen
from Papua New Guinea to the Caribbean,
a single seam that marks the sky indelibly:
Listen for the silent *anawim*.*
There was little I in this island man.

                                    J. McElroy

*Old Testament term for 'marginalised'.

Contents

Contributors

Tim Coles, Ph.D., is Lecturer in Human Geography and University Business Research Fellow at the University of Exeter. His recent research has focused on the position and role of tourism in the restructuring of society and economy, above all in transition states. This has explored some of the conceptual issues surrounding new place imagery, and of the politics and production of difference and destination appeal. He is currently the Honorary Secretary of the Geography of Tourism and Leisure Research Group of the Royal Geographical Society (with IBG) and a member of the editorial board of the journal *Tourism Geographies*.

Dennis Conway received a BA in Geography at Cambridge University (1959–62), a Diploma of Education from Oxford University (1962–3), and an MA and Ph.D. in Geography at the University of Texas at Austin, 1974 and 1976. He joined Indiana University in 1976. Currently, he is Professor of Geography and Latin American and Caribbean Studies at Indiana University, Bloomington, Indiana, USA. He has advanced degrees in geography, training in demography, and field experience conducting surveys in several foreign cultures. He has published widely in four related areas – migration, urbanisation, development and environmental sustainability – with the Caribbean and Nepal as regional specialisations.

David Timothy Duval is a Lecturer in the Department of Tourism at the University of Otago. David received an Advanced BA in Anthropology and Geography and an MA in Anthropology from the University of Manitoba. He holds an MES in Tourism and Development and a Ph.D. from York University. His research interests are in the fields of migration, transnationalism, mobilities and tourism; tourism in island environments; and transport management and planning networks. He has recently published research relating to mobility and return visits among eastern Caribbean migrants living in Toronto.

Gordon Ewing's research focuses on modelling geographical and environmental choice behaviour. A major dimension has been recreation and vacation destination choice. He has studied recreational choice in

Canada, north-eastern USA and Scotland, and vacation choice in the Caribbean. He has contributed to over twenty papers and reports on tourism and recreation. More recently he has published research on choices with environmental implications, including kerbside recycling behaviour, commuting mode choice and vehicle purchase behaviour. Currently he is researching possible links between urban design and travel behaviour. He is currently Chair of the Department of Geography at McGill University.

William C. Found (University Professor, York University) is jointly appointed between the Faculty of Environmental Studies and the Department of Geography. He began his research in the Caribbean while a doctoral student at the University of Florida, and has since conducted research in twenty-four different islands, recently producing twelve documentary films. His teaching includes courses in 'The Geographical Transformation of the Caribbean Islands', and 'Project Planning and Implementation'. He is former Vice President (Academic Affairs) of York, and has held visiting posts at Harvard, the Cuban Academy of Sciences, and Umeå University, Sweden (from which he holds an honorary doctoral degree).

K. Michael Haywood is Professor Emeritus, School of Hospitality and Tourism Management, University of Guelph. As a specialist in the management of hospitality and tourism enterprises, he has published extensively on strategic management topics and issues. His body of work resulted in becoming the second recipient of the John Wiley award for lifetime contributions to the field of hospitality and tourism research. Having resided in the Bahamas during his youth, he maintains an abiding interest in the health and well-being of the tourism industry in the Caribbean region. Michael serves on the editorial boards of numerous industry journals, and provides consultation through his company, The Haywood Group, Inc.

Chandana (Chandi) Jayawardena, Ph.D., is the Academic Director – M.Sc. in Tourism and Hospitality Management, Senior Lecturer in Tourism Management and Research Fellow of the University of the West Indies. He is also the Regional Leader/Editor – Canada and the Caribbean of the Worldwide Hospitality And Tourism Trends (WHATT) and Ambassador – the Caribbean and South America zone of the Hotel and Catering International Management Association (HCIMA). He has held Visiting Professorships in Canada, USA, UK, Guyana, Switzerland and Sri Lanka. Among Chandi's publications are five journal special editions, nine books, nineteen book chapters and thirty-five articles. He has visited twenty-nine Caribbean countries.

Leslie-Ann Jordan is on a New Zealand Commonwealth Scholarship and is currently pursuing a Ph.D. in Tourism at the University of Otago. She holds a B.Sc. in Tourism Management from the University of the West Indies and a Post-graduate Diploma from the University of Otago. Her research interests include tourism development in small island states with special reference to the anglophone Caribbean; tourism planning and development and tourism policy and decision-making. Ms. Jordan has worked as a hotel management intern in the Bahamas and St Kitts and as the Front Office Supervisor at the Trinidad Holiday Inn Hotel. She also worked for a short time with the Tourism and Industrial Development Company of Trinidad and Tobago (TIDCO) as a Tourism Development Officer.

Jerome L. McElroy received his Ph.D. in Economics (1972) from the University of Colorado and has held teaching positions at St John's College in Belize, the University of the Virgin Islands in St Thomas, the University of Notre Dame, and St Mary's College where he is currently Professor of Economics. His early research focused on the economics of tourism, inter-island migration, small-scale agriculture, and planning problems in small economies. More recent work has involved applying the destination life cycle to small islands, developing an index of tourism penetration, and examining the links between tourism and political dependency, and tourism and crime.

Beth Mills (Ph.D., University of California, Davis) has travelled and worked in Grenada and the Grenadines since 1975. She has studied and written about agriculture and land-use change in Grenada and Carriacou since the early 1980s. Her recent work has focused on family land tenure and its meaning to the transnational community of Carriacou. She lives with her two children in Santa Fe, New Mexico, where she works as a community planner and GIS specialist. She returns to the Caribbean to visit friends and colleagues as often as possible.

Simon Milne is Professor of Tourism in the Business Faculty, Auckland University of Technology, where he coordinates the New Zealand Tourism Research Institute. His research focuses on creating stronger links between tourism and surrounding economies. He has considerable experience in economic impact assessment, and in formulating tourism-related economic development strategies. In recent years he has become particularly interested in the use of information and communication technologies to improve the economic performance and sustainability of tourism. Dr Milne has conducted tourism research in the Caribbean, Canada, New Zealand, South Pacific islands, Kenya and Russia.

Janet Henshall Momsen is Professor of Geography in the Department of Human and Community Development at the University of California, Davis. Before coming to California in 1991 she taught in the UK, Canada, Costa Rica and Brazil. Janet has undertaken fieldwork in southern China, Brazil, Costa Rica, Mexico, Sri Lanka and West Africa as well as in the Caribbean. She has degrees in geography and agricultural economics from Oxford, McGill and London Universities. Her publications on Caribbean tourism focus on gender issues and include work for UNED, chapters in various books and the co-edited book *Gender/Tourism/Fun(?)* published in 2002 by Cognizant Communications Press.

Mimi Sheller is a Senior Lecturer in Sociology at Lancaster University, co-director of the Centre for Mobilities Research, and vice-chair of the Society for Caribbean Studies (UK). She is the author of *Democracy after Slavery: Black Publics and Peasant Radicalism in Haiti and Jamaica* (Macmillan, 2000) and *Consuming the Caribbean: From Arawaks to Zombies* (Routledge, 2003), and is co-editor of two forthcoming books, *Uprootings/Regroundings: Questions of Home and Migration* (Berg, 2003) and *Global Places to Play* (Routledge, forthcoming).

Dallen J. Timothy is Associate Professor at Arizona State University and Visiting Professor of Heritage Tourism at Sunderland University, United Kingdom. His current research interests in tourism include politics and political boundaries, planning in developing and peripheral regions, heritage, shopping and consumerism, and destination sustainability. He has authored and edited a wide range of journal articles, books and book chapters dealing with these various issues. He is co-editor of Channel View Publications' Aspects of Tourism book series and works in many editorial capacities with several international journals.

David B. Weaver, Ph.D., is Professor of Tourism and Events Management in the Department of Health, Fitness and Recreation Resources at George Mason University. He is a specialist in ecotourism, sustainable tourism and destination life cycle dynamics, and has authored or co-authored five books and over sixty refereed articles and book chapters on related topics. Dr Weaver is also the editor of *The Encyclopedia of Ecotourism* (CABI Publishing, 2001). He has held previous appointments at Griffith University (Australia) and the University of Regina (Canada).

Paul F. Wilkinson is a Professor in the Faculty of Environmental Studies and the Graduate Program in Geography at York University. Based on experience in Canada, the Caribbean, and Europe, his research focuses on tourism policy and planning, the environmental impacts of tourism and recreation, protected area management, and urban open space planning. He was recently a member of the Panel on the Ecological Integrity of Canada's National Parks.

Robert E. Wood is Professor of Sociology at Rutgers University, Camden Campus. He is the author of *From Marshall Plan to Debt Crisis: Foreign Aid and Development Choices in the World Economy* (University of California Press, 1986) and co-editor, with Michel Picard, of *Tourism, Ethnicity and the State in Asian and Pacific Societies* (University of Hawai'i Press, 1997). He has long been interested in how tourism provides revealing insights into broader structures and processes in society, and has recently been engaged in a study of how the cruise ship industry illuminates fundamental patterns of globalisation.

Preface

Ten years ago, almost to the date, I was involved in archaeological inves-
tigations on the island of St Vincent under the direction of Professor Louis
Allaire, now retired from the University of Manitoba. In the late after-
noons, after all the artifacts (mostly sherds of pottery) were washed and
catalogued, I often found myself hopping in our rented 4 × 4 and exploring
the southern part of the island. I recall marvelling at how few hotels I
would see outside of the main town of Kingstown, and even there they
were relatively few and far between. In fact, my genuine academic interest
in tourism in the Caribbean can be pinpointed to a chance meeting I had
in downtown Kingstown with a family of four, dressed more or less stereo-
typically as mass tourists: freshly-pressed polo shirts, white sneakers and
wide straw hats (perhaps purchased locally – I wasn't entirely certain). In
my consummate effort to be an 'alternative tourist', despite not even being
aware of the existence of such a creature at the time, I was barefoot and
wearing a ratty T-shirt and a pair of rather dirty and faded shorts which,
without a doubt, had seen more prosperous days. In a rather meagre attempt
at 'doing ethnography' (I was, after all, undertaking graduate work in
Anthropology), I asked the family what brought them to St Vincent, especially given that it appeared to be, at least to my non-tourist eyes, such an
undeveloped tourism destination. After all, there were only a handful of
white sand beaches on the island. In answer to this question, which was
undoubtedly perceived as rather intrusive (much like stand-up comedy,
delivery is everything in ethnographic research – I have since become
much more adept), they replied, quite simply, that their travel agent had
recommended it for that very reason: it was unspoiled, lacked the throngs
of people, and was inexpensive in comparison to other more developed
islands nearby, particularly Barbados.

Since that chance meeting, the issue of tourism in the Caribbean has
occupied the forefront of my mind. I watched with great interest as the
Windward Islands were embroiled in a bitter dispute over bananas and
preferential trade arrangements in the 1990s. Entire families, some of whom
I got to know reasonably well, faced economic ruin. Many thought of
getting work in tourist resorts in other islands. One year later, on my first

trip to Antigua, I recall comparing mentally that island with my experience in St Vincent. Antigua was richly developed from a tourism perspective, and there seemed to be tourists everywhere. I was truly hooked and sought to formally understand why such differentiation was taking place. I still do not know the answer.

While I would like to say that this book is the culmination of the quest that these events ultimately sparked, nothing could be further from the truth. In fact, tourism in the Caribbean is, as in most places around the world, an incredibly varied and turbulent phenomenon. Any hope of inductively arriving at a holistic perspective on the subject is probably beyond the grasp of any one individual. As a result, this book should be seen as yet another view of the key issues facing tourism in the region. It is but one version of recent (and past) events that have shaped the tourism sector. It is by no means the final word on the subject, and it is certainly meant to complement the views held by those who really know more about tourism in this region than perhaps anyone else: the residents and 'hosts' of the individual islands.

It is particularly fitting that this book is published roughly ten years after the seminal work of Dennis Gayle and Jon Goodrich appeared (*Tourism Marketing and Management in the Caribbean*, London: Routledge), since for many Caribbeanists working in the field of tourism the Gayle and Goodrich volume provided a benchmark. The genesis of this book came as a result of the realisation that there was a need for an up-to-date, academic synopsis of issues facing the Caribbean region. Beyond the visitor arrival and expenditure statistics, what was needed was a single source in which pertinent issues facing tourism development could be found. Critical questions that are in need of careful consideration include the impact of 11 September 2001, the changing nature of community participation throughout the region, the extent to which ecotourism and other forms of alternative tourism have proliferated, and how tourism development overall has changed the face of economic development in the Caribbean (and vice versa).

It is also particularly fitting that this book is being dedicated to one of the most respected Caribbean scholars of the last century. The work of Klaus de Albuquerque made an immense impression on me personally as a young graduate student. I was fortunate enough to befriend Klaus in 1998 in Antigua and subsequently learn from his vast experience and insights into both tourism and the wider economic framework of the Caribbean as a whole. His passing was troubling, both personally and professionally.

I want to extend my sincere appreciation to all the contributors in this volume for their support and useful guidance throughout the process. Their dedication to this project made it a thoroughly enjoyable undertaking. I am also grateful for my colleagues and friends at the University of Otago, particularly Mike Hall (for his endless encouragement and support), Stephen Boyd, James Higham, Hazel Tucker, Anna Carr, Brent Lovelock,

David Buisson (Dean, School of Business), Frances Cadogan and 'Chicken'. For the most part, they have (largely) indirectly contributed to the development of this book through their own efforts and successes as scholars and, by extension, created the necessary nurturing environment within which this project was developed and operationalised. Thanks also to Kate Feltoe for being a very capable and reliable research assistant.

For their special guidance and inspiration over the years, many thanks to Paul Wilkinson, Dorothy Wilkinson, Bill Found, Jane Couchman and Louis Allaire (who is almost solely responsible for sparking my interest in the Caribbean). Thank you also to the wonderfully supportive and efficient research staff at the Caribbean Tourism Organization head office in St Michael (particularly Adrian McCallister and Gail Clarke) for providing access to recent statistical reports and for their logistical support. Thanks to Polly Pattullo for the encouragement. At Routledge, many thanks to Andrew Mould and Melanie Attridge for their exceptional guidance, patience and support throughout the entire project; thanks to Tom Chandler for his superb copy-editing skills and Nicole Krull for her expert production coordination skills.

Thanks to my parents (Marcia, Brian and Connie) and extended family for their support and encouragement. This book was completed in loving memory of Dr Nancy Benson Hamstra. Finally, I want to especially thank my wife and, indeed, pillar of support, Melissa, for both her presence and tolerance.

David Timothy Duval
Opoho, June 2003

Acknowledgements

Special thanks to:

Kluwer Academic Publishers for kindly allowing the reproduction of Figure 10.2 from Weaver, D. 'Ecotourism in the small island Caribbean', *GeoJournal*, 31(4): 457–65.

Caribbean Tourism Organization for generously allowing access to the *Caribbean Tourism Statistical Report: 2000–2001 Edition* and the *2001–2002 Edition*.

If any unknowing use has been made of copyrighted material, owners should contact the editor via the publishers as every effort has been made to trace owners and to secure permission.

Part I

Trends in Caribbean tourism

1 Trends and circumstances in Caribbean tourism

David Timothy Duval

Introduction

The Caribbean has long been regarded as one of the world's premier travel destinations. While the turbulent nature of global tourism has led to numerous changes in travel patterns, markets and tourist motivations, the extent and scope of tourism in the Caribbean has been substantial. While tourism in the Caribbean is by no means recent (Bryden 1973; Perez 1975; Sealey 1982), periods of economic instability in many island states in the region have effectively enhanced the relative importance of tourism as an alternative economic development strategy. Many island states in the Caribbean are particularly vulnerable to global economic volatility due to a reliance on world markets for various produced goods (Payne and Sutton 2001). Somewhat ironically, the notion of 'smallness', in reference to the economic position of island states in general, and the extent to which the Caribbean functions as a peripheral destination to global flows of economic activity, actually bears little resemblance to the degree to which the region is situated as a key vacation destination for literally millions of foreign travellers. While the Caribbean is economically marginalised, it also plays host to millions of tourists each year.

In light of turbulent economic conditions in many island states in the region, it is not surprising that the scope of the tourism sector in the region is substantial. By the 1990s, tourism in the region generated US$96 billion in expenditures per year and employed some 400,000 people (Gayle and Goodrich 1993). The World Travel and Tourism Council's (WTTC) Tourism Satellite Account (WTTC 2001) indicated that, region-wide, tourism accounted for approximately 2.5 million jobs or 15.5 per cent of total employment in 2001, and contributed 5.8 per cent (or US$9.2 billion) to the region's Gross Domestic Product (GDP). By 2011, tourism is expected to contribute some US$18.7 billion, or 6 per cent of total GDP (WTTC 2001). The significance of tourism in the Caribbean effectively mirrors, and even trumps, the importance and scope of tourism worldwide. For 2001, for example, the World Tourism Organization (WTO) estimated 693 million tourist arrivals worldwide and US$463 billion in international tourism receipts (WTO 2002). While worldwide visitor arrival growth between 1990 and 2000 increased an average of 4.3 per cent, average annual growth in arrivals to the Caribbean increased 4.7 per cent (Caribbean Tourism Organization (CTO) 2002a) (Table 1.1). Despite the economic significance of tourism in the region, however, concerns over the degree of dependency created are often raised (Bryden 1973; Erisman 1983). Holder (1988) has even suggested that tourism in the Caribbean is environmentally dependent given the extent to which it relies on the available natural resource base. Further, Wilkinson (1989) argues that, with the almost inevitability of tourism as a crutch for economic development in certain localities, the risk, or 'folly', of mismanagement and failed integration within other sectors can often be considerable.

By extension, and despite the sizeable economic importance of tourism in the region, the tangible benefits of tourism have come under intense scrutiny (e.g. Archer and Davies 1984; Wilkinson 1989). At issue, in the first instance, is whether the economic benefits of tourism are realised at the local level, as substantial foreign ownership of service providers such as airlines and hotels may often mean a significant degree of leakage (Wilkinson 1989). Further, Wilkinson (1996) questioned the difference between current and real visitor expenditures in the region. Using the data on visitor numbers and expenditures combined with consumer price index shifts for the Bahamas, Wilkinson (1996: 33) noted that: 'In real terms, the Bahamas is only slightly better off in 1993 than it was in 1971 in total visitor expenditures.'

Criticism over tourism is not only limited to economics. The negative biophysical impacts of tourism development have received attention in both

Table 1.1 Worldwide and Caribbean tourist arrivals (1990–2000)

	Worldwide		Caribbean[a]	
	Tourist arrivals (millions)	Change (%)	Tourist arrivals (millions)	Change (%)
1990	457.2	7.2	12.8	6.7
1991	464.0	1.5	13.0	1.6
1992	503.4	8.5	14.0	7.7
1993	519.0	3.1	15.0	7.1
1994	550.5	6.1	15.7	4.7
1995	552.3	0.3	16.2	3.2
1996	599.0	8.5	16.7	3.1
1997	618.2	3.2	17.9	6.9
1998	636.6	3.0	18.2	1.9
1999	652.2	2.5	19.1	4.9
2000	696.7	6.8	20.4	6.9

Source: Caribbean Tourism Organization (2003).

Note
a Includes Cozumel and Cancun.

the academic literature (Weaver 1995; Wilkinson 1989) and among governments and wider regional bodies. Critics cite, for example, pollution and damage to marine environments as a result of mass tourism in the region. Social impacts have also been addressed in the context of the overall community value gained from tourist expenditures (Wilkinson 1999) and with respect to indigenous cultural environments packaged and commoditised for foreign tourists (e.g. Slinger 2000). The question, then, is whether tourism in the Caribbean is, to borrow from Brown (1998), a blight or a blessing.

The Caribbean – physiographic, economic and political aspects

At the outset, and in the spirit of Harrigan's (1974) assertion that it is 'often necessary to have a definition handy', it is important to set forward a geographic definition of the Caribbean that will be followed throughout this book. Those island states considered can perhaps best be referred to as the 'insular Caribbean'. In other words, of interest are those island states stretching from Cuba to Trinidad and Tobago (Figure 1.1). Consequently, the northern coastal countries of the South American mainland (Guyana, Suriname, Venezuela) are outside the purview of this book. Similarly, while Belize, Cozumel and Cancun fall under the purview of the Caribbean

Figure 1.1 Map of the Caribbean

Tourism Organization, they have been broadly excluded from direct consideration in this volume. The reason for excluding these destinations is that recent analyses (e.g. Clancy 2001; Lumsdon and Swift 2001) have already offered adequate coverage.

As defined here, the Caribbean region (Figure 1.1) can be separated into two broader subregions. The Lesser Antilles (the 'eastern Caribbean' or 'southern Caribbean') geographically encompasses the United States and British Virgin Islands in the north to Trinidad and Tobago in the south (Figure 1.2). The Greater Antilles (the 'western Caribbean' or 'northern Caribbean') incorporates Jamaica, Puerto Rico, the island of Hispaniola (incorporating Haiti and the Dominican Republic), the Cayman Islands and Cuba. As Table 1.2 indicates, both the size of the population and the total land area occupied by individual island states in the region vary considerably.

Several geographic anomalies exist in the region. The Bahamas, while situated north of Cuba, are not normally classified as either a Lesser or Greater Antillean island cluster. The Netherlands Antilles, as a political designation, is defined by two geographic groups in the region: one group incorporates the islands of Curaçao and Bonaire (see Figure 1.1) in the southern Caribbean Sea, while the second group is comprised of Saba, Sint Maarten (the Dutch part of the island of St Martin) and Sint Eustatius. Now a kingdom of the Netherlands, Aruba was once part of the Netherlands Antilles but seceded in 1986. Finally, the overseas territory of Bermuda, a UK dependency consisting of a clustering of over 100 coral reefs and small islets, exhibits a climate that is not unlike its Caribbean neighbours, yet it is located almost directly east of the state of North Carolina.

The Lesser Antilles can be subdivided geographically into the Leeward Island group and the Windward Island group. The Leeward group encompasses the British and US Virgin Islands south to, and including, Guadeloupe, while the Windward group is comprised of Dominica, St Lucia, Barbados, St Vincent and the Grenadines, and Grenada. While Trinidad and Tobago are not normally included in the Windward Island characterisation, they are often included in the wider 'Lesser Antilles' characterisation.

The geographic complexity exhibited through subregion designations in the Caribbean is equalled by the political variability found in the region. The Caribbean is home to British Commonwealth members, United States Commonwealth members, British dependencies, autonomous countries within larger kingdoms and independent republics. A number of island states in the Caribbean are independent sovereign nations, while several continue to retain political and economic dependency arrangements with colonial states. Regional political and structural bodies such as the Caribbean Community (CARICOM), the Association of Caribbean States (ACS) and the Organisation of Eastern Caribbean States (OECS) seek to enhance the economic status of the region on the world stage by offering blanket representation in both international and regional affairs. Their

Table 1.2 Selected economic indicators of various Caribbean island states

Official name	Form of government	Population	Land area (sq. km.)
Anguilla	British Dependent Territory	11,915 (est. 1997)	91
Antigua and Barbuda	British Commonwealth with parliamentary democracy	64,362 (est. 1996)	441
Aruba	Separate entity within Kingdom of the Netherlands; autonomy over internal affairs; parliamentary democracy	90,000 (est.)	193
Bahamas	British Commonwealth with parliamentary democracy	83,651 (est. 1995)	13,935
Barbados	British Commonwealth with parliamentary democracy	260,500 (est. 1998)	430
Bermuda	British Dependency	60,000 (est.)	50
British Virgin Islands (BVI)	British Dependency; constitution enables ministerial system headed by Governor	19,000 (1995)	95.4
Cayman Islands	British Crown Colony	32,800 (est.)	260
Cuba	Independent republic; President of the State Council acts as Head of both State and Government	11 million (est. 1996)	114,478
Dominica	British Commonwealth with parliamentary democracy	75,700 (est. 1997)	749
Dominican Republic	Representative democracy; executive power rests with President, who is also Head of Government	7.9 million (1995)	48,442
Grenada	British Commonwealth with parliamentary democracy	99,500 (est. 1997)	311
Guadeloupe	French Overseas Département (DOM)	443,000 (1999)	1,780
Haiti	Independent republic; President is Head of State; Prime Minister is Head of Government	6.5 million (est.)	27,749
Jamaica	British Commonwealth with parliamentary democracy	2.5 million (est. 1995)	10,991
Martinique	French Overseas Département (DOM)	389,000 (1999)	1,079
Montserrat	British colony	3,500 (est. 1997)	102
Netherlands Antilles (Curaçao, Bonaire, St Maarten, St Eustatius, Saba)	Autonomous country within Kingdom of the Netherlands; each island has elected Island Council; parliamentary democracy	190,000 (est. total)	500

Table 1.2 continued

Official name	Form of government	Population	Land area (sq. km.)
Puerto Rico	Commonwealth of the United States	3.7 million (est. 1995)	8,897
The Federation of St Kitts and Nevis	British Commonwealth with parliamentary democracy	42,280 (est. 1997)	269
St Lucia	British Commonwealth with parliamentary democracy	149,621 (est. 1997)	616
St Vincent and the Grenadines	British Commonwealth with parliamentary democracy	111,663 (est. 1997)	389
Trinidad and Tobago	Unitary State, with some autonomy for Tobago	1.3 million (1997)	5,128
Turks and Caicos Islands	British dependency; governed by British Governor	15,000 (est.)	500
United States Virgin Islands	Commonwealth of the United States	105,000 (1995)	

Sources: Compiled from Economic Commission for Latin America and the Caribbean (ECLAC)(1999) and World Statistics and Atlases (http://worldatlas.brinkster.net/).

efforts are crucial, if not difficult, as the economies of the various island states in the region are diverse. Trinidad and Tobago, for example, experienced real GDP growth of 7 per cent in 1999 as a result of strong overall performance in its energy sector (Commonwealth Secretariat 2001), while sugar and bauxite continue to be critical exports for Jamaica. Other economic activities include nutmeg and mace (Grenada), offshore finances (e.g. Aruba, St Lucia) and manufacturing and technology (e.g. Barbados). In 2000, Cuba observed significant growth in its mining sector and the resulting production of petroleum (Economic Commission for Latin America and the Caribbean 2001).

Agriculture continues to be an important source of income for many island states. In the past four to five decades, arrowroot, sugar cane, coconut, indigo, spices and tobacco have been cultivated. While agricultural production continues to be important to the strength of many local economies, its share of the GDP of many island states is falling. In Dominica, for example, agriculture accounted for 24 per cent of GDP in 1985, yet by 1998 it had dropped to 20 per cent. Similarly, agriculture contributed 16 per cent of St Vincent and the Grenadines' GDP in 1985, but by 1998 its share had fallen to 11 per cent (Commonwealth Secretariat 2001). Andreatta (1998) notes that, in anticipation of substantial changes to marketing regimes, non-traditional crops such as root crops, cucumbers, flowers, hot peppers, tomatoes and other non-banana tropic fruits are increasingly being cultivated.

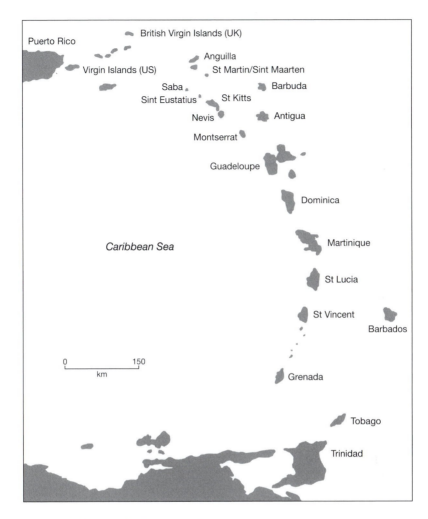

Figure 1.2 Map of the Lesser Antilles

Despite such attempts at agricultural diversification, bananas comprise an important export commodity, particularly in the Windward Islands (Commonwealth Secretariat 2001). In the 1990s, however, the ability of some islands to effectively produce banana crops for export to the European Union was threatened by challenges to the EU's banana import regime (Payne and Sutton 2001). As a result, uncertainty over access to existing and future markets continues to shroud the banana cultivation in the Windwards. As Payne and Sutton (2001: 271) note, the lessons learned from the banana crisis point to the ability, or lack thereof, of small Caribbean states to function in the wider international economic arena:

The Caribbean, especially the smaller countries of the region, here find
themselves particularly disadvantaged: marginalized in the development policy debate, side-lined in the corridors of power and handi-
capped by small size and economic dependence . . . the imperative to
find development alternatives is the central issue facing the Windward
Islands and, by analogy, other parts of the Caribbean with similar situations of domestically desirable, but internationally uncompetitive,
agricultural, extractive, industrial and service sectors.

While part of the decreased emphasis on agricultural production can be
attributed to an increasing amount of attention devoted to the development
of tourism, the reality is that the production of, for example, rice, coffee,
aluminum, bauxite, spices, petroleum and chemical products is often
subject to substantial shifts in demand. Increased competition from other
areas of the world, where the benefits of cheaper labour and larger
economies of scale and scope are significant, has also played a role. For
those reasons, and in the face of economic uncertainty, it is hardly
surprising that many island states have taken notice of the sizeable growth
in the world tourism sector.

Tourism in the Caribbean: some preliminary considerations

Modern conventional mass tourism (i.e. Turner and Ash's (1976) 'golden
hordes') began in the mid-twentieth century in the Caribbean. This form
of tourism can, to a large extent, be characterised by undifferentiated products, origin-packaged holidays, spatially-concentrated planning of facilities,
resorts and activities, and the reliance upon developed markets such as the
United States, Canada and Britain. From a tripographic perspective, mass
tourism in the Caribbean is commonly associated with cruise tourism and
all-inclusive package holidays (largely incorporating resort enclaves
(Freitag 1994)) (Conway 2002; Jayawardena 2002). With the growth in
available leisure time across many developed countries (Shaw and Williams
2002), particularly in the form of patterned, yet temporally repetitive,
holiday distributions, the dominant tourism product in the Caribbean –
which may be considered typical of warm-weather/mass tourism destinations – has been sun, sand and sea.

Several events in the 1950s and 1960s facilitated the development of
conventional mass tourism in the region. The first is the introduction
of jet aircraft, which had a significant impact on the propensity for travel
to the region from key generating markets, such as those in North America
and Europe (Bell 1993). Second, it was during this period that many
island states in the region, especially those associated with the British
Commonwealth, began to move towards political independence. Such independence spawned the desire to seize a certain degree of control over

internal economic development. A conscious focus on tourism meant that some island states could break away from existing colonial dependency arrangements in other economic sectors. One consequence of this, as Bell (1993: 221) noted, was the changing political structure of the region and the introduction of 'sensitive political psyches' that often conjured up suspicions of neo-colonialism as some economic ties to former colonial nations were not severed completely. This was especially felt through tourism (the 'hedonistic face of neocolonialism' (Crick 1989: 322)) because most tourists were from western countries. However, the development of tourism, through neoliberal economic packages (Telfer 2002), was still seen as an opportunity to correct this by taking control of development and management and effectively shaping the tourist; yet as Bell (1993: 220) rightly noted, tourists cannot be 'regulated'.

The third event, which is perhaps more closely related to a broad trend, is the substantial number of migrants leaving the region at the same time as the region itself began to host increasing numbers of visitors (Pattullo 1996). Beginning in the 1950s, but continuing to the present day, many Caribbean nationals migrated (voluntarily, but occasionally out of economic necessity) to the United Kingdom, Canada, New York, Florida and California. The consequence of this has been the creation of numerous Caribbean cultural diasporas around the world. Ironically, these communities now serve as important sources of tourists, as many Caribbean expatriates living abroad frequently travel 'home' to visit family and friends (Duval 2002; see also Stephenson 2002).

The 1960s were characterised by Holder (1993: 21) as 'boom years' for tourism in the region. Growth rates of 10 per cent or more were not uncommon. Bryden (1973: 100) noted increases of over 15 per cent in the years 1961–7 for St Vincent, Grenada, St Lucia, Montserrat, Antigua and the Bahamas. For the same period, the Turks and Caicos, Cayman Islands and Virgin Islands reported a combined increase in arrivals of roughly 24 per cent. The industry management structure at the time, however, was largely in the hands of expatriates, especially large-scale hotel properties. As Sutty (1998) notes, while small-scale hotels and restaurants, largely locally owned and operated, have existed since the early beginnings of tourism in the region, their numbers have always been proportionately small. Moreover, the success of tourism during this period was largely confined to those countries that featured stronger overall economies (e.g. Jamaica, Trinidad and Tobago, and Barbados) (Prime 1976).

According to Bell (1993: 221), by the 1970s tourism began to command 'a measure of respectability', although the negative effects of the 1973 Arab–Israeli war were certainly felt in the region. Local government involvement in the accommodation industry began to grow:

> Some Caribbean governments had deliberately set out to invest in the hotel industry, for what seemed like perfectly valid reasons. During the

1970s, with many hotels operating far below capacity and with major closures imminent, particularly in Jamaica, government was obliged to assume ownership, and in some cases management, of a significant number of hotels so as to protect the jobs of hotel employees.

(Bell 1993: 222)

Despite the problems facing the sector in the 1970s, it was indeed a period of recognition. In terms of the overall economic contribution that tourism activities have made in the Caribbean, approximately US$1.7 billion was generated from foreign exchange as early as 1978 (Holder 1979).

At present, and not unlike other countries in the developing world (Sharpley 2002), many local governments rely on conventional mass forms of tourism to help offset failing manufacturing or agricultural sectors and, consequently, to act as significant sources of foreign exchange. After the positive economic benefits of tourism were first realised in the 1940s (Holder 1993), the subsequent increase in the reliance on tourism for economic success has led to calls for economic and cultural dependency (e.g. Erisman 1983). In fact, in the early 1970s Bryden (1973) warned that the overall economic costs of providing the tourism product may outweigh any benefits received.

While many island states in the Caribbean are associated with a conventional mass tourism profile, the increasing concerns voiced over the unsustainable nature of mass tourism policies, operations and management (e.g. Weaver 2001) have led many governments in the region to consider adopting more sustainable forms of tourism development and management strategies. It is within this context that the Special Committee on Sustainable Tourism of the Association of Caribbean States formally operationalised a Convention on Sustainable Tourism Zone of the Caribbean (STZC) in December 2001. The document (Association of Caribbean States 2002) recognised the importance of sustainable tourism development in the region as sustainability acts as 'the basis for protecting biodiversity, culture and the environment, with the human being at the core of our actions, by virtue of favouring an improved distribution of the benefits derived from tourist development'.

For many small island states or dependencies (SISODs) (Weaver 1995), increased interest in alternative products and management strategies have been used as means to embrace the wildly popular and politically correct mantra of sustainable tourism, although it is important to note that this has not come about at the expense of developing traditional mass forms of tourism. In fact 'alternative tourism' often exists side-by-side with more conventional mass forms in the Caribbean. Indeed, visitors to all-inclusive resorts have opportunities to participate in tours of sensitive natural environments, further blurring the distinction between alternative and mass tourism. The ambiguous nature of alternative tourism, however, requires recognition of the fact that 'alternative' can essentially refer to different forms of investment, development, construction materials, and, more broadly, tourist products. Thus, while Dominica and Montserrat's emphasis on nature tourism and ecotourism are examples of deliberate planning and policy formulation (Weaver 1995), but ultimately reflect the reality that there is a clear lack of '3S' attributes for certain destinations in the region, the emphasis on locally owned accommodation in St Vincent can be held up as a contextually alternative approach to maximising local returns on tourist expenditures.

The emphasis on sustainable, or alternative, tourism is also reflected in the frequent realisation that many forms of tourism may have serious negative impacts on the biophysical environment, the economy and the social and cultural realm of host societies. The implementation of sustainable tourism is of paramount importance for the protection of the fragile ecological (and economic) environments of most microstates in the Caribbean, and there has already been a marked shift in some countries (e.g. Dominica) towards policy and planning frameworks that favour environmentally friendly tourism development.

Whatever the impetus, the growth of alternative tourism in the Caribbean allows for some consideration of the maturing of the industry, to some degree at least, in terms of how it is managed by governments and non-governmental organisations. On the one hand it points to the recognition of new forms of tourist experience that are in demand by visitor segments in key origin markets, while on the other hand it hints at the acceptance of a more responsible management regime incorporating an important economic sector. Examples include the growth in nature-based products (e.g. ecotourism) in island states such as Dominica and culturally derived tourism products such as the annual carnival held in many Commonwealth eastern Caribbean states (e.g. Barbados, Trinidad and Tobago, Grenada, St Vincent, St Lucia, Dominica, St Kitts and Nevis, and Antigua). The increasing emphasis on alternative forms and types of tourism development has certainly not absolved the region of future problems with the sector itself. Fragile environments, once safe from the gaze of tourists, now suffer under the feet of the masses. Additionally, many of the same problems that have plagued tourism in the past still exist.

Magnitude of tourism in the Caribbean

The provision of data on regional visitor arrivals, expenditures and generating markets at the regional level falls largely under the purview of the Caribbean Tourism Organization, although individual island states provide the organisation with specific indicators of tourism sector performance. The compilation of such region-wide data is often problematic, either because some island states chose not to measure one or more indicators or because the actual method of collecting such data varies throughout the region. The consequence of this is that, in some cases, accurate comparisons between island states are difficult. By extension, the aggregate compilation of region-wide data by the Caribbean Tourism Organization includes destinations which may not intuitively be considered as part of the region itself. For example, when reporting visitor arrivals for the region as a whole, the Caribbean Tourism Organization includes visitation to Cancun, Cozumel and Belize as these destinations fall under its official mandate. Table 1.3 outlines some broad trends of tourism for specific island states in the region for the years 1990 and 2000. While the measures by which Caribbean tourism is represented in Table 1.3 are admittedly limited, the data provide at least a cursory insight into the overall magnitude of tourism in the region.

Almost hidden (unintentionally) within the visitation counts presented annually by the Caribbean Tourism Organization is the seasonal nature of tourism in the region. Some destinations, such as Antigua and Barbuda, are less prone to the seasonal effects of visitor arrivals, largely due to the robust nature of tourism in these destinations and the existence of generous airlift access and other connections (i.e. cruise itineraries). Other destinations, however, feature significant seasonal shifts in visitor arrivals. In Bermuda, for example, arrivals in January 2000 were approximately 103,000, while in July and August 2000 they were 157,000 and 164,000 respectively.

Several top line indicators can be used to point to the rapid growth of tourism, even in the past decade. Since 1990 stayover arrivals (i.e. visitors spending more than 24 hours in a destination) have increased 58 per cent, and cruise tourism has grown, on average, by 6.5 per cent each year (CTO 2002a). In 2000 the Caribbean hosted over 30 million visitors, of which 43 per cent were cruise tourists (Table 1.2). Total expenditures for 2000 reached approximately US$17 billion, compared to almost US$7 billion in 1991 (CTO 2002a). The Dominican Republic received the highest amount of visitor spending, although visitor spending per capita is highest in the Turks and Caicos Islands. Examination of expenditure data alone, however, does not always fully explain the state of the tourism sector, nor does it reflect the community benefits that are often expected. For example, Antigua and Barbuda hosted approximately 205,000 visitors in 1991 and registered a total of US$315 million in expenditures (CTO 2002a). By 2000, however, the number of arrivals had increased to 237,000 (an increase of nearly 16 per cent) and the total visitor spend had actually fallen by more than 7 per cent to US$291 million. As well, data for 2000 indicate that stayover tourists spend an average of 10 days in each of St Vincent and the Grenadines and Barbados; however, the per trip spending for Barbados is nearly US$200 higher. Overall, average per trip spending by stayover tourists varies considerably by destination, from US$503 in Anguilla to almost US$1,900 in the Turks and Caicos. For cruise tourists, similar variances exist. In the US Virgin Islands, cruise tourists spend an average

Table 1.3 Selected visitation statistics

	Stayover arrival (thousands)		Cruise arrivals (thousands)		Visitor expenditures (US$ millions)[b]		Average per trip expenditure (US$) (2000)		Average length of stay[e]	
	1990	2000	1990	2000	1990	2000	Stayover[c]	Cruise	1990	2000
Anguilla	31	44	n.a.	n.a.	31	57	505	n.a.	10.6	8.6
Antigua and Barbuda	206	237	227	429	314	291	891	28	n.a.	n.a.
Aruba	433	721	130	490	396	638	n.a.	n.a.	7.8	7.7
Bahamas	1,562	1,596	1,854	2,513	1,193	1,814	1,041	59	5.7f	6.5f
Barbados	432	545	363	533	460	711	1,205	103	10.6	10.1
Bermuda	433	328	113	210	456	431	1,179	210	6.3	6.0
Bonaire	41	51	4	44	24	56	1,021	85	7.0	9.3
British Virgin Islands	160	281	95	188	109	315	934	32	9.3	9.1
Cayman Islands	253	354	362	1,031	276	559	1,262	109	4.9f	n.a.
Cuba	340	1,774	n.a.	n.a.	387	1,857	n.a.	n.a.	8.7	n.a.
Curaçao	208	191	159	309	232	227	1,057	79	7.0	8.2
Dominica	45	70	7	240	28	47	583	27	9.0	n.a.
Dominican Republic	n.a	2,973	n.a.	182	n.a.	2,860	n.a.	n.a.	n.a.	10.0
Grenada	126	129	183	180	42	70	518	17	n.a.	7.2
Guadeloupe	288	807	262	392	234	454	n.a.	n.a.	n.a.	5.2
Haiti	120	140	n.a.	304	46	54	n.a.	n.a.	n.a.	n.a.
Jamaica	841	1,323	386	908	764	1,333	952	80	10.9	10.1
Martinique	282	526	421	286	255	370	677	28	n.a.	13.2
Montserrat	19	10	n.a.	n.a.	10	9	n.a.	n.a.	n.a.	n.a.
Puerto Rico	n.a	3,341	n.a.	1,302	n.a.	2,388	n.a.	n.a.	2.8g	2.5g
Saba	5	9	n.a.	n.a.	n.a.	n.a.	n.a.	n.a.	n.a.	n.a.
St Eustatius	n.a	9a	n.a.	n.a.	n.a.	n.a.	n.a.	n.a.	n.a.	n.a.
St Kitts and Nevis	76	73	34	164	68	58	731d	29d	8.6	9.6
St Lucia	138	270	102	444	173	277	940	52	10.7	9.6
St Maarten	565	432	515	868	310	482	n.a.	n.a.	n.a.	n.a.
St Vincent/Grenadines	54	73	79	86	53	75	1,014	16	9.6	10.0
Trinidad and Tobago	194	398	32	82	101	213	n.a.	n.a.	9.6	n.a.
Turks and Caicos Islands	42	151	n.a.	n.a.	50	285	1,879	n.a.	n.a.	7.5
US Virgin Islands	370	607	1,120	1,768	778	1,157	983	260	4.5	n.a.

Source: Derived from Caribbean Tourism Organization (2002a, 2002c, personal communication).

Notes a Excludes 'Antilleans'. b Includes both stayover and cruise tourists. c Includes all sea arrivals and same-day visitors excluding cruise passengers. d Estimated by the Eastern Caribbean Central Bank. e Intended length of stay in days, unless stated otherwise. f Hotel registrations. g Hotel registrations, includes residents of Puerto Rico.

of US$260, compared to only US$16.2 in St Vincent and the Grenadines (CTO 2002a).

External influences

Not unlike other destinations around the world, Caribbean tourism is particularly vulnerable to external forces. Key among these is, more recently, the impact of the 11 September 2001 hijackings in New York City and Washington, DC. In fact, the months following 11 September 2001 represents the fourth time since 1970 that an actual decline in visitor arrivals to the region has been recorded (Caribbean Tourism Organization 2002b). Previous declines were attributed to the oil crisis in the early 1970s, the global recession in the early 1980s, and the 1991 Gulf War. Like destinations elsewhere, the Caribbean witnessed a slump in arrivals in late 2001 as a result of deflated confidence in travel worldwide. While a small decline in visitor arrivals was initially being realised in mid-2001, the events of 11 September aggravated this trend. Some destinations, such as the Dominican Republic, fared comparatively well, posting a net decline of 6.6 per cent for 2001 over 2000 figures. For others, the aftermath was much more severe. The Bahamas reported a decline of 33 per cent in the last four months of 2001 over 2000 arrival figures. Some destinations, however, rode the strength of their previous winter seasons. The latest estimates from the Caribbean Tourism Organization suggest that several destinations in the region experienced fewer visitor arrivals (CTO 2003) (Table 1.4). Despite a region-wide (including Cozumel and Cancun) drop in tourism receipts of 1.2 per cent, the Caribbean fared well when compared to a global slump of 2.9 per cent (CTO 2003). However, actual arrivals fell by 1.9 per cent in 2001, compared to the worldwide estimate of approximately 0.6 per cent (CTO 2003).

Other external forces that can have consequences for tourism are physiographic and climate-related. Water shortages are common in more arid islands (such as Antigua) and hurricanes wreak havoc on many island states (primarily those in the northern Lesser Antilles and Greater Antilles) and thus severely curtail otherwise active local tourism sectors. St Martin/Sint Maarten is, to some extent, still recovering from Hurricane Luis, which charged through the northern Lesser Antilles in 1995 (*Salt Lake Tribune* 2000). Yet the damage from hurricanes is not restricted to the built environment. Beaches can be stripped of sand or filled with thick layers of sea grass and entire ecosystems can be thrown into disarray. Such resources, obviously important to tourism, can subsequently take years to fully recover.

As the Caribbean region is largely volcanic, the impact of destructive eruptions can devastate an island's economy. In Montserrat, for example, the eruption of the Soufriere Hills volcano, beginning in July 1995, totally destroyed the capital of Plymouth. The eruption was initially responsible

Table 1.4 Stayover arrivals, 2000 and 2001 (thousands)

	2000	2001	Change (%)
Anguilla	43.8	48.0	10
Antigua and Barbuda	236.7	222.1	–6
Aruba	721.2	691.4	–4
Bahamas	1,596.2	1,552.9	–3
Barbados	544.7	507.1	–7
Bermuda	328.3	278.2	–15
Bonaire	51.3	50.4	–2
British Virgin Islands	281.1	295.6	5
Cayman Islands	354.1	334.1	–6
Cuba	1,774.0	1,774.5	0
Curaçao	191.2	204.6	7
Dominica	69.6	66.4	–5
Dominican Republic	2,972.6	2,882.0	–3
Grenada	128.9	123.4	–4
Guadeloupe	807.0	773.4	–4
Haiti	140.5	141.6	1
Jamaica	1,322.7	1,276.5	–3
Martinique	526.3	460.4	–13
Montserrat	10.3	9.8	–5
Puerto Rico	3,341.0	3,608.9	8
Saba	9.1	9.0	–1
St Eustatius	8.9	9.6	8
St Kitts and Nevis	73.1	70.6	–3
St Lucia	269.9	250.1	–7
St Maarten	432.3	402.6	–7
St Vincent/Grenadines	72.9	70.7	–3
Trinidad and Tobago	398.2	383.1	–4
Turks and Caicos Islands	151.4	165.2	9
US Virgin Islands	607.2	592.0	–3

Source: Derived from Caribbean Tourism Organization (2002a, 2003).

for severely restraining visitor arrivals to Montserrat, although many tourists have recently returned for a first-hand glimpse of an active volcano, albeit from a safe distance. Ironically, the volcano itself is now often the object of renewed interest among tourists seeking alternative holiday experiences.

Purpose and scope of the book

As the brief discussion above has highlighted, tourism in the Caribbean is a complex entity, incorporating shifting markets, alternative products and often fluctuating economic returns. Addressing these complexities in a comprehensive manner is difficult at best, although the contributors in this volume have made an attempt to provide specific accounts of the variance and nature of tourism development in the Caribbean. In effect, the goals of this book are twofold. The first is to broadly highlight the central issues in tourism development in the Caribbean region, thus providing a current review of key considerations in Caribbean tourism. Second, the various contributions are designed to provide a useful framework from which future directions in management and development might be situated. In this sense, and taken as a whole, the contributions that follow are intended to act as a forum through which issues facing governments in the Caribbean can be brought forward for discussion.

In the remainder of Part One broad trends in tourism in the region are addressed. In Chapter 2, Sheller utilises an historical approach, informed through rich travel literature, to understanding how Edenic and iconic images of the Caribbean have been constructed. Sheller argues that such images and depictions are at once congruent with broad political and social movements (e.g. Romanticism), and this has been translated and embodied into the 'sexualisation' of the 'exotic' inhabitants of the islands. Following this, in Chapter 3 McElroy outlines the position of the Caribbean in the broader context of island states around the world in the context of increasing levels of competition worldwide. McElroy discusses the nature of such development in the Caribbean specifically through the use of a tourist penetration index, and argues that, for the Caribbean to retain its relatively healthy position as a respected tourism destination, it will need to emphasise quality over quantity through alternative development strategies.

Part Two addresses the key issues facing tourism development in the region. Chapter 4 by Duval and Wilkinson begins by providing a broad overview of tourism in small islands in the context of development and sustainability (social/cultural, environmental and economic). Throughout, Duval and Wilkinson address the complex interrelationships of actors and stakeholders involved in tourism development in the Caribbean. Wilkinson (Chapter 5) probes the policy and planning instruments and various types of government involvement in several island states (Dominica, St Lucia, Cayman Islands, Barbados and the Bahamas). He argues that regional and local organisational structures involved with the development of tourism deserve close examination as they often help shape future expansion efforts in the tourism sector. Consequently, Jordan in Chapter 6 investigates inter-organisational relationships within three twin-island states in the region (Antigua and Barbuda, St Kitts and Nevis, and Trinidad and Tobago), noting that the manner in which tourism is developed in these states is based on their colonial past and their immediate inter-island structural arrangements. The specific conclusions drawn by Jordan, however, have applicability for region-wide governance and management of tourism. Following this, Chapter 7 by Timothy positions tourism in the context of existing supranational economic and political structures (e.g. CARICOM, OECS). Timothy notes that, given the economic importance of tourism, many regionally based supranational organisations, with the exception of the Caribbean Tourism Organization, pay comparatively little attention to tourism. The reasons for this, as Timothy outlines, include, for example, political barriers, policy restrictions, issues of small size and population and the fear of homogenising the tourism product.

Issues involving the management of historic resources and their strong linkage to 'alternative' tourism products are addressed by Found in Chapter 8. As Found points out, the scope of heritage tourism in the region is quite strong, but often requires considerable input and management from government. While perhaps not entirely an 'alternative' tourism product, the increasing popularity of cruise tourism, discussed by Wood in Chapter 9, carries with it equal concern for the management and protection of the natural marine environment, but also raises concerns over the degree of positive economic impacts of cruise tourists. As Wood points out, cruise tourism in the Caribbean is a 'uniquely deterritorialised industry' characterised by foreign ownership, workers and capital.

Reflecting once again on sustainability as applied to tourism, the trend towards ecotourism as a form of sustainable tourism development has been explored in several islands, most notably Dominica (Weaver 1995). Critical, once again, is the management of impacts in natural environments, which is addressed in the context of ecotourism by Weaver in Chapter 10. Weaver questions whether tourism that utilises the often delicate resource base is deliberate or merely circumstantial. As the Caribbean is dominated by small island states, the link between agriculture and tourism provides an excellent example of the interplay between two vital economic sectors. Mills (Chapter 15) examines this interplay from a gendered perspective in her case-study of Carriacou, while Chapter 11 by Conway investigates the extent to which agriculture, coastal zone conservation and tourism can share a sustainable future. Conway argues that, indeed, they can, although directed management is needed. At best, some local production of produce seems to find its way to the tourist's plate.

The extent to which community involvement is incorporated is addressed by Milne and Ewing in Chapter 12. They argue that, by adopting a community perspective for planning and policy endeavours, community values and understandings towards tourism in general are favoured. Management issues in the operation of tourism businesses are addressed by Haywood and Jayawardena in Chapter 13, who provide a synopsis of the interplay between multi-national corporations (MNCs) and the ownership of small accommodation properties, and thus demonstrate the transnational nature of modern economic structures in the region. This theme is also addressed in Chapter 14 by Coles, which features a detailed analysis of resort complexes in the Dominican Republic. Coles highlights the multi-faceted, post-colonial concoction of experience and reflection within the resort product, and despite the mass production of tourist experiences that characterises resort-based mass tourism, 'choices do exist but they are regimented within institutionalised networks of supply and spatial consumption patterns'.

Part Three considers future trends in tourism in the Caribbean. Chapter 16 by Momsen investigates the impact of various externalities in the context of tourism in the region. As Momsen notes, the current characteristics of Caribbean tourism have been moulded by deliberate consumer-oriented strategies by overseas agents as well as changes enacted by local tourism industries. Finally, Chapter 17 concludes with an assessment of past predictions made on Caribbean tourism, and ends by offering several new forecasts in the context of the issues raised throughout.

References

Andreatta, S. (1998) 'Transformation of the agro-food sector: lessons from the Caribbean', *Human Organization* 57: 414–29.

Archer, E.D. and Davies, C.S. (1984) 'Reassessing Third World tourism', *The Tourist Review* 39: 19–23.

Association of Caribbean States (ACS) (2002) *Convention establishing the sustainable tourism zone of the Caribbean*, Association of Caribbean States. Online: <http://www.acs-aec.org/STZC_LEGAL_DOC_AND_INDICATO.DOC> (accessed 31 December 2002).

Bell, J.H. (1993) 'Caribbean tourism in the year 2000', in D.J. Gayle and J.N. Goodrich (eds) *Tourism Marketing and Management in the Caribbean*, London: Routledge.

Brown, F. (1998) *Tourism Reassessed: Blight or Blessing?*, Oxford: Butterworth–Heinemann.

Bryden, J.M. (1973) *Tourism and Development: A Case-Study of the Commonwealth Caribbean*, London: Cambridge University Press.

Caribbean Tourism Organization (CTO) (2002a) *Caribbean Tourism Statistical Report: 2000–2001 Edition*, St Michael: Caribbean Tourism Organization.

—— (2002b) *Caribbean Tourism One Year After 9/11*, St Michael: Caribbean Tourism Organization.

—— (2002c) 'Historic tourism data 1999'. Online: <http://www.onecaribbean.org/information/documentdownload.php?rowid=49> (accessed 29 November 2002).

—— (2003) *Caribbean Tourism Statistical Report: 2001–2002 Edition*, St Michael: Caribbean Tourism Organization.

Clancy, M. (2001) *Exporting Paradise: Tourism and Development in Mexico*, Amsterdam: Elsevier.

Commonwealth Secretariat (2001) *Small States: Economic Review and Basic Statistics*, Annual Series: Sixth volume, Spring 2001, London: Commonwealth Secretariat.

Conway, D. (2002) 'Tourism, agriculture, and the sustainability of terrestrial ecosystems in small islands', in Y. Apostolopoulos and D.J. Gayle (eds) *Island Tourism and Sustainable Development: Caribbean, Pacific and Mediterranean Examples*, Westport: Praeger.

Crick, M. (1989) 'Representations of international tourism in the social sciences: sun, sex, sights, savings and servility', *Annual Review of Anthropology* 18: 307–44.

Duval, D.T. (2002) 'The return visit–return migration connection', in C.M. Hall and A. Williams (eds) *Tourism and Migration: New Relationships between Production and Consumption*, Dordrecht: Kluwer.

Economic Commission for Latin America and the Caribbean (ECLAC) (1999) *Economic Profiles of Caribbean Countries*, LC/CAR/G.572, Port of Spain: ECLAC.

—— (2001) *Summary of Caribbean Economic Performance*, LC/CAR/G.658, Port of Spain: ECLAC.

Erisman, H.M. (1983) 'Tourism and cultural dependency in the West Indies', *Annals of Tourism Research* 10: 337–61.

Freitag, T.G. (1994) 'Enclave tourism development: for whom the benefits roll?', *Annals of Tourism Research* 21: 538–54.

Gayle, D. and J. Goodrich (1993) 'Caribbean tourism marketing, management and development strategies', in D.J. Gayle and J.N. Goodrich (eds) *Tourism Marketing and Management in the Caribbean*, London: Routledge.

Harrigan, N. (1974) 'The legacy of Caribbean history and tourism', *Annals of Tourism Research* 2(1): 13–25.

Holder, J.S. (1979) *The Role of Tourism in Caribbean Development or Buying Time with Tourism in the Caribbean*, Christchurch: Caribbean Tourism Research and Development Centre, Organisation of American States, Inter American Tourism Training Centre.

—— (1988) 'Pattern and impact of tourism on the environment of the Caribbean', *Tourism Management* 9: 119–27.

—— (1993) 'The Caribbean Tourism Organization in historical perspective', in D.J. Gayle and J.N. Goodrich (eds) *Tourism Marketing and Management in the Caribbean*, London: Routlege.

Jayawardena, C. (2002) 'Mastering Caribbean tourism', *International Journal of Contemporary Hospitality Management* 14: 88–93.

Lumsdon, L. and Swift, J. (2001) *Tourism in Latin America*, London: Continuum.

Pattullo, P. (1996) *Last Resorts: The Cost of Tourism in the Caribbean*, London: Cassell.

Payne, A. and Sutton, P. (2001) *Charting Caribbean Development*, London: Macmillan.

Perez, L.A. (1975) 'Tourism in the West Indies', *Journal of Communications* 25: 136–43.

Prime, T.S.S. (1976) *Caribbean Tourism: Profits and Performance Through 1980*, Port of Spain: Key Caribbean Publications Limited.

Salt Lake Tribune (2000) 'There's trouble in Caribbean paradise', 6 May: B7.

Sealey, N.E. (1982) *Tourism in the Caribbean*, London: Hodder & Stoughton.

Sharpley, R. (2002) 'Tourism: a vehicle for development?', in R. Sharpley and D. Telfer (eds) *Tourism and Development: Concepts and Issues*, Clevedon: Channel View Publications.

Shaw, G. and Williams, A.M. (2002) *Critical Issues in Tourism: A Geographical Perspective*, 2nd edn, Oxford: Blackwell.

Slinger, V. (2000) 'Ecotourism in the last indigenous Caribbean community', *Annals of Tourism Research* 27(2): 520–23.

Stephenson, M.L. (2002) 'Travelling to the Ancestral Homelands: the aspirations and experiences of a UK Caribbean community', *Current Issues in Tourism* 5: 378–425.

Sutty, L. (1998) 'Local participation in tourism in the West Indian islands', in E. Laws, B. Faulkner and G. Moscardo (eds) *Embracing and Managing Change in Tourism: International Case Studies*, London: Routledge.

Telfer, D.J. (2002) 'The evolution of tourism and development theory', in R. Sharpley and D.J. Telfer (eds) *Tourism and Development: Concepts and Issues*, Clevedon: Channel View Publications.

Turner, L. and Ash, J. (1976) *The Golden Hordes: International Tourism and the Pleasure Periphery*, New York: St Martin's Press.

Weaver, D.B. (1995) 'Alternative tourism in Montserrat', *Tourism Management* 16: 593–604.

—— (2001) 'Mass tourism and alternative tourism in the Caribbean', in D. Harrison (ed.) *Tourism and the Less Developed World: Issues and Case Studies*, Wallingford: CAB International.

Wilkinson, P.F. (1989) 'Strategies for tourism development in island microstates', *Annals of Tourism Research* 16: 153–77.

—— (1996) 'Graphical images of the Commonwealth Caribbean: the tourist area cycle of evolution', in L.C. Harrison and W. Husbands (eds) *Practicing Responsible Tourism: International Case Studies in Tourism Planning, Policy and Development*, New York: John Wiley & Sons.

—— (1999) 'Caribbean cruise tourism: delusion? Illusion?', *Tourism Geographies* 1(3): 261–82.

World Tourism Organization (WTO) (2002) *Latest Data*. Online: <http://www.world-tourism.org/market_research/facts&figures/ latest_data.htm> (accessed 31 December 2002).

World Travel and Tourism Council (WTTC) (2001) *World Travel and Tourism Council, Year 2001, Tourism Satellite Accounting Research (Caribbean)*, London: World Travel and Tourism Council.

2 Natural hedonism

The invention of Caribbean islands as tropical playgrounds

Mimi Sheller

Introduction

Caribbean tourism is vested in the branding and marketing of Paradise. 'It is the fortune, and the misfortune, of the Caribbean,' argues Polly Pattullo, 'to conjure up the idea of "heaven on earth" or "a little bit of Paradise" in the collective European imagination ... a Garden of Eden before the Fall' (Pattullo 1996: 142). Verdant forests, exotic flora and tropical greenery serve as powerful symbols of the 'Eden' that is imagined before European intrusion. Tobago is a place where you can 'see the islands as Columbus first saw them', for example, while Dominica is described as 'still the primitive garden that Columbus first sighted in 1493. An area of tropical rainforests, flowers of incredible beauty and animals that exist nowhere else in the world'.[1] In this chapter I demonstrate how such imagery picks up on longstanding visual and literary themes in European representations of Caribbean landscapes as microcosms of earthly paradise – including the temptation and corruption that go along with being new Edens. These discursive formations of Caribbean scenery are closely related to the emergence of 'hedonism' as a key set of practices associated with Caribbean tourism. Depictions of Caribbean 'Edenism', I argue, underwrite performances of touristic 'hedonism' by naturalising the region's landscape and its inhabitants as avatars of primitivism, luxuriant corruption, sensual stimulation, ease and availability.[2]

The tropical island has played a crucial part in the history of European literature, philosophy and arts. Richard Grove has shown, for example, how the idea of tropical islands as Edens influenced Daniel Defoe's setting for *Robinson Crusoe* (1719), which is thought to be based on the Caribbean island of Tobago as described in John Poyntz's *The Present Prospect of the Famous and Fertile Islands of Tobago* (1683), and it led to a cult of 'robinsonnades' throughout Europe (Grove 1995: 225–9). From the seventeenth century onwards, Grove argues, 'the tropical environment was increasingly utilised as the symbolic location for the idealised landscapes and aspirations of the Western imagination' (Grove 1995: 3).

This chapter offers a reading of some of the mythic paragons of 'natural' and 'cultivated' scenery in the literature of Caribbean travel over several centuries, showing how themes of luxury and paradise were connected to particular ways of viewing Caribbean people. First I show how the Enlightenment vision of man-made utilisation of a bountiful Nature through orderly cultivation shifted in the nineteenth century to a more Romantic embrace of the luxuriant profusion of tropical nature, through which adventurous bodily immersion and self-discovery could be played out in wild and remote places.[3] Then I analyse how these geographies of risk and desire have become deeply implicated in contemporary practices of touristic hedonism in the region, including sex tourism and the sexual marketing of Caribbean tourism destinations.

Luxury and corruption in island paradise

Early European writing on the Caribbean enshrined the traveller's dream of tropical fecundity and excessive fruitfulness, conjuring up utopian fantasies of sustenance without labour. Such accounts drew on a range of precedents such as the biblical Garden of Eden, the classical garden of the Hesperides and the Renaissance botanic garden. An early seventeenth-century report on St Christopher, for example, emphasised that food was plentiful and required little labour to produce:

> A tree like a Pine, beareth a fruit so great as a Muske Melon, which hath alwayes ripe fruit, flowers or greene fruit, which will refresh two or three men, and very comfortable; Plum trees many, the fruit great and yellow, which but strained into water in foure and twenty houres will be very goode drinke; wilde figge trees there are many . . . all things we there plant doe grow exceedingly, so well as Tobacco; the corne, pease, and beanes, cut but away the stalkes, young sprigs will grow, and so bear fruit for many yeeres together, without any more planting.
>
> (Smith 1819 [1629]: 273)

Coming from a land of four seasons, Europeans had never seen plants that flowered and fruited throughout the year, carrying green fruit, ripe fruit and flowers all at the same time. In discovering the tropical island, 'Paradise had become a realisable geographical reality' (Grove 1995: 51); it was the Garden of Eden before the Fall.

Europeans were eager to grasp this paradise and make it their own. By the early eighteenth century the predominant theme in descriptions of the West Indian colonies was the beauty of cultivated areas. Sir Richard Dutton said of Barbados in 1681 that the dense population and intensive cultivation made it 'one great City adorned with gardens and a most delightful place' (cited in McFarlane 1994: 128). And the French writer Father Charlevoix burbled:

> The country of Barbados has a most beautiful appearance, swelling here and there into gentle hills; shining by the cultivation of every

part, by the verdure of the sugar canes, the bloom and fragrance of the number of orange, lemon, lime and citron trees, the guavas, papas, aloes, and a vast multitude of other elegant and useful plants, that rise intermixed with the houses of the gentlemen which are sown thickly on every part of the island.

(Charlevoix 1766: 319)

As I have traced more fully elsewhere (Sheller 2003), such eighteenth-century depictions of the Caribbean lent support to the institution of slavery by celebrating its capacity to make wild lands productive. They are also the foundation for subsequent developments of the 'tourist gaze' (Urry 1991), which have continued to imbue Caribbean landscapes and scenery with moral meanings.

In describing Antigua, for example, a British traveller in 1804 not only highlights his preference for cultivated scenery, but also comments *negatively* upon the scenes of untouched woods and mountains, 'unrelieved' by human intervention:

the eye traverses a view of one of the fairest and best cultivated tracts of country in the windward islands. It is highly pleasing to a person who has recently come from the woods and mountains of the more southern colonies, to behold so extensive a scene of cleared land. . . . Nothing appears more completely like a garden than the sugar plantation under good cultivation. . . . The green fields of cane . . . were intermixed with provision grounds of yams and eddoes, or the dark and regular parterres of holed land prepared for the reception of the succeeding year's plant-canes. A large windmill on each estate; the planter's dwelling-house and sugar-works, with the negro huts, in their beautiful groves of oranges, plantains, and cocoanut trees, completed a landscape that continually recurred in passing over the island.

(McKinnen 1804: 56–7)

As Mary Louise Pratt suggests in her study of imperial travel writing, the relation of such writers to the landscape is like a viewer to a painting. Gazing through his 'imperial eyes', the landscape is aesthetically composed, attributed with a density of meaning, and fixed by the mastery of the seer over the seen. Such a 'monarch-of-all-I-survey scene', Pratt argues, 'involve[s] particularly explicit interaction between aesthetics and ideology, in what one might call a rhetoric of presence' (Pratt 1992: 204–5). Through the rhetoric of presence the scene is not only recorded, but also ordered and made present to the reader who also masters it.

Yet the European's immersion in the tropical climate and close bodily proximity to slaves carried with it many risks, not only of disease and death, but also of moral corruption and 'tropicalisation'. As Daniel McKinnen observed, 'I am afraid from the mean and disingenuous behaviour of some of the inferior white inhabitants of the town, that the climate, and perhaps their association with the blacks, have not a little relaxed in them the strength and integrity of the British moral character' (McKinnen 1804: 30–1). This kind of moral judgement can also be seen in William Blake's much reproduced illustration of 'A Surinam Planter in his Morning Dress' for John G. Stedman's 1790 narrative (Stedman 1988 [1790]). Here the planter's indulgence in bodily pleasures are represented through his pipe for smoking tobacco and his alcoholic drink being poured by a young bare-breasted slave, described as a 'quaderoon' (and thus suggestive of sexual intercourse between masters and slaves). Dressed in luxury goods from around the world (silk, beaver, fine chintz), his body given over to dissipation, this figure epitomises the European's vulnerability to 'hybridisation' and 'creolisation'. There was a moral danger inherent in the climate itself and in the proximity of 'different' bodies (Young 1995). In the Tropics, it was suggested, the European's active mastery and moral bearing would give way to torpor, self-indulgence, consumption of luxuries and racial degeneracy.

It was not only male travellers who indulged in the sensual fantasies and bodily risks of the West Indies. In a romantic novel set in the French colony of Saint Domingue, but written shortly after the revolution that brought the richest colony in the world to a sticky end, the North American writer Eleanor Sansay conjured up the white Creole woman's luxuries and pleasures:

> I should repose beneath the shade of orange groves; walk on carpets of rose leaves and frenchipone; be fanned to sleep by silent slaves, or have my feet tickled into extacy by the soft hand of a female attendant. Such were the pleasures of the Creole ladies whose time was divided between the bath, the table, the toilette and the lover. What a delightful existence! Thus to pass away life in the arms of voluptuous indolence; to wander over flowery fields of unfading verdure, or through forests of majestic palm-trees, sit by a fountain bursting from a savage rock frequented only by the cooing dove, and indulge in these enchanting solitudes all the reveries of an exalted imagination.
>
> (Sansay 1808: 25)

The luxuriant 'hot-house' of tropical nature, with its 'unfading verdure', casts a moist shadow into this vision of Creole debauchery and self-indulgent luxuriance, a kind of corrupting sink of sin, lethargy and despotism, figured through feminine corruption. In sum, it was through the intertwining tendrils of an Edenic nature, the exercise of mastery, and a proximity to enslaved others that European and North American writers explored the risks and desires of being in the Caribbean. These elements of the imperial gaze have subtly informed the ways in which later tourists came to gaze upon the landscape and experience bodily the pleasures of Caribbean travel.

Entering tropical paradise

In contrast to the eighteenth-century depictions of the Caribbean as a highly cultivated landscape, with a multitude of plantations and horticultural development, in the nineteenth century an emerging artistic appreciation of the sublime allowed for a reinvention of tropical nature in a more romantic genre, particularly in post-emancipation contexts. With the rise of Romanticism, European feeling for certain kinds of landscape began to turn towards a taste for places that appeared to be wild, untamed and untouched by man. A visualisation of primitivism and primal nature increasingly characterise textual constructions of the Caribbean, though this was by no means unique to the Caribbean. As Duncan and Gregory suggest,

> Romanticism marked a post-Enlightenment remapping of the space of representation: it dethroned the sovereignty of Reason and glorified unconstrained impulse, individual expression and the creative spirit. . . . Central to romantic travel was a passion for the wildness of nature, cultural difference, and the desire to be immersed in local colour.
>
> (Duncan and Gregory 1999: 6)

With the abolition of slavery and the decline of the sugar plantations in many of the old colonies, a far more romantic vision of wild scenery and adventure through primitive places began to inform European travel writing.

The English novelist Charles Kingsley popularised this style of romantic tropical scenery in books such as *At Last: A Christmas in the West Indies* (1873 [1871]). Kingsley imagines the first discoverers of the West Indies, amongst 'such a climate, such a soil, such vegetation, such fruits' believing themselves 'to have burst into Fairy-land – to be at the gates of the Earthly Paradise' (Kingsley 1873: 26–7). Here engravings such as 'A Tropic Beach' and 'The High Woods' codified a genre of tropical drawing of remote places thick with trees, festooned with epiphytes, lianas, palms and ferns (Kingsley 1873: 143). In the play of light and dark, foreground and distance, his images often contain a small human figure engaged in an adventure activity such as stalking or fishing, and a wild animal such as a tortoise or monkey. The 'armchair tourist' is invited to join in the heavily illustrated adventures, which found a general readership in outlets like the *Illustrated London News*, thus Kingsley's texts are cited by many later tourists and travel writers as having been influential in their viewing of the Caribbean. Tourism became a mode of moving through tropical landscapes and of experiencing bodily what was already known imaginatively through literature and art.

It is crucial to recognise, however, that forms of viewing 'nature' are always ideological (Pratt 1992; Grove 1995; Poole 1998), and hence also encode relations of inequality between the viewer and the viewed. In the context of the post-emancipation decline of sugar plantations, the myth of the 'natural fertility' of tropical lands became a justification for European and US intervention in the Caribbean. Pulaski Hyatt, the United States Consul in Cuba, adopted the image of natural fertility in order to call for Cuba to be 'seized by an intelligent hand', for example, the United States' occupation of 1898:

> On account of her wonderful garden, fruit, and agricultural resources – impossible of appreciation unless seen – which to the visitor from the unwilling soil and freezing winters of the North seem like a rapturous dream, Cuba has been proudly styled 'The Pearl of the Antilles'. Only the most positive indolence and shiftlessness, and the long-applied withering hand of an oppressive government, have prevented Cuba from being, because of these resources alone, one of the most, if not the most, prolific and profitable spots in the world . . . no opportunity is more promising than this, if seized by an intelligent hand.
>
> (Hyatt and Hyatt 1898: 49–50)

This guide for prospective entrepreneurs posits 'the North' as a place of intelligence from which industrious Americans can more effectively take hold of the 'prolific' and 'profitable' nature of a dream-like Cuba. Such primeval forces and 'virgin' soil supposedly demanded American ingenuity (and capital) to take them in hand, a seizure justified by the island's 'shiftless' and 'indolent' inhabitants, incapable of exploiting its opportunities. Visions of nature, in short, are inseparable from moral hierarchies of civilisation and barbarism, modernity and backwardness.

Following in the footsteps of the explorers, the planters and the armed forces, new possibilities for tourism developed in the late nineteenth century, as fast steam shipping lines originally developed for the fruit trade significantly cut journey times. The ships that brought tourists to the West Indies usually spent less than a day in port, but arrangements could be made to spend a few days in one island, and then rejoin the cruise. The traditional inns and guesthouses of the main towns, with their bawdy reputation (many famed as brothels), were now joined by more 'respectable' large hotels built especially for the new tourist trade, such as the Myrtle Bank Hotel in Jamaica. The Caribbean voyage was promoted as a picturesque and healthful escape from winter weather. Now the literature of descriptive travel was joined by a new genre written specifically for the tourist market.

William Agnew Paton wrote an account of his 'Voyage to the Caribbees' by steamer from New York in 1888. Arriving in Antigua, he travelled along the roads of the island to see the scenery:

> At intervals along the road we passed darkies of every age, of both sexes, on their way to or from town, carrying baskets of fruits and

vegetables; we heard some of them singing, but as we approached they stepped aside to make way for us, and watched us in silence, always ready and delighted to return our greetings. Close to some of the negro shanties were little gardens planted with potatoes, yams, pea-bushes, arrowroot, and the like.

(Paton 1888: 66)

As in earlier travel literature the scenery is narrated in terms of the author's movement through it. His terminology reflects the everyday racism of the northern United States, while his idea of going 'down the islands' and viewing the friendly 'darkies' suggests a sense of proprietorship and being at home as he moves through this landscape. Contrary to the vast efforts that went into making these Caribbean landscapes hospitable to man, he envisioned Dominica as a natural Eden: 'Food is abundant, living is cheap, the island is not overcrowded; therefore the darkies have an easy time, as no one needs go hungry at any time of the year – no one, at least, who will walk into the woods, where are wild fruits and vegetables to be had at no more trouble to the would-be eater than to put forth his hand and pluck' (Paton 1888: 95). Thus the myth of 'easy' living in the Caribbean arose out of a notion of tropical fecundity, and was accompanied by assumptions that the 'natives' were available to serve tourists, offer friendly greetings, and become part of the scenery.

E.A. Hastings Jay wrote of his 'Four Months Cruising in the West Indies' in 1900 in terms of the scenery's relation to earlier literary representations of the tropics:

Here, for the first time, was the tropical beach! How often, from child-hood, I had tried to picture it from Kingsley's vivid descriptions or the histories of the early explorers. There were the cocoa-nut palms, with clusters of green cocoa-nuts growing all along the sea-line out of the soft white sand, with beautiful rainbow colours in the water as it moved lazily backwards and forwards, glittering in the brilliant sunlight.

(Hastings Jay 1900: 34–5)

For such travellers the entire archive of travel writing informed their experience of the Caribbean; 'real' experiences were always just a reflection of the vivid Caribbean in their imaginations. In Trinidad, Jay immersed himself in the storybook forests where he could enjoy an uninterrupted view of the natives from a safe distance:

masses of creepers cover even the tallest trees, climbing the trunks and spreading over the branches, then falling in festoons to the ground. ... Sitting on a rock at the side of the water, I gazed long upon the scene before me. Some coolies were bathing in a beautiful pool at

the bend of the river, their bronze colouring making a fine contrast
to the green of the forest behind them. . . . They chatter to each other
in a jargon which is unintelligible . . . there suddenly appears a little
band of boys and girls. Some of them have a scanty covering of cloth
or cotton, others are as nature made them. They all plunge in just as
they are into the cool water . . . I tore myself away at last, eager to
explore farther into the heart of the forest.

(Hastings Jay 1900: 84–6)

Here Kingsleyesque landscapes are experienced through viewing 'bronze'
bodies against the greenery and gazing upon scantily clad children playing
in water. The tourist's eager eyes seek out Caribbean bodies, and especially bared skin. The unintelligible 'jargon' allows him to feel a distance
from these woodland nymphs, who are naturalised as part of the scenery,
and the viewer imagines himself as an explorer discovering the 'heart' of
the tropical forest.

Susan de Forest Day, another visitor from New York, in 1897, had
similar motives of rediscovering a childhood fantasy, though sometimes
reality did not live up to expectations:

We longed to see the palm trees and sugar cane, to eat the luscious
fruits and to float over summer seas, basking in the warm tropical sun,
while the trade wind softly fanned our brows. . . . Without any real
reason we at first find St Thomas a trifle disappointing. . . . Perhaps
those bare, barren, rugged mountains, whose counterparts we had seen
time and again in our own everyday America, did not come up to the
ideal we had formed of the wealth and luxuriance of tropical vegetation – an ideal almost unconsciously derived from the old geographies
of our childish days in which the picture of a dense jungle, with
serpents gracefully festooned from tree to tree and a monkey in one
corner, always was the symbol of the torrid zone.

(Day 1899: 8, 28–9)

Thus earlier literary and visual representations of the Caribbean such as
Kingsley's deeply informed the tourist's desire to visit the Caribbean,
mapped in their 'unconscious' before ever setting foot there. It is this
fantasy 'torrid zone' that was unceasingly packaged and sold for northern
consumers. One early guidebook describes Jamaica as 'the New Riviera':
'now the mere mention of the island's name conjures up the vision of a
blue sky over a blue sea, in which is set a beautiful island of luxurious
vegetation and lovely scenery; fragrant with the odour of spices and
flowers, with an atmosphere refreshed by invigorating sea breezes'
(Johnston 1903: 10). Tourists thus entered the Caribbean with a set of pre-
conceptions about the natural scenery and expectations of bodily recreation
and pleasure.

Embodiments of the exotic

How, then, was a vision of nature turned into a set of expectations about Caribbean people and the services that could be received from them? In touring 'through' the islands, moving from one to another, the ideas of ease, luxury and relaxation were crucial. The tourist immersed his or her body in a tropical experience of sights, scents and tastes in which nature was understood to be more bountiful, more colourful, with more flowers, exotic fruits and leafy greenery. In these Edenic places where others laboured and living would be easy, tourists are encouraged to believe that they can engage guiltlessly in sensuous abandon and bodily pleasures. But just as in the earliest accounts of the Caribbean plantations, those who enter this tropical paradise put their own bodies at risk of moral corruption. Insofar as the familiar imageries of palm-fringed beaches or verdant exotic rainforests used in Caribbean tourism promotion conjure up a fantasy of 'pleasure islands', these fantasies of tropical nature are closely allied with transgression of moral codes through gluttony, intoxication and sexual encounters with exotic 'others', practices which can be collectively referred to as 'hedonism'.

With the development of the package holiday and the cruise-ship industry it became increasingly easy for tourists to get to the Caribbean and experience its charms in a pre-packaged form in which fantasy Caribbean landscapes are ineluctably linked to fantasies of encounters with Caribbean people.[4] Many of the earlier travel writers cited here, through their desires, rhetoric and imagery, shaped the itineraries, perceptions and 'performances' of the tourists who followed in their footsteps.[5] I argue that the 'naturalisation' of the social and economic inequalities of the contemporary tourist economy occurs via three steps: the objectification of Caribbean people as part of the natural landscape, the equation of that landscape (and hence those who people it) with sexuality and corruption, and finally the marketing of the Caribbean via imagined geographies of tropical enticement and sexual availability. In these Ur-tropics 'the inhabitants are identified with the natural world of luxuriant growth and fecundity' (Root 1998: 62), and the adventure of entering the tropics (for the white tourist) becomes a sexual adventure with 'dark' Others.

Again we can begin with Kingsley, whose work served as a travel guide for many who followed in his footsteps. He writes that the 'general coarseness' of the young 'Negro' women in Port-of-Spain, Trinidad, in the 1870s 'shocks' the stranger:

> It must be remembered that this is a seaport town; and one in which the licence usual in such places on both sides of the Atlantic is aggravated by the superabundant animal vigour and the perfect independence of the younger women. It is a painful subject. I shall touch it in these pages as seldom and as lightly as I can.
>
> (Kingsley 1873: 88)

By attributing the practice of prostitution to the 'animal vigour' of the 'masculine' 'Negro women' of the port city, he subtly links his romantic natural scenery of primal forests and wild shores to an implied primitivism of the inhabitants. Moreover, in skirting around this 'painful' (yet prurient) subject Kingsley effectively entices his readers to read on. His imagery of the Caribbean as a slightly risky and risqué 'Earthly Paradise' informs much subsequent writing on the touristic encounter.

Tourists have long objectified Caribbean men, women and children by viewing their bodies as part of the tropical landscape. In 1932, for example, John van Dyke wrote 'sketches' of Jamaica, in which the people form a crucial part of 'the picture':

> Seen along the Jamaica roads, under the broken sunlight filtered through palm and bamboo, the black is decidedly picturesque. He has the fine line and movement of an animal, the dark skin born of a tropical sun, and the female of the species comes in to help out the picture with the glow of bright clothing. Male and female after their kind they belong to the landscape as much as the waving palm or the flowering bougainvillea or the gay hibiscus. They are exotic, tropical, indigenous, and fit in the picture perfectly, keeping their place without the slightest note of discord.
>
> (Van Dyke 1932: 23–4)

Van Dyke's dehumanisation or animalisation of West Indians served a new American imperialism, which was grasping the 'renaturalised' Caribbean islands through both adventure tourism and military ventures.

Objectification of 'naturalised' black bodies became quite explicitly linked to sexual interest in them especially with the spread of sex tourism in the post-World War II period (cf. Enloe 1989 on the connections between bananas, beaches and bases). Travelling on a banana boat from Bristol to Dominica in 1952, for example, James Pope-Hennessy writes of his arrival in St Lucia that 'boys and girls stand about in the streets of Castries, drinking rum when they can afford it, and vaguely offering themselves or their young sisters, or little brothers, to the crews of passing ships' (Pope-Hennessy 1954: 21). The well-known Irish travel writer Patrick Leigh Fermor also animalises and sexualises the black male bodies seen on his travels:

> The young men are nearly all beautifully built, and look their best when they are working without their shirts, and displaying magnificent shoulders and torsos that taper down to flat stomachs and phenomenally narrow waists. Their bodies have the symmetry and perfection of machines. Muscles and joints melt smoothly into each other under skin which shines like a seal's or an otter's.
>
> (Fermor 1955 [1950]: 15)

It becomes clear in his account, and others, that some of these lithe bodies (including those of children) are available for sexual hire. Both Pope-Hennessy (1954) and Fermor describe the 'Saga-boys' (or 'Swagger-boys') of Trinidad as being involved in prostitution or pimping, linked to the prevalence of US naval ships in the region during the war. Here I want to focus on how this past informs the present.

By the 1950s, cities such as Kingston, Jamaica, and Havana, Cuba, enjoyed reputations as the epitomes of tropical decadence and pleasure. Havana became known as the 'brothel of the Caribbean' (Pattullo 1996: 90) and had up to 10,000 sex workers (Schwartz 1997: 122). This history contributed to the resurrection of sex tourism in Cuba in the 1990s (Rundle 2001), as Fidel Castro reopened the country to the international tourist market, and Havana was resurrected as a travel destination trading on the crumbling colonial past, the 1950s American cars, the macho aura of writer Ernest Hemingway, and the promise of sex. Despite the continuing US embargo against Cuba, the surreal overlay of an economy based on dollars onto a population still living on pesos has produced a marketing dream: the close yet unattainable, the illicit embrace of the taboo island. One typical recent book is illustrated with copious pictures of 'brown' women in skimpy cabaret costumes. The text easily slides from the *belle époque* days when Havana was 'the Paris of the Caribbean' to a tawdry invitation to enjoy Havana's twenty-first-century 'smells of tobacco, rum and lovely girls' (Lechthaler 1997: 75, 92).

The sexualisation of 'exotic' bodies has become a standard tool of Caribbean tourist promotion, and feeds into the development of sex tourism in the region (Clift and Carter 2000; Kempadoo 1999). As Jacqueline Sánchez Taylor has shown, sex tourism packages Caribbean people as 'embodied commodities' by turning the long history of sexual exploitation under colonial rule into a 'lived colonial fantasy' available for the mass tourist consumer. 'A key component of sex tourism', she argues, 'is the objectification of a sexualized, racialized "Other", available at a low price' (Sánchez Taylor 2000: 42). Thus both tourists and travel writers report unflinchingly on prostitution by disguising it as 'holiday romance'. Describing the Dominican Republic, for example, one journalist claims that 'the local economy and the tourists seem to coexist tolerably . . . [with] local girls keen to acquire a big-spending, sun-inflamed lover for a week' (Sweeney 1999: 1). The global economic inequalities, gender inequalities and racial order underlying such transactions are thus 'naturalised', as having a 'lover for a week' becomes part of the promise of the 'all-inclusive' experience (for women tourists as much as for men).

Furthermore, sexualised strategies for entering the extremely competitive international market for tourism are promoted by many Caribbean states, with suggestive marketing slogans such as 'It's better in the Bahamas'. As Jacqui Alexander argues, sexual legislation that polices 'erotic autonomy' mobilises the Bahamian population as 'loyal sexual citizens to service heterosexuality, tourism, and the nation simultaneously':

> The state institutionalization of economic viability through hetero-sexual sex for pleasure links important economic and psychic elements for both the imperial tourist (the invisible subject of colonial law) and for a presumably 'servile' population whom the state is bent on rena-tivizing. . . . The state actively socializes loyal heterosexual citizens into tourism, its primary strategy of economic modernization [,] by sexualizing them and positioning them as commodities.
>
> (Alexander 1997: 67–9, 90)

Thus a loyal citizenry advances the national economic agenda by appearing as cheerful natives, available and ready to service the needs of tourists. The state's orchestration of the pleasurable experience of being in the Bahamas reinforces the sexual hedonism of tourism by making it appear a 'natural' part of Caribbean culture.

For writers who 'continue the imperial narrative', Alexander argues, 'nature figures as raw material for American (European) creative expan-siveness, nature is positioned to collude in phantasmic representations of Black people, the very rhetorical strategies that state and private corpora-tions utilise to market Bahamians and the Bahamas to the rest of the world' (Alexander 1997: 93–4). In St Lucia, for example, the sexual nature of the Caribbean is written onto the land itself. Here the former owner of the 'luxury' island of Mustique, Colin Tennant, operates a bar known as 'Bang Between the Pitons', situated in the symbolic 'cleavage' between the two breast-shaped hills (Baker 2000: 15). Resorts in this area specialise in eco-tourism, emphasising the sensual pleasures of waterfalls, hot springs and rainforest immersion, but the beachfront hotels are not far away. Such developments, whatever their 'ecological' credentials, are nevertheless tapping into the currents of 'Edenism' that inform the sex-tourist economy.

Jamaica, too, is described as a 'pleasure island', though occupying an ambivalent place as both a site of enjoyable escapism and the dangerous terrain of drugs, criminals and violence (Aitkenhead 2000). As one British journalist describes it: 'The first time I went to Jamaica, I didn't know much about the place beyond a vague impression of pirates, palm trees, Noël Coward, ganja and beneath that a sense of intensity, a lurking volup-tuous danger' (Jenkins 2001: 6). It is this frisson of danger that adds to the allure of the place, and attracts those seeking extreme pleasures at resorts such as the infamous Hedonism II. Hedonistic practices of holiday abandon today serve to mark 'the islands' as places differing from the tourist's point of origin. The West Indies are inscribed as 'resorts' beyond civilisation, utopian/dystopian places where the normal rules of civility can be suspended. The deep layering and reiteration of such representations of the Caribbean has reinforced an imaginary geography in which it becomes a carnivalistic site for hedonistic consumption of illicit substances (raunchy dancing, sex with 'black' or 'mulatto' others, smoking ganja).

Hints of Oriental luxury, self-indulgence and corruption are also inscribed onto Caribbean landscapes and into the fantasies of tourists.[6] There is often an explicit 'orientalization' (Said 1991[1978]) of the Caribbean, indicated by Eastern imagery such as Richard Branson's Necker Island, described as a 'Balinese playground in the Caribbean . . . when you own a private island, history, geography and just about everything else are whatever you choose to make them' (Mallalieu 2001: 2). Or Mustique, often described as a 'pleasure island' for the rich and famous, where prospective visitors are instructed, 'Indulge yourself' and enjoy 'a taste for luxury' in the buildings designed by British theatre set designer Oliver Messel, 'built along Japanese lines or in the spirit of Bali, with names like Moongate, Obsidian and Sleeping Dragon' (Pietrasik 2001: 12). Such Orientalist fantasies of excess feed into unequal relations of power between tourists and workers in the service economy.

Expectations of the kinds of services the tourist can expect to receive from local people depend on locating the Caribbean as a 'natural Paradise'. One all-inclusive resort in St Lucia, for example, is described as 'a microcosm, a sanitised parallel universe where going abroad involves no stress because it involves no difficulty, where everyone's pleasant and speaks your language and whatever you desire will be yours' (Heathcote 1999: 4). An article in a British Airways magazine concludes that in Jamaica 'you'll be in the nearest thing we have on earth to the Garden of Eden, and to make it even better, it's after Eve tempted Adam with the apple'. In this new Eden tourists can indulge in the temptations of primal nature and hedonistic abandon to its associated sexual temptations without guilt; the laws and strictures of morality (and self-discipline) are temporarily suspended in this fantasy Jamaica, vested in Hedonism as much as Edenism. The transgression of moral boundaries serves to reinforce the constitution of geographies of difference that define the North as 'civilised' and the Caribbean as 'unreal', like 'going back in time'. These touristic performances reflect a long history of the inscription of corruption onto the landscapes and inhabitants of these 'pleasure islands'.

Notes

1 From itinerary of the Noble Caledonian Ltd 'West Indies: Hidden Treasures' 14-night cruise on the *Levant*, 8th to 23rd February, 2002, as advertised in *The Financial Times*.

2 This chapter draws in part on *Consuming the Caribbean: From Arawaks to Zombies* (Sheller 2003), research for which was funded by the British Academy, the Arts and Humanities Research Board, and the Faculty of Social Sciences, Lancaster University.

3 Such shifts in European perception and representation of landscapes have been explored more widely in relation to Northern travellers in Central and South

America (Manthorne 1989; Poole 1998), but have not yet been tracked in relation to the Caribbean. For a more detailed discussion see Sheller (2003: ch. 2).
4 For an interesting discussion of similar tropes of sexual encounter in travel
writers' depictions of the indigenous Carib people of Dominica see Hulme
(2000).
5 On tourism as a 'performative' practice see Macnaghten and Urry (2001) and
Sheller and Urry (forthcoming).
6 See Sheller (2003: ch. 4) for a more detailed discussion of Orientalism in the
Caribbean.

References

Aitkenhead, D. (2000) 'Pleasure island', *Guardian,* Travel Section, 30 November, pp.
2–3.
Alexander, M.J. (1997) 'Erotic autonomy as a politics of decolonization: an anatomy
of feminist and state practices in the Bahamas tourist economy', in M.J. Alexander
and C.T. Mohanty (eds) *Feminist Genealogies, Colonial Legacies, Democratic
Futures*, New York: Routledge.
Baker, L. (2000) 'Sleeping Beauty', *Guardian*, Travel Section, 6 May, pp. 14–15.
Charlevoix, P. (1766) *A Voyage to North America; undertaken by command of the
present King of France, Containing . . . A Description and Natural History of the
Islands of the West Indies belonging to the different powers of Europe*, 2 vols.,
Dublin: John Exshaw and James Potts.
Clift, S. and Carter, S. (eds) (2000) *Tourism and Sex: Culture, Commerce and
Coercion*, London and New York: Pinter.
Day, Susan de Forest (1899) *The Cruise of the Scythian in the West Indies*, London,
New York and Chicago: F. Tennyson Neely.
Duncan, J. and Gregory, D. (eds) (1999) *Writes of Passage: Reading Travel Writing*,
London and New York: Routledge.
Enloe, C. (1989) *Bananas, Beaches and Bases: Making Feminist Sense of
International Politics*, London: Pandora.
Fermor, P.L. (1955 [1950]) *The Traveller's Tree: A Journey Through the Caribbean
Islands*, Reprint, London: John Murray.
Grove, R. (1995) *Green Imperialism: Colonial Expansion, Tropical Island Edens
and the Origins of Environmentalism, 1600–1860*, Cambridge: Cambridge
University Press.
Hastings Jay, E.A. (1900) *A Glimpse of the Tropics, Or, Four Months Cruising in
the West Indies*, London: Sampson Low, Marston & Co.
Heathcote, E. (1999) 'Oh no, another day in paradise', *Independent on Sunday*,
Travel Section, 30 May, p. 4.
Hulme, P. (2000) *Remnants of Conquest: The Island Carib and their Visitors,
1877–1998*, Oxford and New York: Oxford University Press.
Hyatt, P. and Hyatt, J. (1898) *Cuba: Its Resources and Opportunities*, New York:
J.S. Ogilvie Publishing Co.
Jenkins, A. (2001) 'Lust for Life', *Guardian,* Travel Section, 17 February, p. 6.
Johnston, J. (1903) *Jamaica: . . . The New Riviera, A Pictorial Description of the
Island and its Attractions,* London: Cassell & Co.
Kempadoo, K. (ed.) (1999) *Sun, Sex and Gold: Tourism and Sex Work in the
Caribbean,* Lanham and Oxford: Rowman & Littlefield.

Kingsley, C. (1873 [1871]) *At Last: A Christmas in the West Indies*, London: Macmillan & Co.

Lechthaler, E. (1997) *Rum Drinks and Havanas: Cuba Classics*, New York: Abbeville Press.

McFarlane, A. (1994) *The British in the Americas, 1480–1815*, London and New York: Longman.

McKinnen, D. (1804) *A Tour Through the British West Indies in the Years 1802 and 1803, giving a particular account of the Bahama Islands*, London: J. White & R. Taylor.

Macnaghten, P. and Urry, J. (eds) (2001) *Bodies of Nature*, London: Sage.

Mallalieu, B. (2001) 'Island Special', *Guardian*, Travel Section, 6 January, p. 2.

Manthorne, K. (1989) *Tropical Renaissance: North American Artists Exploring Latin America, 1839–1879*, Washington, DC: Smithsonian Institution Press.

Paton, W.A. (1888) *Down the Islands: A Voyage to the Caribbees*, London: Kegan Paul, Trench & Co.

Pattullo, P. (1996) *Last Resorts: The Cost of Tourism in the Caribbean*, London: Cassell.

Pietrasik, A. (2001) 'Pleasure island', *Guardian*, Travel Section, 11 August, p. 12.

Poole, D. (1998) 'Landscape and the imperial subject: US images of the Andes, 1859–1930', in G. Joseph, C. Legrand and R. Salvatore (eds) *Close Encounters of Empire*, Durham, NC and London: Duke University Press.

Pope-Hennessy, J. (1954) *The Baths of Absalom: A Footnote to Froude*, London: Allan Wingate.

Pratt, M.L. (1992) *Imperial Eyes: Travel Writing and Transculturation*, London: Routledge.

Root, Deborah (1998) *Cannibal Culture: Art, Appropriation and the Commodification of Difference*, Boulder and Oxford: Westview Press.

Rundle, M.L. (2001) 'Tourism, social change and *Jineterismo* in contemporary Cuba', Paper presented at the 25th Annual Conference of the Society for Caribbean Studies, 2–4 July, University of Nottingham.

Said, E. (1991 [1978]) *Orientalism*, Harmondsworth: Penguin.

Sánchez Taylor, J. (2000) 'Tourism and "embodied" commodities: sex tourism in the Caribbean', in S. Clift and S. Carter (eds) *Tourism and Sex: Culture, Commerce and Coercion*, London and New York: Pinter.

Sansay, E. (1808) *Secret History, or the Horrors of St Domingo*, Philadelphia, n.p.

Schwartz, R. (1997) *Pleasure Island: Tourism and Temptation in Cuba*, Lincoln: University of Nebraska Press.

Sheller, M. (2003) *Consuming the Caribbean: From Arawaks to Zombies*, London and New York: Routledge.

Sheller, M. and Urry, J. (eds) (Routledge: forthcoming) *Global Places to Play*.

Smith, J. (1819 [1629]) 'The beginning and proceedings of the new plantation of St Christopher by Captain Warner', in *The True Travels, Adventures and Observations of Captain John Smith in Europe, Asia, Africke and America, beginning about the yeere 1593, and continued to this present 1629*, 2 vols., Richmond, VA: Franklin Press.

Stedman, J.G. (1988 [1790]) *Narrative of a Five Years' Expedition Against the Revolted Negroes of Surinam*, ed., introd. and notes by Richard and Sally Price, Baltimore and London: Johns Hopkins University Press.

Sweeney, P. (1999) 'The secret life of the Merengue Island', *Independent on Sunday*, Travel Section, 10 January, p. 1.

Urry, J. (1991) *The Tourist Gaze: Leisure and Travel in Contemporary Societies*, London: Sage.

Van Dyke, J. (1932) *In the West Indies: Sketches and Studies in Tropic Seas and Islands*, New York and London: Charles Scribner's Sons.

Young, R. (1995) *Colonial Desire: Hybridity in Theory, Culture and Race*, New York and London: Routledge.

3 Global perspectives of Caribbean tourism

Jerome L. McElroy

Introduction

Global forces have historically left indelible imprints on small countries. This has been especially true for island societies in the past half century. In the postwar era, their political economy has been largely defined by two worldwide forces: the juggernaut of decolonisation and the global spread of international tourism. In the first case, since 1960 over two dozen islands have achieved full political independence (Central Intelligence Agency 2002). As the terms of trade stagnated for colonial plantation staples, local elites who replaced metropolitan leadership used their new-found autonomy to create tax havens and other non-traditional activities like offshore finance and ship registry (Baldacchino and Milne 2000). Most enduring success was achieved by capitalising on their comparative advantage in recreational amenities. Like hand in glove, the rapid transformation of tourism into the world's largest industry – accounting for 12, 10, 9 and 8 per cent respectively of global exports, Gross Domestic Product, investment and employment (World Travel and Tourism Council [WTTC] 2002) – coincided with the restructuring of island economies away from traditional exports like sugar and copra towards mass tourism. These forces transformed the insular landscape across the Caribbean, Mediterranean and Northern Pacific into the pleasure periphery of North America, Europe and Japan respectively (Turner and Ash 1976).

In the Caribbean, a host of historical and contemporary forces have shaped the gestation of tourism. In contrast to remote Pacific, Indian Ocean and African islands, the geographic proximity of the Antilles to the dominant origin markets in North America (Europe to a lesser extent) has significantly enhanced their low-cost appeal. In addition, centuries of core–periphery commerce have facilitated the islands' access to foreign capital for hotel investment and to a steady supply of imported food, beverage and luxury gifts essential to satisfy affluent metropolitan vacation demand. Long-term political ties have also played a role and continue to do so since sixteen of the twenty-nine islands remain dependencies. The capital base of the tourism industry, the airports and docks and main island roadways, has largely been created by aid finance. More recently, in the late 1970s US policy-makers provided tax incentives for US corporations holding offshore conventions in the region. Moreover, the 1990 revision of the Caribbean Basin Initiative (CBI) increased the duty-free gift allowance of US citizens returning from the Caribbean from US$400 to US$600 and in the US territories (Puerto Rico and US Virgin Islands) from US$800 to US$1,200 (Coyle 1993). Finally, domestic economic policy has been influential. Island governments, habituated to a colonial tradition of high-volume, low-value-added monocultural exports, provided a receptive pro-market environment with generous tax incentives to establish the similar development of mass tourism. The confluence of all these forces with rising western affluence and the advent of paid vacations, package tours and jet technology combined to create the tropical island getaway as the North American vacation (or cruise) of choice.

Geography also played a part in the spread of tourism across the region. Because of proximity to the affluent US Atlantic seaboard, tourism historically evolved from north to south in four waves. In the late nineteenth and early twentieth centuries, a small stream of wealthy Americans established tourism in the Greater Antilles, primarily Cuba and Jamaica, and also Bermuda. The take-off into mass tourism took place in the 1950s and 1960s with the appearance of jet travel and the US embargo of Cuba. As a result, the Bahamas, Puerto Rico and the US Virgin Islands (USVI) in the north and Aruba and Barbados in the south became popular international resort destinations (Seward and Spinrad 1982). During the late 1970s and 1980s, a third wave of even more rapid growth washed across the archipelago. In the Greater Antilles, it included the Dominican Republic and the resurgence of Cuba as well as the British Virgin Islands (BVI) and the Cayman Islands. A final pulse engulfed the rest of the Lesser Antilles spreading south from the Leeward Islands to the Windward Islands, from St Maarten and Antigua through Guadeloupe, Martinique and St Lucia to Trinidad and Curaçao. The Turks and Caicos Islands represent the most recent mass tourism destination, achieving 100,000 stayover visitors by the late 1990s.

The cyclic pace of this postwar expansion was also affected by a variety of external and internal factors. According to Bell (1993), in the 1950s and 1960s tourism did not attract wholesale government support because it was mainly controlled by external investors using black labour to service affluent whites and appeared to be neo-colonialism in modern guise. It also 'attracted criticism from a vocal, newly independent intelligentsia, and from some international development agencies, because of sociocultural factors' (Holder 1993: 21). Growth slowed in the early 1970s because of volatility caused by the oil crisis and socialist experiments (e.g. Grenada, Guyana and Jamaica) in the region. Since then, however, the secular decline in agriculture, continuing instability in mining, the relatively high cost of manufacturing, and the failure of CARICOM (Caribbean Community) to meet enhanced trade and investment expectations have fostered a more favourable climate for tourism promotion and growth has surged.

At the threshold of the new millennium, the economic significance of tourism has intensified in a region buffeted by the damaging contours of globalisation. These include the export of manufacturing (textiles) jobs to Mexico through NAFTA (North American Free Trade Agreement), the on-going phasing out of banana production for export through the consolidation of the European Union, and a drastic falloff in US aid since 1990. Most recently they include the aftermath of terrorist attacks in the US, the slowdown in global growth and uncertainty from the threat of another war in the Middle East. In addition, dependence on tourism as the primary growth engine is occurring at a time when the non-sustainability of past mass tourism practice has become apparent. Condominiums, hotels and roadwork on steep hillsides have damaged forests and watersheds, causing erosion, silting over streams and wetlands and polluting lagoons (McElroy and de Albuquerque 1998). Mangrove forests and salt ponds have been destroyed by the placement of large-scale resorts, marinas and infrastructure along shorelines, depleting endemic species, archaeological artifacts and reef systems already weakened by sand mining, yacht anchoring and sewage dumping (Wilkinson 1989). In short, just as centuries of deforestation for sugar culture 'resulted in the environmental devastation of island interiors ... mass tourism itself has had a particularly noticeable direct impact on highly vulnerable coastal environments, such as beaches, coral reefs and mangrove forests' (Weaver 1998: 189).

The Caribbean in the world

Pressure to grow Caribbean tourism is also increasing in an era of intensifying global competition. According to Harrison (2001), the past decade has witnessed a decline in traditional European and North American leaders and the rise of Less Developed Country (LDC) destinations. Noteworthy are rising market shares for East Asia, the Pacific and Africa. Some of the most rapid growth is taking place in eastern European economies in transition and especially in south-east Asia. In the latter case, growth has been fuelled by a combination of four related factors (Hitchcock *et al.* 1993): (1) the increasing availability of attractive packages and deep discounting, (2) the shift away from more familiar Mediterranean and Aegean destinations towards the more exotic, (3) the rising importance of newly industrialised countries (Hong Kong, Singapore, South Korea, Taiwan) as origin markets, and (4) aggressive promotion by ASEAN (Association of South-East Asian Nations) countries (Indonesia, Malaysia, Philippines, Thailand).

Despite these trends, tourism in the insular Caribbean (excluding Bermuda) has out-performed the industry worldwide, and the region has increased market share. For example, between 1970 and 2000 the number of Caribbean stayover arrivals increased nearly five times from roughly 3.5 million to 17.2 million (Caribbean Tourism Organization [CTO] 2002;

US Department of Commerce 1993). Average compound growth rose 5.2 per cent per year versus 4.9 per cent for world arrivals over the same period. Annual growth was fastest during the 1980s (5.9 per cent) and slowest during the 1990s (4.4 per cent), perhaps a reflection both of tourism's maturity as well as of growing international competition. Over the three decades, the insular Caribbean's share of total stayover arrivals rose slightly from 2.23 per cent in 1970 to 2.46 per cent in 2000.

As a result of a generation of mass tourism development, the Caribbean has become the most tourist-penetrated region in the world. This is indicated by the Tourism Satellite Accounts (TSA) which aggregate direct and indirect effects of visitor spending to determine the industry's economywide impact. According to recent World Travel and Tourism Council Estimates (2001), tourism accounts for roughly 17 per cent of total Caribbean Gross Domestic Product (GDP) in contrast to the next most dependent regions: 12 per cent for North America, Europe and Oceania. Second, tourism accounts for over 21 per cent of all Caribbean capital formation while comparable figures for Oceania (13 per cent) and North America/Europe (10 per cent) are significantly lower. Third, regional tourism accounts for nearly 20 per cent of exports in contrast to 15 per cent for Oceania and 7–8 per cent for North America/Europe. Finally, tourism accounts for roughly 16 per cent of total Caribbean employment compared to 12 per cent or less for the other regions. In addition, these TSA figures are likely too low since they exclude six islands in the estimates: Bonaire, Montserrat, Saba, St Eustatius, St Maarten and the Turks and Caicos. The World Travel and Tourism Council forecasts that the Caribbean will continue to lead the world in economic importance over the next decade (WTTC 2002).

Tourism Penetration Index

A snapshot of the comparative performance of Caribbean destinations (and the region indirectly) in the context of worldwide tourism can be developed by applying the Tourism Penetration Index (TPI) to a subset of small islands across all major oceanic basins. This technique attempts to comprehensively measure tourism development in small islands (roughly one million population or less) with the use of three standard and easily accessible impact indicators: (1) per capita visitor spending to measure economic impact; (2) average daily visitor density [(stayovers × average stay + one-day excursionists) divided by the host population × 365] per 1,000 population to measure socio-cultural impact; and (3) hotel rooms per square kilometre to measure tourism's impact on the insular landscape and fragile ecology. The advantage of these indicators as constituted in the TPI is that they represent an integrated comprehensive measure of overall tourism development, but because of their highly aggregative nature, they remain rough approximations. As such the TPI cannot account for geographic or seasonal concentration of visitation nor capture a destination's tourism style or long-term experience with and/or adaptation to tourism (McElroy 2002).

To construct the index, standardised indices based on the three indicators were calculated by taking the value of each variable for each destination, subtracting the minimum of that value for the whole sample, and then dividing the result by the sample maximum minus the sample minimum according to the formula:

$$(X - X\text{min}) / (X\text{max} - X\text{min})$$

The overall TPI scores and destination rankings were then calculated as the unweighted average of the three standardised impact indices (McElroy and de Albuquerque 1998).

The interpretation of the results is based loosely on Butler's (1980) destination (or product) lifecycle model that argues that tourism is dynamic and successful destinations normally pass through a regular sequence of growth stages that parallel the S-shaped logistic curve. More specifically, the interpretation is informed by an abbreviated three-stage version of the lifecycle comprising low-density, intermediate and high-density levels or stages (de Albuquerque and McElroy 1992). The overriding assumption is that Butler's evolutionary theory provides a useful framework for understanding the general processes and patterns of tourism development (Hovinen 2002).

To operationalise the index, a sample of forty-seven small islands, with populations of roughly one million or less, and for which relatively complete data were available, was selected. To ensure uniformity, all data were taken from standard sources: the tourism data from the *Compendium of Tourism Statistics* (WTO 2002), and the population and area figures from *The World Factbook* (Central Intelligence Agency 2001). The resulting sample includes twenty-three islands in the Caribbean, fourteen in the Pacific, five in the Indian Ocean, two in the Atlantic (Bermuda and Cape Verde), two in the Mediterranean (Malta and Cyprus) and Bahrain in the Persian Gulf. Three have slightly more than one million inhabitants (Hawai'i, Mauritius and Trinidad), and all are less than 20,000 square kilometres in area except the Solomons.

Results

Table 3.1 presents the basic data and Table 3.2[1] records the three impact variables, their standardised indices and their combined TPI scores and destination rankings. Although highly aggregative and only a rudimentary first approximation, the TPI loosely ranks the forty-seven islands from most (St Maarten) to least (Solomons) penetrated. More important than individual destination rankings, however, are the island clusters revealed by the TPI and informed by a sense of the literature and historical observation. The sample divides roughly into three distinct levels of tourism

Table 3.1 Selected indicators of small islands[a]

	Land area (km²)	Population (thousands)	Tourists (thousands)	Day visitors (thousands)	Length of stay (nights)	Rooms	Total spending ($US million)
Anguilla	91	12	44	68	8.6	1,067	55
Antigua	440	67	232	357	7.0	3,185	290
Aruba	193	70	721	490	8.6	7,783	638
Bahamas	10,070	298	1,596	2,608	5.4	14,701	1,814
Bahrain	620	645	2,420	1,449	2.3	6,766	469
Barbados	430	275	545	533	10.1	6,456	711
Bermuda	50	64	328	210	6.2	3,339	431
Bonaire	311	12	51	43	9.3	1,050	87
British Virgin Islands	150	21	281	188	9.4	1,637	315
Cape Verde	4,030	405	83	–	7.0	2,391	33[b]
Cayman Islands	260	36	354	1,031	6.5	4,318	440
Comoros	2,170	596	24	–	7[b]	389	15
Cook Islands	240	21	73	–	11.0	783	36
Curaçao	544	147	191	309	8.5	2,768	227
Cyprus	9,240	763	2,686	313	11.0	43,363	1,894
Dominica	750	71	69	240	9.2	890	47
Fiji	18,270	844	294	10	8.5	5,283	195
French Polynesia	3,660	254	252	25	12.1	3,357	470[b]
Grenada	340	89	129	187	7.2	1,822	67
Guadeloupe	1,706	431	623	392	5.2	8,136	418
Guam	541	158	1,288	9	3.0[b]	10,110	1,908
Hawai'i[d]	16,760	1,212	6,738	–	8.7	71,480	11,133
Kiribati[d]	717	94	2	4	18.0	436	3
Maldives	300	311	467	–	8.4	8,329	344
Malta	320	395	1,216	171	8.4	20,344	614
Marshalls	181	71	5	–	5.6	305	4

Martinique	1,060	415	526	290	13.2	6,766	302
Mauritius	1,850	1,190	656	22	10.4	8,657	542
Montserrat[c]	100	8	10	–	7.0[b]	243	9
New Caledonia	18,575	205	110	50	16.0	2,401	110
Marianas	477	75	517	12	3.6	4,551	656[b]
Palau	458	19	58	–	6.0[f]	973	47[b]
Réunion	2,500	733	430	–	15.7	2,719	276
Saba[c]	13	2	9	15	7.0[b]	87	18[b]
St Eustatius	21	2	19[b]	6[b]	7.0[b]	63	17[b]
St Kitts	269	39	84	140	8.7	1,754	70
St Lucia	610	158	270	457	9.6	4,428	277
St Maarten	41	36	432	868	5.0	3,065	482
St Vincent	340	116	73	183	10.6	1,747	74
Samoa	2,850	179	88	–	7.6[e]	763	40
Seychelles	455	80	130	10	10.4	2,479	117[b]
Solomons[c]	27,540	480	21	–	13.0	860	7
Tonga	718	104	35	8	17.0[b]	642	7
Trinidad	5,130	1,170	399	82	10.0[b]	4,532	210
Turks and Caicos	430	18	152	–	7.5	2,023	285
US Virgin Islands	349	122	607	1,870	4.5	4,997	1,157
Vanuatu	14,760	193	58	48	7.6	1,060	58

Sources: Derived from World Tourism Organization (2002); Central Intelligence Agency (2001).

Notes
a Population data generally for 2001. Tourism data are for 2000 unless otherwise indicated.
b Author's estimate.
c 1999.
d 1998.
e 1997.
f 1996.

Table 3.2 Construction of the Tourism Penetration Index, destination scores and rankings

	Spend per population (US$)	Density (thousands)[a]	Rooms per km²	[Impact indices[b]]			TPI scores[c]
				Spending	Density	Rooms	
Most tourism developed							
St Maarten	13,389	230	74.8	0.8455	0.6223	1.0000	0.8226
British Virgin Islands	15,000	369	10.9	0.9473	1.0000	0.1457	0.6977
Aruba	9,114	262	40.3	0.5752	0.7092	0.5388	0.6077
Cayman Islands	12,222	254	21.0	0.7717	0.6875	0.2807	0.5760
Bermuda	6,734	96	66.8	0.4248	0.2582	0.8930	0.5253
Turks and Caicos	15,833	174	4.7	1.0000	0.4701	0.0628	0.5110
Guam	12,706	67	18.7	0.8023	0.1793	0.2500	0.4105
Malta	1,554	72	63.6	0.0973	0.1929	0.8503	0.3802
St Eustatius	8,500	190	3.0	0.5364	0.5136	0.0401	0.3633
US Virgin Islands	9,484	103	14.3	0.5986	0.2772	0.1912	0.3557
Hawai'i	9,186	133	4.3	0.5798	0.3587	0.0575	0.3320
Saba	9,000	107	6.7	0.5680	0.2880	0.0899	0.3153
Marianas	8,747	68	9.5	0.5520	0.1793	0.1270	0.2861
AVERAGE	10,113	156	26.1	0.6384	0.4410	0.3482	0.4756
Intermediate tourism developed							
Bonaire	7,250	118	3.4	0.4574	0.3179	0.0455	0.2736
Anguilla	4,583	102	11.7	0.2888	0.2745	0.1564	0.2399
Bahamas	6,087	103	1.5	0.3839	0.2772	0.0201	0.2271
Antigua	4,328	81	7.2	0.2727	0.2174	0.0963	0.1955
Maldives	1,106	35	27.8	0.0690	0.0924	0.3717	0.1777
Barbados	2,585	60	15.0	0.1625	0.1603	0.2005	0.1744
Cyprus	2,482	107	4.7	0.1560	0.2880	0.0628	0.1689
Cook Islands	1,714	105	3.3	0.1074	0.2826	0.0441	0.1447
St Kitts	1,795	61	6.5	0.1125	0.1630	0.0869	0.1208
St Lucia	1,753	53	7.3	0.1099	0.1413	0.0976	0.1163

Palau	2,474	50	2.1	0.1555	0.1332	0.0281	0.1056
Seychelles	1,463	47	5.5	0.0915	0.1250	0.0735	0.0967
Bahrain	727	30	10.9	0.0450	0.0788	0.1457	0.0900
Curaçao	1,544	36	5.1	0.0967	0.0951	0.0682	0.0867
Martinique	728	48	6.4	0.0451	0.1277	0.0856	0.0861
Polynesia	1,850	33	0.9	0.1160	0.0870	0.0120	0.0717
Grenada	752	34	5.4	0.0466	0.0897	0.0722	0.0695
Guadeloupe	968	23	4.8	0.0602	0.0598	0.0642	0.0614
St Vincent	638	23	5.1	0.0394	0.0598	0.0682	0.0558
Montserrat	1,125	24	2.4	0.0702	0.0625	0.0321	0.0549
Dominica	662	34	1.2	0.0409	0.0897	0.0160	0.0489
AVERAGE	2,220	57	6.6	0.1393	0.1544	0.0880	0.1269
Least tourism developed							
Mauritius	455	11	4.7	0.0278	0.0272	0.0628	0.0393
Réunion	376	25	1.1	0.0228	0.0652	0.0147	0.0342
New Caledonia	536	24	0.1	0.0329	0.0625	0.0013	0.0322
Tonga	67	16	0.9	0.0033	0.0408	0.0120	0.0187
Trinidad	180	10	0.9	0.0104	0.0245	0.0120	0.0156
Samoa	223	10	0.3	0.0131	0.0245	0.0040	0.0139
Fiji	231	8	0.3	0.0137	0.0190	0.0040	0.0122
Vanuatu	301	7	0.1	0.0181	0.0163	0.0013	0.0119
Marshalls	56	1	1.7	0.0026	0.0000	0.0227	0.0084
Cape Verde	81	4	0.6	0.0042	0.0082	0.0080	0.0068
Kiribati	32	1	0.6	0.0011	0.0000	0.0080	0.0030
Comoros	25	1	0.2	0.0010	0.0000	0.0027	0.0012
Solomons	15	2	0.0	0.0000	0.0027	0.0000	0.0009
AVERAGE	198	9	0.8	0.0116	0.0223	0.0118	0.0152

Sources: Derived from World Tourism Organization (2002); Central Intelligence Agency (2001).

Notes

a Calculated as: [(Tourists * Stay) + Day]/[(Population * 365)*1,000].
b Calculated as: (Indicator Value − Minimum)/(Maximum − Minimum).
c Unweighted Average of the three impact indices.

development (least, intermediate, most) that correspond broadly to Butler's lifecycle trajectory of increasing visitation, facility scale and environmental impact. The Caribbean dominates the most developed group comprising the top six (including Bermuda) in this sub-sample and 70 per cent of all most developed destinations. The Caribbean also comprises five of the top six in the intermediate group. This strong small-island international showing is noteworthy since four of the region's tourism leaders (Cuba, Dominican Republic, Jamaica and Puerto Rico), which account for over half of all visitor activity in the Caribbean, were excluded from the analysis because of their large size. On the other hand, the more remote and less developed South Pacific and Indian Ocean destinations dominate the least tourism-penetrated group.

With some exceptions, the most developed islands define a large slice of the pleasure periphery and include thirteen mature destinations with an average per capita visitor spending over $10,000 and an average daily visitor density approaching 160 tourists per 1,000 residents. Visitors on average represent the rough equivalent of a 16 per cent increase in the daily population. Their landscapes are crowded with over twenty-five rooms per square kilometre. This group includes three clusters: (1) five traditional resort islands comprising Bermuda, British Virgins (BVI), Hawai'i, Malta and US Virgins (USVI); (2) six more recent mass tourism destinations of Aruba, Caymans, Guam, Marianas, St Maarten and Turks/Caicos; and (3) two small Dutch Antilles, Saba and St Eustatius, known for their dive tourism. Although the TPI does capture the heavily built character of Bermuda, Malta and St Maarten with over sixty rooms per square kilometre, as a land-based indicator it overstates the tourism impact in the BVI and the two tiny Dutch Antilles that cater principally to marine tourism. These latter two should more appropriately be considered intermediate destinations.

As a group, many of these most developed destinations share a relatively unique profile. According to the literature (McElroy and de Albuquerque 1992) and recent Caribbean Tourism Organization data (2002), these mature, affluent areas, excluding Saba and St Eustatius, are characterised by relatively slow population, visitor and room growth rates. Their market is by and large dominated by shorter-staying visitors (6.6 nights on average) with strong preference for hotels, large-scale (100+ rooms) facilities and man-made attractions. They also exhibit the highest levels of hotel occupancy, promotional spending, overseas marketing offices and employment and (for the Caribbean) cruise passenger traffic. As an indicator of their integration into the global tourist economy, they tend to display the lowest degree of seasonality through special year-round packages (e.g. honeymoon weekends, conventions, carnivals, regattas). They also tend to exhibit a relatively high degree of man-made attractions (e.g. casinos, golf courses, conventioneering). Many of these established destinations are among the most frequently cited in the literature for tourism-induced ecosystem damage, marine pollution, overcrowding, host tensions and declining vacation quality (Jenner and Smith 1992; Beller *et al.* 1990; Towle 1985).

In addition to their climate and natural assets, these popular resort areas share geographic proximity to their major respective origin markets and long-standing and continuing commercial relationships. Historically, many functioned as colonial nodes in centre–periphery trade, and these links have fostered the inflow of postwar hotel and other investment. Moreover, all except Malta are political dependencies or states (e.g. Hawai'i) with the benefit of enduring legal and fiscal ties. In several cases this has nourished access to aid-financed transport and communications infrastructure and other advantages conducive to tourism growth. These include for the US Territories (Guam, Marianas, USVI), for example, ease of travel for American visitors (no passports, same language and currency) and special tax and duty-free gift/liquor purchase allowances not available to their politically independent island neighbours.

In contrast to the high-density group, the least developed destinations contain thirteen primarily Pacific, Indian and African islands at the early stage of the resort cycle. As a group they average less than $200 in per capita visitor spending, one hotel per square kilometre, and contribute only a 1 per cent increment to the daily population. Many are considerably larger in population and area than the mature destinations and possess more abundant agricultural and mineral resources and more diversified economies. They cluster into three heterogeneous sub-groups. At the bottom are several low-income remote Pacific (Marshalls, Kiribati, Solomons, Vanuatu) and African (Cape Verde, Comoros) outposts with limited tourism development and characterised by heavy dependence on emigration, remittances, aid and public employment, i.e. the contours of so-called MIRAB (Migration, Remittances, Aid and Bureaucracy) societies (Bertram and Watters 1986). Towards the top are four destinations with highly developed tourist regions masked by the TPI: Fiji, Mauritius, New Caledonia and Réunion. In the middle are islands advancing along the cycle: Trinidad experiencing 10 per cent stayover growth since the mid-1990s (CTO 2002) and Samoa achieving increasing international recognition as an ecotourism destination.

Many least developed islands exhibit characteristics of Butler's early tourism stages: small-scale facilities and infrastructures, limited visitor and room growth, and less-penetrated ecosystems and cultures than the mature islands. They tend to spend least on promotion and have the lowest share of hotel rooms in large facilities. They also have the highest ratio of regional inter-island visitors and enjoy the longest average length of visitor stay (eleven nights), i.e. up to two weeks in New Caledonia, Kiribati, Réunion, Solomons and Tonga. Their slower progress up the resort cycle stems from a variety of factors. These include geographic isolation and limited infrastructure which have inhibited the expansion of direct air access to metropolitan origin markets. In contrast to the Caribbean, many also possess a less extensive colonial and commercial history with the West which has slowed their socio-economic modernisation and attractiveness for foreign investment. Because of their diversified economies, policy-makers have felt somewhat less pressure to aggressively promote mass tourism. Finally, in some cases (e.g. Comoros, Fiji), political instability has hindered development.

The twenty-one intermediate destinations are the most dynamic and heterogeneous. Two-thirds are Caribbean islands. As a group their average TPI scores fall cleanly between the most and least developed. For example, average per capita visitor spending is approximately $2,200, average daily density is fifty-seven visitors per 1,000 residents, and average rooms per square kilometre is seven. Their average length of visitor stay is 8.6 nights. In many cases, these islands are characterised by very rapid visitor growth and hotel and infrastructure construction. In contrast to the most developed destinations, they tend to have higher rates of seasonality and lower levels of promotional spending. In terms of lifecycle progression, all have considerable tourism experience, display increasing facility scale and socio-environmental impact, and tend to differ in size, economic structure and tourism style. Clustered at the top are five developed Caribbean destinations plus Maldives and Cyprus. Two TPI anomalies are apparent. Bonaire is likely over-ranked because of its emphasis on dive tourism. On the other hand, Bahamas is undoubtedly under-ranked and should more accurately be considered a mature destination because of its high-density concentration of activity in the Freeport–Nassau area. However, its intermediate TPI score results from the archipelago's large land area and the lack of development in the out islands.

In the middle are a large number of diversified island economies with smaller tourism sectors in various stages of transition from dependence on traditional exports. In the Caribbean, there are St Lucia and Grenada (bananas) plus St Kitts/Nevis (sugar), Curaçao (petroleum) and the French dependencies of Guadeloupe and Martinique, replacing sugar with banana exports to France. In the Pacific, there are French Polynesia (farming), the MIRAB economy of Cook Islands, heavily dependent on remittances and subsidies from New Zealand, and Palau, a TOURAB (Tourism and Bureaucracy: see Apostolopoulos and Gayle 2002) international dive destination in transition from subsistence production and public employment. In the Indian Ocean, there are Maldives (fishing) and Seychelles (tuna). At the bottom are two recent graduates from low-density status, Dominica and St Vincent and the Grenadines, and Montserrat. The former have become increasingly recognised as ecotourism destinations with Dominica promoting itself as the 'Native Island of the Caribbean' and encouraging 'small-scale facilities, local ownership, and an emphasis on environmental, historical and cultural attractions' (Weaver 2001: 168). The TPI ranking for Montserrat, a popular North American retirement resort, has declined since 1995 when devastating eruptions of the Soufriere Hills volcano rendered over half the island uninhabitable and spawned widespread emigration.

One of the major weaknesses of the TPI is its propensity to underestimate tourism impacts in larger islands. In the Caribbean case, for example, this excludes all of the Greater Antilles which account for the lion's share of tourism: Cuba, Dominican Republic, Jamaica and Puerto Rico. Together in 2000 these four (plus Haiti) accounted for over 50 per cent of the 16.9 million stayovers to the insular Caribbean (excluding Bermuda), over 50 per cent of the $16.4 billion in visitor spending, and nearly 60 per cent of the 207,400 rooms (World Tourism Organization 2002). However, were they included in the present TPI analysis, they would be positioned in the early low-density stage of the lifecycle with per capita spending between $150 (Cuba) and $625 (Puerto Rico), daily densities between five (Cuba) and fifteen (Jamaica) visitors per 1,000 population, and an average of one room per square kilometre. All of these destinations, however, have intensely developed tourist zones the aggregative all-island TPI masks: Veradero and Cayo Coco in Cuba, the north Puerto Plata coast in the Dominican Republic, Montego Bay and Ocho Rios in Jamaica, and the Condado area in Puerto Rico.

To remedy this TPI shortcoming, Padilla and McElroy (2003) developed regional indicators for the Dominican Republic, a good test case because it represents one of the fastest-growing destinations in the Caribbean and presently accounts for over 17 per cent of all insular stayovers and tourist expenditures. The Dominican Republic also maintains the largest facility infrastructure containing 25 per cent of all insular rooms available. Provisional results indicate sharply different regional stages of tourism development. For example, Punta Cana on the east coast resembles islands at the top end of the lifecycle with average spending of $8,000 per capita and daily visitor densities of over 200 per 1,000 population. The Puerto Plata/Semana region on the north coast resembles intermediate destinations with per capita spending of $2,300 and sixty daily visitors while the two south-east regions (Santo Domingo and La Romana/San Pedro) are similar to least penetrated islands. Analogous results are expected for the other three dominant destinations when regional data become available.

New directions

To maintain its position in the global tourist economy, the Caribbean will have to devise alternative strategies to traditional mass tourism promotion that emphasise quality over quantity. This is particularly important in a region characterised by relatively high travel costs and room rates, high labour and utility costs, and low profits and income multipliers (Tewarie 2002). As a first approximation, the TPI can function as an early warning signal indirectly identifying threats to sustainability along the three major stages of the resort cycle. For mature destinations, the key challenge is to sustain vacation quality. This will require restoring damage and preventing encroachment in fragile areas, better managing visitation through space and time, and expanding average visitor stay and quality by designing nature, culture and heritage attractions. Bermuda's experience is instructive. Stagnation and complaints of saturation in the 1980s prompted a long-term policy reassessment that resulted in controls on bed capacity, cruise ships and construction design. Since then tourism has stabilised and natural and heritage amenities have been protected (McElroy 2001).

For highly developed intermediate destinations, the key issue is controlling growth within insular socio-economic and ecosystem absorptive capacities. This will require containing further renewable resource losses, sequencing large developments over long time horizons, and providing residents not only with decision-making participation but also with a viable financial stake in the industry. This last can be achieved by using incentives to foster small-scale, local, labour-intensive businesses, local purchases by hotels/restaurants and developers, and targeting regional tourists whose spending per dollar has greater domestic impact (Sinclair and Vokes 1993). A 10 per cent increase in the tourism income multiplier maintains economic contribution, with roughly 10 per cent fewer visitors and socio-environmental impact. Here the experience of Antigua is useful. Between 1975 and 1990, total visitors doubled twice and more coastal ecosystems and habitats were damaged than in all previous history (de Albuquerque and McElroy 1995). Since 1990 stayover visitors have been volatile and/or flat. In a reverse of the Bermuda case, Antigua's failure has been largely due to a tradition of environmental neglect, marginal citizen participation, and a persistent policy preference favouring short-term economic gains over long-term environmental stability.

For less-developed, intermediate and low-density destinations, the key challenge is achieving commercial viability. Ideally this demands a long-term participatory planning process to identify unique assets, to construct infrastructure necessary for sustainability accessing these assets, and to devise a compatible destination identity to promote. This is a time-consuming task but such islands have room to manoeuvre because of their early position in the resort cycle and their less exploited natural and cultural amenities. They also have the greatest opportunity for developing ecotourism attractions. Some of the more successful examples include the Hol Chan Marine Reserve and Chao Creek Terrestrial Reserve on the Belizean mainland, the renowned marine parks in Bonaire and Saba, the expansion of heritage sites in St Lucia and Cayman Brac, the USVI National Park in St John. Clearly the best example of a deliberate ecotourism destination is Dominica, promoting a combination of its riverine ecology, volcanic assets and Carib culture (see also Weaver, Chapter 10). However, managers must remain constantly vigilant lest success breed pressure for unsustainable visitation at such delicate sites. According to France and Wheeler (1995: 68), they

> should not be seduced by the myth of green tourism, which may be little more than a growth policy masquerading behind the pseudo

respectability of a green mantle. The critical factor in the 'success' of any ecotourism development is effective control of numbers.

Other challenges

There are a number of other external and internal challenges the Caribbean must confront to retain its premier position in global tourism. A brief sampling of the more serious deserves mention. First, in the wake of 11 September and the employment losses and eroded profits from deep discounting, the region must ensure adequate transport security infrastructure is in place to avoid further pull-outs like the discontinuation of two British cruise ships to Trinidad and Tobago (*Jamaica Gleaner* 2003). Second and related, 2002 was one of the worst years in airline history as evidenced by bankruptcy threats from mainland carriers and substantial losses suffered by major regional carriers like Air Jamaica, BWIA and LIAT. Since air transport is the bone marrow of the long-haul tourism typifying the Caribbean, concerted efforts must be made to explore long-term partnerships with both mainland and island airlines to consolidate service and stabilise visitor access to the region.

Third, although their overall economic impact is considerably lower than stayovers, cruise arrivals represent the fastest growing segment of the industry and will soon rival the hotel sector in bed/berth capacity. Because of the security, comfort, high satisfaction and vacation variety they offer, cruise ships have become very competitive with land-based tourism. To maximise their contribution, better partnerships must be established between cruise lines and island governments to reduce the impact of their sewage and other debris on the delicate marine environment (Ocean Conservancy 2002). In addition, their regional impact can be enhanced with greater on-land purchases from island suppliers and more cooperative 'fly–cruise' packages that promote overnight hotel stays as is presently under way in Bermuda, Jamaica and elsewhere. Fourth, for their part, regional decision-makers not only need to appreciate the increasing significance of tourism in their policy-making, but also to recognise the long-term value of cooperatively developing uniform region-wide promotion, incentive and regulatory regimes to reduce destination competition with foreign (e.g. air, cruise, hotel) suppliers. The recent mounting of the first major Caribbean marketing campaign ('Life Needs the Caribbean') is a promising step (Harewood 2002).

Finally, given the importance of visitor safety, crime is perhaps the most systemic internal threat to Caribbean tourism. Periodic reports of cruise ship cancellations because of robberies against passengers and crew underline the problem (Schladen 2002). Although the nexus between crime and tourism is becoming clearer – lucrative targets lacking precaution in unfamiliar surroundings less likely to report and/or return for trial (de Albuquerque and McElroy 1999a) – the relationship is complicated by the increasing significance of the region as a primary narco-traffic route for cocaine transshipment from Colombia to the US (Drug Enforcement Administration 2001). Although violent crime is highest in many of the most mature destinations (de Albuquerque and McElroy 1999b), much is linked to turf wars among rival posses (gangs) as well as the deportation of Caribbean criminals from the metropolis: 'they leave our islands as high school criminals and return to us as post graduates' (Commonwealth Secretariat 1997: 105). Whatever the source of rising criminality, in the intermediate run Caribbean leaders must redouble efforts to strengthen enforcement with both local and extra-regional resources, and in the longer run address the more daunting task of creating a more egalitarian tourism in which all social strata share a viable economic stake and the motivation to sustain it.

Note

1   The data presented in these tables may differ somewhat from Duval (Chapter 1) due to the utilisation of different sources. The reader is reminded that there are multiple sources of travel and tourism data pertaining to the Caribbean.

References

Apostolopoulos, Y. and Gayle, D.J. (2002) 'From MIRAB to TOURAB? Searching for sustainable development in the maritime Caribbean, Pacific, and Mediterranean', in Y. Apostolopoulos and D.J. Gayle (eds) *Island Tourism and Sustainable Development: Caribbean, Pacific, and Mediterranean Examples*, Westport: Praeger.

Baldacchino, G. and Milne, D. (2000) *Lessons from the Political Economy of Small Islands: The Resourcefulness of Jurisdiction*, New York: St Martin's Press.

Bell, J.H. (1993) 'Caribbean tourism in the year 2000', in D.J. Gayle and J.N. Goodrich (eds) *Tourism Marketing and Management in the Caribbean*, London: Routledge.

Beller, W., d'Ayala, P. and Hein, P. (1990) *Sustainable Development and Environmental Management of Small Islands*, Paris: UNESCO-Parthenon.

Bertram, G. and Watters, R.F. (1986) 'The MIRAB process: earlier analysis in context', *Pacific Viewpoint* 27: 47–59.

Butler, R.W. (1980) 'The concept of a tourist area cycle of evolution: implications for management of resources', *Canadian Geographer* 24: 5–12.

Caribbean Tourism Organization (CTO) (2002) *Caribbean Tourism Statistical Report (2000–2001 Edition)*, St Michael, Barbados: Caribbean Tourism Organization.

Central Intelligence Agency (2002, 2001) *The World Factbook*, Washington, DC: CIA. Online: //www.odci.gov/cia/publications/factbook/index.html> (accessed 22 January 2003).

Commonwealth Secretariat (1997) *A Future for Small States: Overcoming Vulnerability*, London: Commonwealth Secretariat.

Coyle, M.D. (1993) 'US trade policy and tourism development in the Caribbean', in D.J. Gayle and J.N. Goodrich (eds) *Tourism Marketing and Management in the Caribbean*, London: Routledge.

de Albuquerque, K. and McElroy, J.L. (1992) 'Caribbean small-island tourism styles and sustainable strategies', *Environmental Management* 16: 619–32.

de Albuquerque, K. and McElroy, J.L. (1995) 'Antigua and Barbuda: A Legacy of Environmental Degradation, Policy Failure and Coastal Decline', Washington, DC: Supplementary Paper No. 5, USAID, EPAT/MUCIA.

de Albuquerque, K. and McElroy, J.L. (1999a) 'Tourism and crime in the Caribbean', *Annals of Tourism Research* 26: 968–81.

de Albuquerque, K. and McElroy, J.L. (1999b) 'A longitudinal study of serious crime in the Caribbean', *Caribbean Journal of Criminology and Social Psychology* 4: 32–70.

Drug Enforcement Administration (2001) 'The drug trade in the Caribbean: a threat assessment', Washington, DC: US Department of Justice, Intelligence Division.

France, L. and Wheeler, B. (1995) 'Sustainable tourism in the Caribbean', in D. Barker and D.F.M. McGregor (eds) *Environment and Development in the Caribbean: Geographical Perspectives*, Jamaica: University of West Indies Press.

Harewood, C. (2002) 'Trust selling region as one-talk tourism', *Barbados Daily Nation,* 12 September. Online: http//www.nationnews.com/StoryView.cfm? Record=29024&Section=Business&Current=2002%2D09%2D12%2000%3A00% 3A00 (accessed 12 September 2002).

Harrison, D. (2001) 'Less developed countries and tourism: the overall pattern', in D. Harrison (ed.) *Tourism and the Less Developed World: Issues and Case Studies*, Wallingford, UK: CABI Publishing.

Hitchcock, M., King, V.T. and Parnwell, M.J.G. (1993) *Tourism in South-East Asia*, London: Routledge.

Holder, J.S. (1993) 'Caribbean Tourism Organization in historical perspective', in D.J. Gayle and J.N. Goodrich (eds) *Tourism Marketing and Management in the Caribbean*, London: Routledge.

Hovinen, G. (2002) 'Revisiting the destination lifecycle model', *Annals of Tourism Research* 29: 209–30.

Jamaica Gleaner (2003) 'Cruise ships pull out of T & T', 21 January. Online: <http: //www.jamaica-gleaner.com> (accessed 21 January 2003).

Jenner, P. and Smith, C. (1992) *The Tourism Industry and the Environment*, London: Economist Intelligence Unit, Special Report No. 2453.

McElroy, J.L. (2001) 'Island tourism: a development strategy for biodiversity', *Insula* 10: 21–2.

—— (2002) 'The impact of tourism in small islands: a global comparison', in F. di Castri and V. Balaji (eds) *Tourism, Biodiversity and Information*, Leiden: Backhuys Publishers.

McElroy, J.L. and Albuquerque, K. (1992) 'An integrated sustainable ecotourism for small Caribbean islands', Bloomington: Indiana University, Center for Global Change and World Peace, Occasional Paper No. 8, Series on Environment and Development.

McElroy, J.L. and de Albuquerque, K. (1998) 'Tourism penetration index in small Caribbean islands', *Annals of Tourism Research* 25: 145–68.

Ocean Conservancy (2002) 'Cruise control: a hard look at the cruise industry and its impacts on the marine environment', *(ENN) Environmental News Network*, 29 May. Online: <wysiwg: //55/http://www.enn.com/direct(display-release.asp? id=7007> (accessed 29 May 2002).

Padilla, A. and McElroy, J.L. (2003) 'The TPI in larger Caribbean islands: the case of the Dominican Republic', paper presented at the 28th Caribbean Studies Conference, Belize City, Belize: 26–31 May.

Schladen, M. (2002) 'Cruise line to visit DR on dates of canceled St Croix calls', *The Virgin Islands Daily News*, 17 December. Online: <http://www.dynatek.com/wadabagei/index2.htm> (accessed 17 December 2002).

Seward, S.B. and Spinrad, B.K. (1982) *Tourism in the Caribbean*, Ottawa: International Development Research Centre.

Sinclair, M.T. and Vokes, R.W.A. (1993) 'The economics of tourism in Asia and the Pacific', in M. Hitchcock, V. King and M. Parnwell (eds) *Tourism in South-East Asia*, London: Routledge.

Tewarie, B. (2002) 'The development of a sustainable tourism sector in the Caribbean', in Y. Apostolopoulos and D.J. Gayle (eds) *Island Tourism and Sustainable Development: Caribbean, Pacific and Mediterranean Experiences*, London: Praeger.

Towle, E. (1985) 'The island microcosm', in J.R. Clark (ed.) *Coastal Resources Management: Development Case Studies*, Washington, DC: National Park Service.

Turner, L. and Ash, J. (1976) *The Golden Hordes: International Tourism and the Pleasure Periphery*, New York: St Martin's Press.

US Department of Commerce (1993) *Tourism in the Caribbean Basin*, Washington, DC: US Travel and Tourism Administration.

Weaver, D.B. (1998) *Ecotourism in the Less Developed World*, Wallingford, UK: CABI Publishing.

—— (2001) 'Mass tourism and alternative tourism in the Caribbean', in D. Harrison (ed.) *Tourism and the Less Developed World: Issues and Case Studies*, Wallingford, UK: CABI Publishing.

Wilkinson, P. (1989) 'Strategies for tourism development in island microstates', *Annals of Tourism Research* 16: 153–77.

World Travel and Tourism Council (WTTC) (2002, 2001) *The Travel and Tourism Economy 2002: Special End of Year Update*, London: World Travel and Tourism Council. Online: <http://www.wttc.org> (accessed 23 January 2003).

World Tourism Organization (WTO) (2002) *Compendium Tourism Statistics, 2002 Edition*, Madrid: World Tourism Organization.

Part II

Tourism development in the Caribbean

4 Tourism development in the Caribbean

Meaning and influences

David Timothy Duval and
Paul F. Wilkinson

Introduction

The purpose of this chapter is to provide a broad overview of tourism development in the Caribbean. As the region is comprised of small island states (or island microstates (Wilkinson 1989)), the nature of tourism in such environments is discussed with the goal of considering the natural, social and economic constraints facing island environments worldwide. The context of tourism development in the Caribbean is presented based on three broad themes: the meaning of tourism development through the lens of both neo-colonialism and dependency theories; tourism in small islands; and factors in Caribbean tourism development, particularly in terms of sustainable development.

Tourism has come to dominate much of the economy in the Caribbean. As Mather and Todd (1993: 11) noted, 'There is probably no other region in the world in which tourism as a source of income employment, hard currency earnings and economic growth has greater importance than in the Caribbean.' Tourism in the Caribbean has effectively grown in parallel to worldwide patterns. Using data from the Caribbean Tourism Organization (CTO) (which therefore includes destinations such as Cancun and Cozumel in Mexico), there was a 383 per cent increase in stayover arrivals in the Caribbean from 4.2 million in 1970 to 20.3 million in 2000, compared to a 337 per cent increase worldwide from 159.7 million to 698.8 million in the same period (CTO 2002: 23). Similarly, there was a 465 per cent increase in tourist receipts in the Caribbean from US$3.5 billion in 1980 to US$19.8 billion in 2000, compared to a 352 per cent increase world-wide from US$105.3 billion to US$476 billion in the same period (CTO 2002: 26).

The role of tourism as a potential economic force in the region was first voiced strongly by the Anglo-American Caribbean Commission in 1945 in its examination of potential post-war development alternatives. The debate was later fuelled by the Tripartite Economic Survey (Great Britain Ministry of Overseas Development 1966), financed by the governments of Canada, United Kingdom and United States, which argued that tourism and light industry were the types of labour-intensive industries on which the small islands of the Caribbean could base their economies. This view was supported by the Zinder (1969) report, which had a great deal of influence on international aid agency policies. Zinder, however, was subsequently severely criticised by many authors because of what were considered to be overly optimistic cost–benefit ratios and economic multipliers (e.g. Bryden 1973; Levitt and Gulati 1970) and overly-low leakage rates (e.g. Britton 1977; Perez 1973–4). Such criticisms have had a substantial impact upon opinion throughout the development literature through the extension of their conclusions to other islands. There are few reliable and consistent studies of these issues, but it appears that (a) a large proportion of tourist revenues (even after multinational corporations take out their costs and profits) is lost through the purchase of foreign products (e.g. liquor, food, hotel equipment) and through profits being exported; and (b) the multiplier effect on the rest of the economy is not as high as is often predicted or reported (Wilkinson 1989: 164).

The growth of Caribbean tourism is largely related to the presence in a very small area of a range of tourism destinations as interesting and as diverse as any other region in the world, complemented by a favourable climate, areas of exceptional beauty, and diverse cultures:

> Amid this beauty, and despite its glamorous image as one of the world's primary playgrounds for the (relatively) rich, however, it is easy to overlook the fact that many of the Caribbean territories are poor, over-populated, underdeveloped countries in which population growth generally exceeds the rate of growth in employment. There are some islands in which children still grow up poor and hungry and, at its worst (in countries such as Haiti and in some of the rural parts of the small Windward Islands in particular), the region presents development problems every bit as intractable as those in Africa. Because of the small scale, however, many of these problems remain hidden and the casual visitor may get little or no impression of the difficulties faced by many of the region's inhabitants.
>
> (Mather and Todd 1993: 9)

Such poverty is not, however, universal in the Caribbean. Some islands rank among the wealthiest societies worldwide (e.g. Cayman Islands' GDP per capita of US$18,215 in 2000 (CTO 2002: 279, 289)), while others (e.g. Anguilla at US$5,044) have in recent years risen from poverty to increasing prosperity. For both of these examples, the main source of that prosperity has been tourism. Nevertheless, while not faced with the dire problems confronting Haiti (approximately US$34), other states remain quite poor (e.g. Jamaica: US$2,414; Dominica: US$2,196). They too, however, are seeing tourism becoming increasingly more important in their economies.

In fact, a common characteristic of most Caribbean states – rich or poor – is an important and growing tourism sector, with its impacts becoming a major factor in not only the economic environment, but also the social and biophysical environments. Moreover, as traditional economic activities (e.g. plantation agriculture, fishing, mining) wane, it seems likely that tourism and other service activities (e.g. offshore banking and insurance) will become even more predominant in the region.

Tourism is not a recent phenomenon in the Caribbean, but it was only with the advent of mass tourism in the late 1950s and of inexpensive jet air travel in the early 1960s that tourism swiftly and dramatically became a major economic activity in the region. In the 1960s, the great optimism for the potential of tourism led to the rapid expansion in (largely foreign) investment in tourism facilities, generous financial incentives by national governments (many of whom were newly independent of their 'Northern' colonial masters or looking towards imminent independence) wanting to attract foreign investment, and government provision of infrastructure (financed in large part by foreign aid). Growth in visitor arrivals and expenditures – which, although dangerously simplistic, are the most frequently-used tourism statistics worldwide – continued into the 1970s. The first energy 'crisis' of 1973, worldwide recession, and some strong nationalistic expressions (which many tourists perceived to have racial overtones) led, however, to a temporary decline in the middle part of the decade. As Bell (1993: 222) aptly notes in his assessment of the early 1970s situation, 'Tourism stalled.'

After 1975, however, there was renewed optimism about the vitality of the sector, as a result of such factors as improved economic conditions in North America, better marketing and promotion, the opening of new markets (e.g. Europe, favoured by a softening in currency exchange rates (Bell 1993)), and increasingly successful promotion of summer (low-season) vacations. The second energy 'crisis' of 1979 caused another temporary slump, but the 1980s ushered in a period of overall growth for the region (contrasted by serious problems for particular destinations) – halted only briefly by global uncertainty related to the Gulf War of 1991 and more recently to the events of 11 September 2001. (The latter resulted in a decline of stayover visitors to the Caribbean in 2001 of 4 per cent over the previous year (Caribbean Tourism Organization 2002: 13).)

The meaning of development

At the outset, it is important to distinguish what is meant by the term 'development'. In many respects the term is commonly seen as synonymous with planning for wider social 'betterment' or 'progress', which can often be problematic in the context of tourism (Sharpley 2002). However, development as it is used here is meant to convey a more situational approach. In this sense, it is meant to portray the activities surrounding the existing economic function of, in this case, tourism development.

The meaning of 'development' remains either obscure or contested in much of the tourism literature (Wilkinson 1997: 26). The main point of contention is that of distinguishing economic growth from development (Gayle and Goodrich 1993). Lewis (1955) states clearly that economic growth implies the constant creation of enhanced capacity to produce wealth. Measured in terms of GDP levels and rates of increase, such growth is generated by product and service innovation, employment, export expansion, and increased investment. Uphoff and Ilchman (1972), in contrast, argue that development emphasises productivity and distribution, or the creation of optimal capacity to challenge human abilities, as well as to satisfy human needs and desires, over time.

Building on the latter conception, a Lasswellian (1958) political economy view of development is adopted here, following Smith's (1977: 207) argument for development as 'welfare improvement . . . [which] means a better state of affairs, with respect to who gets what where'. In that sense, therefore, the major goals of development are sustenance of life, self-esteem and freedom – each broadly defined (Goulet 1993). Similarly, writers such as Seers (1979) and Bromley and Bromley (1982) adopt a set of socially and politically oriented criteria of development: improvement, modernisation, increasing welfare and the enhancement of the quality of life.

When the mainstream development literature deals with tourism, it tends to focus on the relationships between tourism and development in terms of the 'impacts of tourism' (Pearce 1989). Pearce's contention remains true: the development literature generally ignores tourism, except for criticising it as a tool of continued colonialisation. Erisman (1983: 339) argues that such harsh comments against tourism are symptomatic of the hostility which, among others, many Caribbean nationalists may ascribe towards the externally controlled tourism sector. To this end, Perez (1975: 1) states that

> The travel industry in the Caribbean may very well represent the latest development in the historical evolution of the neocolonial context of the West Indian socio-economic experience. Through tourism, developed metropolitan centers . . . in collaboration with West Indian elites, have delivered the Caribbean archipelago to another regimen of monoculture.

Heavily influenced by Latin American *dependencia* theorists (e.g. Cardosa 1972; Frank 1981, 1972), these critics have been mainly concerned with tourism's economic ramifications (particularly increased dependency on the United States). Dependence is 'a situation in which a certain group of countries have their economy conditioned by the development and expansion of another economy to which the former is subject' (Dos Santos 1972: 71–2). In the tourism literature, the most commonly adopted form of dependency theory is that of the plantation model (e.g. Bianchi 2002; Weaver 1988; see also Hall 1994), which uses an analogy of tourism being similar in effect to monocrop export agriculture. Bianchi (2002: 270) argues that a wider pattern of land-use associated with tourism very much replicates previous systems of a 'colonial space-economy'.

While studies of tourism development worldwide have examined the tangible benefits and negative impacts, very few have addressed the approach to tourism from a political economy perspective (cf. Bianchi 2002). A benefit of a political economy approach is the focusing of attention on the formative political, economic, social and institutional processes which influence societal organisation (Potter and Binns 1988: 279). Such an approach goes beyond the usual concerns of tourism research to include a consideration of institutions, notably those related to tourism policy and planning:

> the character and performance of national institutions are specific to each country's economic history, social composition, and class structure. These factors are only on the level of appearances of underlying, less observable social mechanisms. It is important to go beyond these appearances. The political economy of tourism encompasses the ways in which the industry manifests the division of labour, class relations, ideological content, and social distributions specific to a social formation.
>
> (Britton 1987: 171)

In effect, the political economy approach attempts to go beyond the 'developmentalist paradigm' (Browlett 1980) which has been the focus of much tourism research; in that paradigm, 'Researchers interpreted their findings through the lens of apolitical, ahistorical social-science perspectives' (Britton 1987: 185). In that sense, this approach is influenced by underdevelopment theory and neo-Marxian social science which call for consideration of the underlying political, ideological, economic and class dynamics that structure small developing countries – thus requiring an historical approach to study the evolution of those dynamics.

Tourism in small islands

While it is not entirely clear why tourists choose island environments as holiday destinations (Harrison 2001), Baum (1997) suggests that, for some tourists, the lure of such environments is indeed special and rather unique in comparison to other forms or types of destinations. The means of transport for travelling to and from an island destination, particularly the requirement of travel by plane or boat, represents one aspect that is, for Baum (1997: 21), equally matched by 'the feeling of separateness, of being cut off from the mainland'. Royle (2001: 193) argues that there exists a romantic perception of island vacations: 'Where better for a tourist to get away from it all physically as well as spiritually than to go to a separate piece of land?' For Harrison (2001: 9), islands are 'warm and sensuous' and offer 'stressed-out visitors the much-needed opportunity to relax, escape, recharge their batteries, and generally appreciate a way of life that, sadly, has been lost in the too-busy commercial environment of the globalizing, postindustrial Western world'. What is interesting is that Lowenthal (1992: 18) remarked earlier that there was a time when islands were

> held up as ideal laboratories of nature and human culture. Boundedness and isolation made them more self-contained than continental realms; their unique cultures, like their endemic species, were supposedly protected from undue contamination by the outside world. It is now hard to imagine how anyone could have believed this.

The popularity of island destinations worldwide leads to consideration of a number of related problems stemming from this success. The first is the nature of islandness itself and the subsequent economic implications, what Ryan (2001) has remarked as being a 'case for marginalities' that this brings. For example, Wilkinson (1997: 27–8) summarises Connell's (1988) inventory of various constraints that hinder wider economic development in small island states:

- no advantages from economies of scale (which are reduced further by fragmentation)
- a limited range of resources
- a narrowly specialised economy, based historically primarily on agricultural commodities
- small, open economies with minimal ability to influence terms of trade or to manage and control their own economies
- limited ability to adjust to changes in the international economic environment
- dependence for key services on external institutions such as universities, regional training facilities, banking institutions
- a narrow range of local skills and problems of matching local skills and jobs (often exacerbated by a brain and skill drain)
- a small GDP (hence problems of establishing import-substitution industries) yet alongside considerable overseas economic investment in key sectors of the economy, and especially commerce
- high transport, infrastructure and administration costs
- cultural domination by metropolitan countries
- geographical constraints (e.g. soils, climate)
- vulnerability to natural hazards
- vulnerability to externally influenced illegal activities.

However, the degree to which islands are more vulnerable than mainland states is not always clear. One view holds that islands are especially vulnerable because of limited natural resources that often, although not always, preclude the degree of economic diversification and successes that other nations enjoy. In other words, a limited natural resource base is quite often a function of restricted size (Stabler and Goodall 1996). Another view, however, is that islands are no more vulnerable than bounded nations. Harrison (1996) suggests that, to juxtapose islands and mainland states on the basis of geography, political status and population density is not necessarily enough to conclude that islands are any more susceptible to environmental degradation, either from tourism or other intrusive industries.

Tourism is often seen as a panacea for solving development problems in island states (McElroy and de Albuquerque 1996; Wilkinson 1989). However, with the development of tourism in such spatially restricted environments comes a variety of problems. Socially, tourism development – especially the mass variety – can result in changes to social structures among residents. This is particularly the case with island tourism development in developing countries or where overall economic development has hitherto been somewhat depressed. Problems of 'socio-cultural spillovers include casino gambling, cheap commercialisation of folklore and historical attractions, rising levels of crime and the erosion of local social values in tandem with the admiration of foreign lifestyles' (Lockhart 1997: 4). Of course, such problems are not limited to islands, but it is the insularity and geographic spatial concentration of both tourism development and existing settlement that can often amplify the interaction between tourist and local.

Significant consideration is often given to the role(s) of regulation and management in addressing limits of capacity. In other words, with the high degree of environmental sensitivity that is often attributed to island environments, the question of such an environment's physical ability to withstand a set number of visitors often comes into question. However, as Johnson and Thomas (1996) rightly argue, problematic in such approaches is how capacity is ultimately measured. Environmental overrun is just as much a function of limitations through political legislation or the enforcement of guidelines governing, for example, solid waste management as it is the number of visitors staying on a particular stretch of coast on a particular island (Johnson and Thomas 1996; see also Conlin 1995; Edwards 1996).

Broadly speaking, and from an economic perspective, the move towards the creation or expansion of tourism within many island states can often merely shift the burden of dependence on larger metropoles from one form of production (e.g. agriculture) to another (e.g. experiences). Although external support in the form of aid and grants for agricultural production in island states still occurs, many have chosen a route through which tourism may ultimately off-set such dependence. Such a move is hardly surprising, especially given that tourism can often contribute well over half of the total GDP for smaller island states. For example, in 2000, visitor expenditure as a percentage of GDP was 63.8 per cent for St Lucia, 63.4 per cent for Antigua and Barbuda, and 56.7 per cent for the United States Virgin Islands (CTO 2002: 261). This is augmented in those situations where islands have few, if any, economic alternatives. In fact, where natural resources and the overall potential for economic diversification are limited, tourism can become critical for the survival of local economies (Apostolopoulos and Gayle 2002a; Harrison 2001).

Compounded upon these issues of economic costs and benefits is the fact that resident population levels do not necessarily exhibit stability. Island states that rely on institutionalised economic linkages with dominant nations have been referred to as MIRAB societies (Migration, Remittances, Aid and Bureaucracy) (Bertram and Watters 1985), which are characterised by substantial out-migration to the larger parent state, supportive development schemes and investment projects, large bureaucracies, and often substantial transfer payments. In an effort to diversify the economic base of such dependent states, many are turning to tourism. For Guthunz and von Krosigk (1996), MIRAB is becoming TouRAB (see also Apostolopoulos and Gayle 2002a). Unfortunately, the development of tourism may not always reverse the wider trends of out-migration.

Factors in Caribbean tourism development

Given the considerations discussed above with reference to tourism in island environments, particularly the nature of economic dependence, environmental sustainability, and the degree to which governments have embarked upon often aggressive development strategies for local tourism sectors, of interest here is how such issues have been realised in the Caribbean. On the whole, the nature of tourism development in the Caribbean seems to be characterised by a delicate balance between internal interests and priorities, the economic realities facing many island states, and the external, often uncontrollable, nature of tourism worldwide.

Understanding the nature of tourism development can hardly be done through the examination of visitor arrivals. The distinct physiographic differences in the islands mask stark differences in the nature and type of tourism on offer. Some island states have uniquely different tourism products. Dominica, for example, has successfully positioned itself as a nature tourism destination, although chronologically the extent of tourism development there has been somewhat recent. In contrast, Barbados and Antigua have engaged in tourism development initiative longer and offer more conventional mass forms (e.g. beach holiday resorts) of tourism.

Elements influencing the rate and form of tourism development in the region include the nature of tourism development and the concept of sustainable tourism development. Each of these is discussed below.

The nature of tourism development in the Caribbean

The nature of tourism development necessarily involves the interaction and interplay between actors, stakeholders, tourists and the wider economic and political systems. Tourism in the Caribbean has been perpetuated by an uneasy relationship with local populations and governments. On the one hand, the benefits of tourism have long been used to offset declines in agricultural production and the shrinking or disappearance of overseas markets, while on the other hand it has spurred suggestions that tourism is a new form of slavery and dependency.

A recent discussion draft put forward by the Caribbean Group for Cooperation in Economic Development within the World Bank suggests that, with the exception of Cuba, 'most Caribbean countries have not articulated targets for tourism development' (Bonnick 2000: 47; see also Hinch 1990). Further, Bonnick (2000: 49) notes that while past efforts of tourism development were primarily limited to the provision of taxes for investment purposes and general marketing activities, in the future, government will be required to:

- act more dynamically to attract private investors
- assist in assembling land and other resources
- involve the people in planning and development
- identify and programme complementary infrastructure and utility developments
- establish appropriate zoning laws and ensure adequate land titling and management.

The role of tourism in the economic development of the Caribbean seems certain: the West Indian Royal Commission in 1945 recommended tourism as a way to 'create employment, diversify the economy, promote infrastructural development and help maintain the balance of payments of the colonies' (Dann 1996: 118). It has only been during the past few decades that some of the more significant complications arising as a result of conventional mass tourism development strategies have been truly recognised (Bell 1993; Butler 1980; Davies 1996).

Many islands, such as the Dominican Republic and the Bahamas, have traditionally (perhaps even consciously) engaged in conventional mass tourism sectors by offering little choice in activities to guests (largely as a result of limited attractions or services), thus leaving the sector particularly vulnerable to severe and potentially damaging market fluctuations (based on such factors as seasonality and market targets) and even collapse.

Wilkinson (1989) suggests that an increase in tourism dependency often occurs with an increase in the modernisation of existing tourism sectors (e.g. airport improvements in the eastern Caribbean Commonwealth islands such as Barbados and St Lucia in the 1970s and 1980s financed by Canadian aid programmes). Following this, it is not uncommon for the local economies of developing nations to be heavily dependent on both the flow of revenue generated by tourism sectors and the revenue generated by foreign currency exchange from tourist dollars. Unfortunately, this arrangement is reminiscent of the systems of economic development that have been established in the region since colonisation.

As noted above, the argument can certainly be forwarded that tourism functions as an extension of dependency theory (Pattullo 1996) and more or less operates within a neo-colonial atmosphere. Baldacchino (1996: 161) has stressed that many developing nations are unable to control external tourism services such as markets, tour operators and airlines. Thus, this contribution generates a substantial loss in real revenue generation. The increasing popularity of the Caribbean as a tourism destination is testimony to this, a fact conceded by Pattullo (1996), who indicates that such relationships are quite common throughout the Caribbean. The central question, therefore, should not only involve the identification of options that are decidedly different from the conventional mass tourism sector model, but should also encompass issues with respect to economic, social, environmental and cultural sustainability in an effort to highlight alternative forms of economic development. In this sense, alternative tourism, as a concept, seeks to counter the claim that tourism in general highlights 'existing inequalities, economic problems and social tensions' (Britton 1989: 93).

Sustainable development, tourism and island environments

In the academic literature, a considerable amount of attention has been devoted to the sustainable development of tourism in island environments (see e.g. Apostolopoulos and Gayle 2002b; Briguglio *et al.* 1996a, 1996b; Conlin and Baum 1995; Lockhart and Drakakis-Smith 1997). Because sustainability as a concept is built on the respectful management and use of resources in order to ensure continued management and use, as applied to tourism it is often coordinated in response to its negative impacts. The constricted environments within which tourism operates in island environments leads to considerations of, for example, waste management, harmful outdoor recreational activities (McElroy and de Albuquerque 2002) and, in some cases, the social angle of sustainability such as the perceived negative connotations of, for example, the commoditisation of culture (Ayres 2002). However measured, the reality is that tourism in island states offers the potential for unbalanced environmental relationships. McElroy and de Albuquerque (1991) identify five aspects of such imbalances:

•   the actual size of the wider tourism sector which, in island environments, often means interactions with constricted, environmentally sensitive areas;

- by extension, the degree and size of the tourism sector in such environments (which can lead to substantial burdening of the local infrastructure);
- resource use, the consequential strain on such resources, fluctuates with the seasonal nature of visitor arrivals and their use of the environment;
- policies which often (though not always) tend to favour an outright increase in visitor arrivals as opposed to those which favour increased expenditures, minimise leakages and/or enhancing multipliers; and
- the fact that the carrying capacity of island states is often, in some localities, well above that which the physical and social environment can endure.

There is, of course, another meaning for development, in the sense of *tourism development*, a sector-specific term which might be narrowly defined as the provision or enhancement of facilities and services to meet the needs of tourists (Pearce 1989). Pearce argues that, in a broader sense, tourism can also be seen as a means of development, the path to achieve some end state or condition. In this light, the impacts of tourism are re-examined in terms of how these impacts contribute to national, regional or community development and how the state of development in any area depends on the way tourism there has developed. In these terms, therefore, development can be considered as both a process and a state.

In general, the literature agrees with Smith's (1983) contention that there is never a single, right answer about the desirable level of tourism development or the capacity of a destination region to host tourists. Rather, limits and capacities are best obtained by policy reviews and evaluation while development is allowed to proceed slowly and incrementally: 'Only through cautious, deliberate evaluation of on-going development and policy applications can intelligent decisions be made regarding what is best for host and guest' (Smith 1983: 182). Such a conclusion leads into the current debate on blending tourism with sustainable development in the Brundtland sense (World Commission on Environment and Development (WCED) 1987).

The concept of sustainability is so vague that, as Butcher (2003: 27) notes, 'the term can be moulded to fit one's preference' or 'can be used in a variety of circumstances by a variety of people to convey a variety of meanings'. Given the lack of a widely accepted definition of 'sustainable tourism', the Brundtland (WCED 1987) definition of sustainable development is presented as a beginning point: development that meets the needs of the present generation without compromising the ability of future generations to meet their own needs. The concept has been the subject of much debate, but it is not the purpose here to enter that debate.

Butler (1993: 29) provides a 'working' definition of sustainable development in the context of tourism:

tourism which is developed and maintained in an area (community, environment) in such a manner and at such a scale that it remains viable over an indefinite period and does not degrade or alter the environment (human and physical) in which it exists to such a degree that it prohibits the successful development and wellbeing of other activities and processes.

Butler (1993: 29) argues, however, that this is not the same as 'sustainable tourism', which is 'tourism which is in a form which can maintain its viability in an area for an indefinite period of time'. Butler states that the notion of tourism being an evolutionary process is widely held; similarly, it is also broadly recognised that tourism is extremely dynamic and that the processes and impacts associated with it are susceptible to change. This element of change is a crucial factor, particularly when tourism is being considered in the context of sustainable development because sustainable development implies some degree of stability and permanence, at least in the very long-term view. Butler (1993: 30), therefore, argues that this does not blend well with a highly dynamic and constantly changing phenomenon such as tourism. For this reason, the view taken here is that sustainable development in terms of tourism is an heuristic concept (i.e. an organising concept or a 'guiding fiction'), rather than an operational one (Wilkinson 1997: 31).

Concern over the sustainability of the tourism sector in the Caribbean has grown over the past several decades. Much of this concern coincides with the rapid scale of development and successes attributed to tourism in the region. What is interesting is that the level and scale of this concern, however, is not merely limited to the conservation and protection of those natural resources which are central to many visitor experiences. On a more macro level, there is the recognition of the need for the sector to be economically sustainable. While governments and region-wide organisations such as the Caribbean Tourism Organization (CTO) naturally want to ensure that the environmental sustainability of the region is encouraged, they should be forgiven for equally wanting to see the region prosper from visitor expenditures. In fact, this reflects one of the paradoxes of the sustainability debate within tourism. Indeed, measuring the success of the sector is often more than just numbers. According to the Prime Minister of the Bahamas, Perry Christie, 'That is "ego-tourism" which, by and large, has been the hallmark of tourism in our region for years. That form of tourism has more to do with how our efforts make us feel as opposed to how they make our constituents and our visitors feel' (*Guyana Chronicle Online*, 13 November 2002).

However characterised, sustainability can be examined from three perspectives: environmental, social/cultural and economic. While these are separated here for the sake of convenience, they are in effect quite closely interrelated. As well, an emphasis on one aspect does not preclude a positive or even negative impact on another. For example, by focusing on the preservation of cultural traits and social values, such elements may have at their core the preservation of the natural environment.

Environmental sustainability

There is widespread recognition in the Caribbean that environmental conservation is critical to the future success of tourism, largely because the very attractions visited by tourists are often non-renewable and are, in effect, much of the reason why tourists choose particular destinations. When they are renewable, the intrusion of external elements (e.g. the effects of tourism activities) often throws the equilibrium of such resources into disarray. As such the management of resources becomes a critical issue (Cushnahan 2001; Henderson 2000).

One example of attempts to maintain environmental sustainability is that of marine protected areas (MPAs). For instance, Jamaica's National Environment Planning Agency (NEPA) is responsible for the management of the country's five MPAs. While NEPA has been restructured with more emphasis on monitoring, education and enforcement, three of the MPAs (Ocho Rios Marine Park, Palisadoes and Port Royal Protected Area and Portland Bight Protected Area) are rated in terms of level of management as having 'None', one (Montego Bay Marine Park) as 'Low-Moderate' and only Negril Marine Park as 'Moderate' (Wilkinson 2002: 294). As a result, NEPA is turning to non-governmental organisations (NGOs) to provide active management and fund-raising. For example, operation of the Negril Marine Park, which became a protected area in 1995, has recently been handed over to the Negril Coral Reef Preservation Society (NCRPS). NCRPS will be required to monitor impacts in the park, which could extend to visitor use through a user-fee system (*Jamaica Observer*, 27 October 2002).

In contrast, the Saba National Marine Park has long been deemed a case of effective MPA management. Established in 1987, it circles the entire island of Saba from the high-water mark to a depth of 60 metres (Wilkinson 2002: 334–5). It is managed by the Saba Conservation Foundation, a local NGO, using a multiple-use zoning plan with mooring buoys in place to prevent anchor damage from dive boats. Financially self-supporting through visitor fees, souvenir sales and donations, the park employs a marine park manager, an assistant manager and a visitor centre attendant.

Alternative tourism is often seen as a viable option in the face of 'insular mass tourism' (e.g. Davies 1996; Duval 1998; McElroy and de Albuquerque 1996). A key question, however, remains over the true alternative nature of such development. Heritage tourism puts fragile micro-environments, such as historic forts, at considerable risk if proper management frameworks are not established (see Found, Chapter 8). Cultural resources promoted and packaged for (alternative) tourist consumption equally put social and cultural structures under the lens of curious foreigners. Alternative forms of tourism can be alternative at almost any level. A shift in the target market for a particular product can be promoted as a shift towards alternative tourism through specific marketing. Low-rise construction of accommodation units in coastal environments may certainly be seen to be alternative in the sense that they are mindful of the visual pollution manifested elsewhere in island environments, but if sewage discharge remains untreated as it is sent into the Caribbean Sea, the true effect of 'alternative' remains in doubt.

Social/cultural sustainability

The wider context of social and cultural sustainability can be set as the broader environment within which hosts and guests share similar resources and space. The nature of social and cultural sustainability, then, is very much positioned as a process of negotiation (Tucker 2003). For Mowforth and Munt (1997: 106), social/cultural sustainability speaks to the ability of a community to almost withstand the presence of tourists, and to even continue to function without 'disharmony'. Such an argument is not without foundation as tourism has the potential to open previously non-existent social divisions (e.g. increasing differences between the beneficiaries of tourism and those who are marginalised by it) or to exacerbate already existing ones (e.g. the creation of spatial ghettoes, either for the tourists or those excluded from tourism) (Mowforth and Munt 1997: 107–9). Stonich *et al.* (1995) speak of several such divisions, including reduced access for local people to needed natural resources, escalating prices and increased outside ownership of local resources. In terms of cultures, aspects which could be changed as a result of the influence of tourists 'with different habits, styles, customs and means of exchange' include 'the relationships within that society, the mores of interaction, the styles of life, the customs and traditions' (Mowforth and Munt 1997: 109).

Britton (1977) recognised the damaging effects of industrialised tourism in his case-study of St Vincent, suggesting that resort-based development antagonises relations between host and guest. Boxill (2002) discusses the various layers of social impacts from tourism, particularly employment and housing, the latter of which can be characterised by luxurious hotels that are often proximal to local residents living in poverty. Rising crime rates in certain tourist areas, coupled with the increasing use of illicit drugs, is not uncommon (de Albuquerque and McElroy 1999). Most importantly, perhaps, and as noted by Boxill (2002), is the nature of accommodation provision in the region. With the advent of enclave-based tourism, where guests receive often exemplary hospitality, local accommodation providers can often hardly compete. As Pratt (2002) observes, the use of education to promote and engage in socially sustainable tourism development is critical.

There have been attempts in the Caribbean to alleviate some of these problems. For example, as early as 1988, Bermuda limited the number of cruise-ship arrivals following concerns of overcrowding; cruise ships are also banned on Sundays for religious reasons (Conlin 1995). Bermuda has also recently created the African Diaspora tour, which not only celebrates their history but also gives African-Americans – a major niche market still largely untapped by the Caribbean – another reason to visit (*Guyana Chronicle Online*, 13 November 2002). As well, Slinger's (2000) examination of the 'Carib' of Dominica draws attention to the indigenous form of cultural tourism (although Slinger positions it as ecotourism), and under-developed countries are increasingly pushing forward with tourism-development projects where the central product is the social relations, customs, traditions and life-styles of local residents. In fact, the line that separates 'hosts' from 'guests' becomes intentionally blurred as destinations offering authentic cultural experiences (read as cultural or ethnic tourism) look for more ways to encourage guests to become part of the local culture.

The Government of St Lucia recognised that tourism had come to dominate the national economy, but few of its benefits trickled down to the segments of the population with smaller annual incomes. As a result, it initiated the St Lucia Heritage Tourism Programme in 1998 with the following mission:

> To establish heritage tourism as a viable and sustainable component of St Lucia's tourism product by facilitating a process of education, capacity building, product development, marketing, credit access and the promotion of environmental and cultural protection for the benefit of host communities and St Lucians.
>
> (Renard 2001: 2)

The programme's objectives are as follows:

- To develop the island's tourism product, thus enriching the visitor experience through the provision of unique, authentic and natural/cultural visitor activities.
- To enhance St Lucia's image in the market place as a 'green' destination, with a unique blend of attractions, and types of accommodation.
- To diversify and decentralise the tourist product and benefits, resulting in integration of rural communities island-wide into the tourism industry, providing jobs, and a sense of participation in and ownership of the industry.
- To contribute to the sustainable management of the island's natural and cultural resources.

(Renard 2001: 2)

The Programme is involved in a variety of activities, including product development and management, marketing, capacity building, awareness and communication, and policy and programming. Renard (2001) provides a detailed analysis of the first phase of the Programme, including case studies concerning participatory planning in a village; creation of a marketing brand, 'Heritage Tours, Explore St Lucia', and authorisation for its use by tours around the country; locally based tourism in a village; and development of a heritage tourism site. The Programme appears to have been successful and its second phase began in 2002 (St Lucia Heritage Tourism Programme 2003).

Economic sustainability

The goal of economic sustainability is to ensure long-term returns on investments. In the context of tourism, local governments have a vested interest in maintaining visitor numbers in order to ensure a prosperous sector. Government revenues from taxation remain high (Table 4.1); despite the call for the reduction of taxation it can often lead to artificially inflated prices (Palmer 1993). The economic sustainability of a particular destination may be at odds with the environmental or even social dimensions of sustainability. For Mowforth and Munt (1997: 111), economic sustainability is

> a level of economic gain from the activity sufficient either to cover the cost of any special measures taken to cater for the tourist and to mitigate the effects of the tourist's presence or to offer an income appropriate to the inconvenience caused to the local community visited – without violating any of the other conditions – or both.

They go on to argue that economic sustainability should not compete with other aspects of sustainability, nor is it the only condition of sustainability; however, 'the question of who gains financially and who loses financially often sets the power and control issue in sharper and more immediate focus than all other facets of sustainability'.

Where tourism accounts for a significant proportion of a nation's GDP, which is often the case in the Caribbean (Table 4.1), maintaining an active tourism sector is often vital. Tewarie (2002) suggests that a variety of issues can affect the degree of economic sustainability of tourism in the Caribbean, including, for example, increasing competition (from both other destinations and cruise ships), low retention of revenue from tourism receipts due to high costs of imports, occupancy rates and other high costs associated with accommodations, and investment incentives that do not emphasise local building materials.

Table 4.1 Selected development indicators for Caribbean tourism, 2000 data

	Visitor expenditure (% of GDP)	Visitor expenditure per capita (US$)	Rooms per thousand population	Government revenue from hotel occupancy tax (US$ millions)
Anguilla	83	4,198	79	1.73
Antigua and Barbuda	63	4,054	44	8.15
Aruba	32	6,766	80	10.50
Bahamas	44	5,948	45	10.00
Barbados	33	2,659	24	0.20
Bermuda	20	6,885	53	n.a.
Bonaire	n.a.	4,133	78	n.a.
British Virgin Islands	49	15,834	82	4.40
Cayman Islands	n.a.	13,672	131	11.60
Cuba	11	13,672	3	n.a.
Curaçao	n.a.	1,638	21	4.14
Dominica	28	619	12	0.23
Dominican Republic	14	387	7	9.50
Grenada	28	688	18	n.a.
Guadeloupe	n.a.	1,075	19	n.a.
Haiti	n.a.	7	n.a.	n.a.
Jamaica	21	513	9	n.a.
Martinique	n.a.	946	22	n.a.
Montserrat	43	1,698	50	0.03
Puerto Rico	4	625	3	n.a.
Saba	n.a.	n.a.	5	n.a.
St Eustatius	n.a.	n.a.	31	n.a.
St Kitts and Nevis	28	1,441	43	1.72
St Lucia	64	1,774	29	7.86
St Maarten	n.a.	11,297	83	n.a.
St Vincent/Grenadines	33	672	16	1.91
Trinidad and Tobago	4	164	4	n.a.
Turks and Caicos Islands	n.a.	18,237	130	n.a.
US Virgin Islands	57	10,652	46	12.60

Source: Derived from Caribbean Tourism Organization (2002).

Conclusion

It is clear that tourism has followed a rather uneven pace of development in the Caribbean. This is partially explained by the unevenness of those economic sectors in which it is heralded as a necessary form of economic development, especially given some of the realities that many island states face as they become players on the world stage. It is also partially explained by the diversity of the tourism product that is available. To some extent, conscious decisions have been, and continue to be, made with respect to the overall direction of the tourism product in the region; but there is the alternative reality that some destinations face with respect to tourism being developed as an exercise designed primarily to accommodate lacklustre performance of other economic sectors.

This chapter has, albeit briefly, sought to provide an overview of key issues facing tourism development in the region. Perhaps most critical are the more regulatory and government-issued policy and planning frameworks that chart the course of tourism development. Coupled with this is the degree of sustainability that is both desired and achieved, although it is recognised that the measurement of such sustainability is indeed problematic. Shifting markets (a consequence of world events and changes in perception of the overall product) are concerns, as is the degree of political instability in the region that prevents accurate and meaningful planning for destructive fluctuations.

References

Apostolopoulos, Y. and Gayle, D.J. (2002a) 'From MIRAB to TOURAB? Searching for sustainable development in the maritime Caribbean, Pacific, and Mediterranean', in Y. Apostolopoulos and D.J. Gayle (eds.) *Island Tourism and Sustainable Development: Caribbean, Pacific, and Mediterranean Examples*, Westport: Praeger.

Apostolopoulos, Y. and Gayle, D.J. (eds) (2002b) *Island Tourism and Sustainable Development: Caribbean, Pacific, and Mediterranean Examples*, Westport: Praeger.

Ayres, R. (2002) 'Cultural tourism in small-island states: contradictions and ambiguities', in Y. Apostolopoulos and D.J. Gayle (eds) *Island Tourism and Sustainable Development: Caribbean, Pacific, and Mediterranean Examples*, Westport: Praeger.

Baldacchino, G. (1996) 'Visitor management and the sustainable tourism agenda', in L. Briguglio, B. Archer, J. Jafari and G. Wall (eds) *Sustainable Tourism in Islands and Small States: Issues and Policies*, New York: Pinter Press.

Baum, T. (1997) 'The fascination of islands: a tourist perspective', in D.G. Lockhart and D. Drakakis-Smith (eds) *Island Tourism: Trends and Prospects*, London: Pinter.

Bell, J.H. (1993) 'Caribbean tourism in the year 2000', in D.J. Gayle and J.N. Goodrich (eds) *Tourism Marketing and Management in the Caribbean*, London: Routledge.

Bertram, I.G. and Watters, R.F. (1985) 'The MIRAB economy in South Pacific microstates', *Pacific Viewpoint* 26: 497–519.

Bianchi, R. (2002) 'Towards a new political economy of global tourism', in R. Sharpley and D. Telfer (eds) *Tourism and Development: Concepts and Issues*, Clevedon: Channel View Publications.

Bonnick, G. (2000) *Toward a Caribbean Vision 2020: a regional perspective on development, challenges, opportunities and strategies for the next two decades (discussion draft)*, Caribbean Group for Cooperation in Economic Development (World Bank).

Boxill, I. (2002) 'Challenges of academic research in Caribbean tourism', in C. Jayawardena (ed.) *Tourism Hospitality Education and Training in the Caribbean*, Barbados: The University of the West Indies Press.

Briguglio, L., Archer, B., Jafari, J. and Wall, G. (eds) (1996a) *Sustainable Tourism in Islands and Small States: Issues and Policies*, New York: Pinter Press.

Briguglio, L., Butler, R., Harrison, D. and Filho, W.L. (eds) (1996b) *Sustainable Tourism in Islands and Small States: Case Studies*, New York: Pinter Press.

Britton, R. (1977) 'Making tourism more supportive of small state development: the case of St Vincent', *Annals of Tourism Research* 4(5): 268–78.

Britton, S. (1987) 'Tourism in small developing countries: development issues and research needs', in S. Britton and W. Clarke (eds) *Ambiguous Alternative: Tourism in Small Developing Countries*, Suva, Fiji: University of the South Pacific Press.

—— (1989) 'Tourism, dependency and development: a mode of analysis', in T.V. Singh, H.L. Theuns and F.M. Go (eds) *Towards Appropriate Tourism: The Case of Developing Countries*, Frankfurt am Main: Peter Lang.

Bromley, R.D.F. and Bromley, R. (1982) *South American Development: A Geographical Introduction*, Cambridge: Cambridge University Press.

Browlett, J. (1980) 'Development, the diffusionist paradigm and geography', *Progress in Human Geography* 4(1): 57–80.

Bryden, J. (1973) *Tourism and Development: A Case-Study of the Commonwealth Caribbean*, Cambridge: Cambridge University Press.

Butcher, J. (2003) *The Moralisation of Tourism: Sun, Sand . . . and Saving the World?*, London: Routledge.

Butler, R. (1980) 'The concept of a tourist area cycle of evolution: implications for management of resources', *Canadian Geographer* 24: 5–12.

—— (1993) 'Tourism – an evolutionary process', in J. G. Nelson, R. Butler and G. Wall (eds) *Tourism and Sustainable Development: Monitoring, Planning, Managing*, Publication Series No. 37, Waterloo: Department of Geography, University of Waterloo.

Cardosa, F.H. (1972) 'Dependency and development in Latin America', *New Left Review* 74(4): 83–95.

Caribbean Tourism Organization (CTO) (2002) *Caribbean Tourism Statistical Report: 2000–2001 Edition*, St Michael, Barbados: Caribbean Tourism Organization.

Conlin, M.V. (1995) 'Rejuvenation planning for island tourism: the Bermuda example', in M.V. Conlin and T. Baum (eds) *Island Tourism: Management Principles and Practice*, Chichester: Wiley.

Conlin, M.V. and Baum, T. (eds) (1995) *Island Tourism: Management Principles and Practice*, Chichester: Wiley.

Connell, J. (1988) *Sovereignty and Survival: Island Microstates in the Third World*, Research Monograph No. 3, Sydney: Department of Geography, University of Sydney.

Cushnahan, G. (2001) 'Resource use and tourism on a small Indonesian island', *Tourism Recreation Research* 26(3): 25–31.

Dann, G.M.S. (1996) 'Socio-cultural issues in St Lucian tourism', in L. Briguglio, R. Butler, D. Harrison and W.L. Filho (eds) *Sustainable Tourism in Islands and Small States: Case Studies*, New York: Pinter Press.

Davies, B. (1996) 'Island states and the problems of mass tourism', in L. Briguglio, B. Archer, J. Jafari and G. Wall (eds) *Sustainable Tourism in Islands and Small States: Issues and Policies*, New York: Pinter Press.

de Albuquerque, K. and McElroy, J. (1999) 'Tourism and crime in the Caribbean', *Annals of Tourism Research* 26(4): 968–84.

Dos Santos, T. (1972) 'Dependence and the international system', in J. Cockcroft, A.G. Frank and D.L. Johnson (eds) *Dependence and Underdevelopment: Latin America's Political Economy*, New York: Anchor.

Duval, D.T. (1998) 'Alternative tourism on St. Vincent', *Caribbean Geography* 9: 44–57.

Edwards, J. (1996) 'Visitor management and the sustainable tourism agenda', in L. Briguglio, B. Archer, J. Jafari and G. Wall (eds) *Sustainable Tourism in Islands and Small States: Issues and Policies*, New York: Pinter Press.

Erisman, H.M. (1983) 'Tourism and cultural dependency in the West Indies', *Annals of Tourism Research* 10(3): 337–61.

Frank, A.G. (1972) *Lumpen-Bourgeoisie and Lumpen-Development: Dependence, Class and Politics in Latin America*, New York: Monthly Review Press.

—— (1981) 'The development of underdevelopment', in M. Smith, R. Little and M. Schackleton (eds) *Perspectives on World Politics*, London: Croom Helm.

Gayle, D.J. and Goodrich, J.N. (1993) 'Caribbean tourism marketing, management and development strategies', in D.J. Gayle and J.N. Goodrich (eds) *Tourism Marketing and Management in the Caribbean*, London: Routledge.

Goulet, R. (1993) 'Society, policy and policy analysis – review and critique of a theoretical debate?', Unpublished paper, York University, Toronto.

Great Britain Ministry of Overseas Development (1966) *Tripartite Economic Survey*, London: Her Majesty's Stationery Office.

Guthunz, U. and von Krosigk, F. (1996) 'Tourism development in small island states: from MIRAB to TouRAB?', in L. Briguglio, B. Archer, J. Jafari and G. Wall (eds) *Sustainable Tourism in Islands and Small States: Issues and Policies,* New York: Pinter Press.

Guyana Chronicle Online (2002) www.guyanachronicle.com/news.html (accessed 13 November 2002).

Hall, C.M. (1994) 'Is tourism still the plantation economy of the South Pacific? The case of Fiji', *Tourism Recreation Research* 19: 41–8.

Harrison, D. (1996) 'Sustainability and tourism: reflections from a muddy pool', in L. Briguglio, B. Archer, J. Jafari and G. Wall (eds) *Sustainable Tourism in Islands and Small States: Issues and Policies*, New York: Pinter Press.

—— (2001) 'Islands, image and tourism', *Tourism Recreation Research* 26: 9–14.

Henderson, J.C. (2000) 'Managing tourism in small islands: the case of Pulau Ubin, Singapore', *Journal of Sustainable Tourism* 8(3): 250–62.

Hinch, T.D. (1990) 'The Cuban tourism industry: its re-emergence and future', *Tourism Management* 11(3): 214–26.

Jamaica Observer (2002) http://www.jamaicaobserver.com/magazines/Environment/html/20021020T230000–0500_33965_OBS_NCRPS_TAKES_OVER_NEGRIL_MARINE_PARK.asp

Johnson, P. and Thomas, B. (1996) 'Tourism capacity: a critique', in L. Briguglio, B. Archer, J. Jafari and G. Wall (eds) *Sustainable Tourism in Islands and Small States: Issues and Policies*, New York: Pinter Press.

Lasswell, H. (1958) *Politics: Who Gets What, When, How?*, New York: Meridian.

Levitt, K. and Gulati, I. (1970) 'Income effect of tourist spending: mystification multiplied – a critical comment on the Zinder Report', *Social and Economic Studies* 19(3): 326–43.

Lewis, C.A. (1955) *The Theory of Economic Growth*, London: Allen and Unwin.

Lockhart, D.G. (1997) 'Islands and tourism: an overview', in D.G. Lockhart and D. Drakakis-Smith (eds) *Island Tourism: Trends and Prospects*, London: Pinter Press.

Lockhart, D.G. and Drakakis-Smith, D. (eds) (1997) *Island Tourism: Trends and Prospects*, London: Pinter Press.

Lowenthal, D. (1992) 'Small tropical islands: a general overview', in H. Hintjens and M.D.D. Newitt (eds) *The Political Economy of Small Tropical Islands: The Importance of Being Small*, Exeter: University of Exeter Press.

McElroy, J. and de Albuquerque, K. (1991) 'Tourism styles and policy responses in the open economy-closed environment context', in N.P. Girvan and D. Simmons (eds) *Caribbean Ecology and Economics*, St Michael, Barbados: Caribbean Conservation Association.

—— (1996) 'Sustainable alternatives to insular mass tourism: recent theory and practice', in L. Briguglio, B. Archer, J. Jafari and G. Wall (eds) *Sustainable Tourism in Islands and Small States: Issues and Policies*, London: Pinter Press.

—— (2002) 'Problems for managing sustainable tourism in small islands', in Y. Apostolopoulos and D.J. Gayle (eds) *Island Tourism and Sustainable Development: Caribbean, Pacific, and Mediterranean Examples*, Westport: Praeger.

Mather, S. and Todd, G. (1993) *Tourism in the Caribbean*, Special Report No. 455, London: The Economist Intelligence Unit.

Mowforth, M. and Munt, I. (1997) *Tourism and Sustainability: New Tourism in the Third World*, London: Routledge.

Palmer, R. (1993) 'Tourism and taxes: the case of Barbados', in in D.J. Gayle and J.N. Goodrich (eds) *Tourism Marketing and Management in the Caribbean*, London: Routledge.

Pattullo, P. (1996) *Last Resorts: The Cost of Tourism in the Caribbean*, London: Cassell.

Pearce, D.G. (1989) *Tourist Development*, Harlow: Longman.

Perez, L.A. (1973–4) 'Aspects of underdevelopment: tourism in the West Indies', *Science and Society* 37: 478–80.

—— (1975) 'Tourism in the West Indies', *Journal of Communications* 25: 136–43.

Potter, R.B. and J.A. Binns (1988) 'Power, politics and society', in M. Pacione (ed.) *The Geography of the Third World: Progress and Prospect*, London: Routledge.

Pratt, G.A. (2002) 'Sustainable tourism development in the Caribbean: the role of education', in C. Jayawardena (ed.) *Tourism Hospitality Education and Training in the Caribbean*, Barbados: The University of the West Indies Press.

Renard, Y. (2001) 'Practical strategies for pro-poor tourism: a case-study of the St Lucia Heritage Tourism Programme', *PPT Working Paper No. 7*, London: Pro-Poor Tourism Project (Overseas Development Institute, International Institute for Environment and Development, and Centre for Responsible Tourism at the University of Greenwich).

Royle, S.A. (2001) *A Geography of Islands: Small Island Insularity*, London: Routledge.

Ryan, C. (2001) 'Tourism in the South Pacific – a case of marginalities', *Tourism Recreation Research* 26(3): 43–9.

St Lucia Heritage Tourism Programme (2003) 'The St Lucia Heritage Tourism Programme', www.stluciaheritage.com.

Seers, D. (1979) 'The meaning of development', in D. Lehmann (ed.) *Development Theory: Four Critical Studies*, London: Cass.

Sharpley, R. (2002) 'Tourism: a vehicle for development?', in R. Sharpley and D. Telfer (eds) *Tourism and Development: Concepts and Issues*, Clevedon: Channel View Publications.

Slinger, V. (2000) 'Ecotourism in the last indigenous Caribbean community', *Annals of Tourism Research* 27(2): 520–3.

Smith, D.M. (1977) *Human Geography: A Welfare Approach*, London: Arnold.

Smith, S.L.J. (1983) *Recreation Geography*, London: Longman.

Stabler, M.J. and Goodall, B. (1996) 'Environmental auditing in planning for sustainable island tourism', in L. Briguglio, B. Archer, J. Jafari and G. Wall (eds) *Sustainable Tourism in Islands and Small States: Issues and Policies*, New York: Pinter Press.

Stonich, S., Sorenson, J. and Hundt, A. (1995) 'Ethnicity, class and gender in tourism developments: the case of the Bay Islands, Honduras', *Journal of Sustainable Tourism* 3(1): 1–28.

Tewarie, B. (2002) 'The development of a sustainable tourism sector in the Caribbean,' in Y. Apostolopoulos and D.J. Gayle (eds) *Island Tourism and Sustainable Development: Caribbean, Pacific, and Mediterranean Examples*, Westport: Praeger.

Tucker, H. (2003) *Living with Tourism: Negotiating Identities in a Turkish Village*, London: Routledge.

Uphoff, N. and Ilchman, W. (1972) *The Political Economy of Development*, Princeton: Princeton University Press.

Weaver, D.B. (1988) 'The evolution of a "plantation" tourism landscape on the Caribbean island of Antigua', *Tijdschrift voor Economische en Sociale Geografie* 79: 319–31.

Wilkinson, C. (ed.) (2002) *Status of Coral Reefs of the World: 2002*, Townsville: Australian Institute of Marine Science.

Wilkinson, P. (1989) 'Strategies for tourism development in island microstates', *Annals of Tourism Research* 16: 153–77.

—— (1997) *Tourism Policy and Planning: Case Studies from the Commonwealth Caribbean*, Elmsford, NY: Cognizant Communications.

World Commission on Environment and Development (WCED) (1987) *Our Common Future*, Oxford: Oxford University Press.

Zinder and Associates (1969) 'The future of tourism in the eastern Caribbean', Washington, DC: United States Agency for International Development.

5 Caribbean tourism policy and planning

Paul F. Wilkinson

Introduction

'Globalisation' is a term under which prevailing models of social, economic and political organisation have popularly been collected, although the term itself has only seen widespread use since the 1980s (Wilkinson 2000). Within the context of western society, there have been three waves of globalisation: colonisation, development (in the sense of capitalist development) and what is frequently termed neoliberal capitalism. This third wave affects virtually all aspects of economic activity that are related to free-market economies: communications, transportation, agriculture, manufacturing, etc.

In one sense, tourism has long been 'global' or international in that tourists, as the demand side of the tourism system, have for many centuries visited other countries. As tourism demand increased, however, with growing prosperity in the developed nations following World War II and more particularly with the advent of jet airplanes in the early 1960s, the boom in mass tourism both led to and was facilitated by increased involvement by multinational corporations (MNCs) in the supply side of the tourism system. Modern international tourism has come, therefore, to be dominated in corporate terms by MNCs, including airlines, hotels, tour wholesalers, tour operators, travel agents and car rental companies.

One result of the globalisation of tourism is that destination nations see much of the economic benefits of tourism lost through high economic leakages and low economic multipliers, to the advantage of MNCs and the nations from which the majority of tourists originate. Moreover, because of the large amounts of capital required to develop a major tourism sector and of the frequent absence of available local capital, developing countries are usually forced to allow MNCs to dominate their tourism sectors. It has often been argued, therefore, that the governments of developing countries – especially small island states – are virtually powerless in controlling the shape and direction of their economies – particularly tourism – when faced with the global economy and the enormous strength of MNCs.

This chapter will examine the contrary argument, that such governments can exert a high degree of control over their tourism development through effective policy and planning. Following a discussion of the basic nature of tourism policy and planning, examples will be drawn from island states of the Commonwealth Caribbean. An analysis of whether these states have been effectively involved in tourism policy and planning will conclude the chapter.

Tourism policy and planning

Until recently, little attention has been paid in the tourism literature to the analysis of tourism policy and its subsequent implementation through tourism planning. An outstanding exception to this lack of research on tourism policy and planning is political scientist Linda Richter, who (1989: 11) argues that 'where tourism succeeds or fails is largely a function of political and administrative action and is not a function of economic or business expertise.'

The situation, however, is changing. For example, in their revised model of destination competitiveness and sustainability, Ritchie and Crouch (2000: 2) now include tourism policy as a separate, major element:

> Tourism Policy, which we define as
> *A set of regulations, rules, guidelines, directives, and development/ promotion objectives and strategies within which the collective and individual decisions directly affecting tourism development and the daily activities within a destination are taken,*
> seeks primarily to create an environment within which tourism can flourish in an adaptive sustainable manner.

Similarly, Clancy (1999: 7) argues that

> the fundamental development debate has now come to center on domestic politics and especially the key role played by the state in affecting development patterns. It may be argued, in fact, that state actors and policies have become the primary, if not solitary set of independent variables that explain successes and failures to the exclusion of other factors.

This chapter will use Acerenza's (1985: 60) definition of *tourism policy*: 'the complex of tourism related decisions which, integrated harmoniously with the national policy for development, determines the orientation of the sector, and the action to be taken.' He sees tourism policy as providing the broad guidelines which shape the development of the sector, while the development strategy constitutes the means by which resources are used to meet defined objectives.

Although the general planning literature provides few references to *tourism planning*, Inskeep (1988) argues that it is a sufficiently specialised and different form which demands special attention on its own. Getz (1987:

3) defines it as 'A process, based on research and evaluation, which seeks to optimise the potential contribution of tourism to human welfare and environmental quality.' Mathieson and Wall (1982: 186) provide a very succinct statement on its main objectives: 'to ensure that opportunities are available for tourists to gain enjoyable and satisfying experiences and, at the same time, to provide a means for improving the way of life of residents of destination areas.'

It is widely argued that local control and the extent of policy and planning practices are two key factors in the political arrangement of tourism destination. Jenkins and Henry (1982: 506), for example, hypothesise that 'For each developing country, the degree of active involvement by government in the tourist sector will reflect the importance of tourism in the economy.' As measures of tourism's economic importance, they (1982: 506) use contribution to Gross Domestic Product (GDP) and national income, foreign exchange earnings, employment and income generated, and contribution to government revenues: 'Using these four indicators, one would expect government to intervene actively in the tourist sector either when tourism is of major economic significance or where government follows a system of centrally-planned economic activity.'

Jenkins and Henry (1982: 501) define *active involvement* as 'a deliberate action by government, introduced to favour the tourism sector'. It implies not only a recognition by government of the specific needs of the tourism sector, but also of the necessity for government's operational participation to attain stated objectives. There are two types of active involvement:

- *managerial*: a government not only sets out tourism objectives (e.g. in a tourism development plan), but also induces organisational and legislative support to attain them (e.g. tourism investment incentives legislation)
- *developmental*: a government undertakes an operational role in tourism, either for ideological reasons or because of the inability or unwillingness of the private sector to become involved in tourism (e.g. government financing or ownership of hotels).

In contrast, *passive involvement* occurs when a 'government undertakes an action that may have implications for tourism, but is not specifically intended to favour or influence tourism' (Jenkins and Henry 1982: 501). There are two types of passive involvement:

- *mandatory*: legislation relates to the country as a whole and is not intended to discriminate in favour of the tourism sector, although it may have implications for tourism (e.g. investment incentives legislation)
- *supportive*: a government does not deliberately inhibit the development of tourism, nor does it encourage it (e.g. approving a private-sector 'national' tourist board).

Arguing that 'it is probable that most governments would be involved in policy-making in each area, if only in a passive role', Jenkins and Henry (1982: 506) discuss five general 'areas of concern' in the formulation of tourism policy: foreign exchange earnings, foreign investment, employment in tourism, land-use policies and air transport and tourism. 'Even a necessarily brief consideration of each area will indicate the need for active government involvement' in order to avoid what Jenkins and Henry (1982: 506) describe as a progression that has become all too familiar in developing countries:

> Unspoiled place with unique character attracts tourists; new buildings and amenities necessary to house tourists bring about change; more tourists produce more change; loss of initial attractive character becomes an element responsible for departure of tourists; and final result is economic, social and financial disaster.

This progression suggests that potential conflicts between short-term benefits and longer-term objectives have not been recognised or anticipated and that, to achieve these objectives, it will be necessary to monitor and control the nature and pace of tourism development: 'As in most government decisions, evaluation [of means to achieve the objectives] should be seen in terms of political economy rather than mere economic rationale' (Jenkins and Henry 1982: 506).

Despite the almost continuous growth in visitor arrivals and expenditures, numerous decision-makers and researchers have expressed concern about Caribbean tourism for many years, citing problems related to such factors as obsolete facilities, overcrowding, environmental impacts, and 'the need to apply long-range planning strategies based on accurate market and product information' (Spinrad *et al.* 1982: 15). The point about strategies – or the lack of them – was clearly stated long ago by Tinsley (1979: 310) who contends that the state of the Caribbean tourism sector up to that point was 'the result of twenty-five years of non-planning'. (Tinsley's conception of 'planning' is much more general than that adopted above and includes both policy and planning.) The questions arise as to whether Tinsley accurately describes the situation prior to 1979 and whether he accurately describes the situation since 1979.

This chapter will examine whether the governments of Caribbean islands can exert a high degree of control over their tourism development through effective policy and planning. The situation up to the 1990s in five island states will be examined briefly, in increasing order of the size of their tourism sectors: Dominica, St Lucia, Cayman Islands, Barbados and the Bahamas (see Wilkinson 1997 for detailed analyses). It should be noted that all of the islands have various pieces of tourism-related legislation (e.g. hotel acts) and destination marketing organisations or national tourism boards; because of space limitations, these will not be discussed here.

Dominica

Since the Government of the Commonwealth of Dominica (GOCD) quickly and wisely rejected the Shankland Cox (1971) report's call for mass tourism based on highly unrealistic assumptions about the projected growth in the number of tourists both to the region and to Dominica, Dominica's approach to tourism policy has been relatively successful. There has been slow but steady growth in numbers of tourists and tourist expenditure (with the exception of downturns caused by political problems in the 1970s and hurricane damage in 1979 and 1980), reaching 69,578 stayover tourists, 239,796 cruise passengers, and tourism expenditures of US$47.3 million (representing 28.2 per cent of GDP) in 2000 (Caribbean Tourism Organization (CTO) 2002: 157 and 261). Such success is a result of a realistic and pragmatic recognition of the country's lack of suitability for mass tourism (largely related to the absence of white sand beaches and an international airport) and its potential to attract a small but growing number of tourists who mainly are interested in its natural environment. A 1975 United Nations Development Programme-sponsored report (Kastarlak 1975) clearly delineated the limitations of Dominica's tourism development potential and recommended that the focus not be on the development of major beach-oriented hotels, but on lower-scale developments aimed at the island's environment and its attraction to specialised market segments. This success has been fuelled by a marketing programme which, while relatively low-key and modest in scale, presents a clear and honest picture.

In the area of tourism planning, Dominica has not been as successful as on the policy front. There are notable exceptions, particularly two national parks, but their exceptionality is mainly in their potential, which will require careful management and substantial improvements in related infrastructure to become major tourism magnets and to avoid serious environmental degradation – the universal problem of use versus preservation of fragile natural environments. Land-use and site-development planning do exist with respect to particular hotel development proposals, but that is a reactive process and not one of deliberate government involvement in development.

Early attempts at tourism planning were undertaken, but not approved by the Government. For example, a *National Structure Plan for 1976–1990* (GOCD 1976) saw tourism as playing a modest role in the country's economy in view of the realities of the regional market and available resources. It proposed a dispersed pattern of development: beach-oriented hotels, locally owned cluster hotels, modest-scale resort subdivisions, development of historic and cultural areas, etc. None of these proposals were implemented, largely due to the lack of finances and of tourist demand. However, the plan's concept of Dominica as a nature-oriented tourism destination remains as a guiding force to the present. A subsequent national structure plan (GOCD 1985) made similar proposals, but it too was never approved, for similar reasons.

The picture remains different on the policy side. In a widely distributed document, the Government (GOCD 1987) clearly states that its tourism policy is to optimise the sector's contribution to the national economy through improved marketing, revived interest in cruise ships, linkages with agriculture and related sectors, and protection of natural resources. This policy was supported by a subsequent marketing plan (Giersch 1987) which emphasised a combination of natural, cultural and historic environments, rather than beach-oriented facilities. This report remains the basis of the country's marketing focus – a focus that must be judged a success given the increase in tourist and cruise-ship arrivals since it was formulated.

The Government finally accepted and implemented a tourism sector plan (Tourism Planning and Research Associates 1991) which clearly designates the basic framework of its tourism policy: a focus on the 'nature island' image, small-scale developments, integration with the rural community, strong economic linkages to other sectors, orientation on conservation, and tourism at a level within the 'absorptive capacity' of the island. As a result of the 1987 policy and this plan, Weaver (1993, 1991) concludes that Dominica is the Caribbean destination most closely associated with comprehensive ecotourism and that it has developed a 'deliberate' form of ecotourism based on conscious policy direction, as opposed to other destinations with 'circumstantial' ecotourism, or the mere appearance of alternative tourism.

It is difficult to criticise the Government for taking such an approach of low involvement in tourism planning, given its emphasis on agriculture as the backbone of the economy and its weak financial position. Nevertheless, economic diversification is a goal and, because of the internal and external problems with agriculture and manufacturing, tourism appears to be the only viable economic alternative. The question, however, of who would fund increased tourism development remains. It is Government policy to encourage small-scale development and local participation, yet local capital and expertise is limited. Moreover, involvement from foreign investors on anything beyond a modest scale is unlikely without a major improvement in air access. Finally, international aid agencies are no longer interested in the tourism sector or international airport construction.

It appears reasonable that the Government should continue its current realistic and pragmatic policy which fosters steady but incremental growth in tourist numbers and expenditures through improved and targeted marketing focused on the island's environment, coupled with improved services (e.g. roads, information, tour guides) and upgraded accommodations. Moreover, if a marketing goal is to attract more tourists from the upper end of the market, then greater attention has to be given to the quality of existing and new accommodations, in terms of location, design and services, largely through the strengthening of the planning process.

St Lucia

With five important exceptions, the history of tourism policy and planning in St Lucia has been characterised by a 'hands-off' philosophy in which the Government has focused on dealing with specific private-sector (usually foreign) development proposals on a project-by-project incremental basis – in effect, site-specific land-use planning. The results have been mixed. On the one hand, the number of tourists and the level of expenditures have slowly but continuously risen, reaching 269,850 stayover tourists, 443,551 cruise passengers, and US$277 million in tourism expenditures (63.8 per cent of GDP) in 2000 (CTO 2002: 214, 261). Also, new developments and development proposals have steadily come on line, including several major luxury and all-inclusive resorts. On the other hand, a transition to high-density tourism – which potentially could cause severe environmental stress – seems almost inevitable unless the Government takes a more pro-active role in tourism policy and planning. Such an emphasis on high-density tourism could cause local dissatisfaction, because there is strong evidence that the St Lucian people want more local involvement in the sector; for example, Dann (1992) finds that three-quarters of his sample of residents want local people to have more say in tourism decision-making.

The five exceptions indicate the Government's willingness to consider alternative forms of involvement in the tourism sector. The first two exceptions concerned direct government participation in tourism development and operation. In the Rodney Bay Development beginning in the 1970s, the Government was an active partner in a scheme to develop a major tourism zone (which was only partially completed).

The second instance was reactive, in which the Government became involved in the operation of several financially troubled hotels in the mid-1970s. Most notably, it took over the bankrupt Halcyon Days Hotel and ran it while searching for an international hotel chain to take it over. The ensuing debate on gambling led to a major policy stance based on strong cultural convictions of the public. Given the weak fiscal position of the Government, however, future direct participation in the tourism sector seems unlikely, a policy position stated in the 1991 National Tourism Policy (Government of St Lucia 1989). The third exception was the 1977 National Development Plan (Government of St Lucia 1977), which was never adopted officially, but which set the framework for tourism zoning that remains today. A new national planning agenda is, however, needed to avoid the current and potential environmental problems related to tourism.

Fourth, the 1991 National Tourism Policy seems to provide a strong direction to guide the future style, form and location of tourism development. The question remains, however, as to how and whether such a policy can be integrated with a land-use planning process which is focused on particular development proposals, rather than the overall structure of the tourism sector. For example, while the policy calls for an emphasis on small hotels with a high degree of local participation, there seem to be few resources available to foster such development and, indeed, innovation in general in the tourism sector. Conversely, the Government continues to approve large-scale hotels owned by MNCs. (The current Government announced that a new tourism policy and plan would be in place by late 2002, but neither has been released to date.) Finally, the Government's support of the St Lucia Heritage Tourism Programme (St Lucia Heritage Tourism Programme 2003) suggests that there is an increasing awareness that the concentration of tourism development in the north-west coastal area does not promote the dispersal of the benefits of tourism to the entire country, particularly to rural communities hard hit by the downturn in banana exports.

The danger is clear: mass tourism characterised by large numbers of tourists, but tourist expenditures with high leakage rates and low multipliers – and increased environmental stress. Clearly, a reconciliation of policy and practice is in order; otherwise, serious environmental degradation is likely.

Cayman Islands

The Cayman Islands have shown remarkable growth in their tourism sector in the past three decades, reaching 354,087 stayover tourists, 1,030,857 cruise passengers, and US$559.2 million in tourism expenditures (79.1 per cent of GDP) in 2000 (CTO 2002: 145, 261). In a statement that is still applicable, Weaver (1990: 13) quotes a former Caymanian politician responsible for tourism: 'This remarkable surge of visitors to our shores is not a sudden or random happening. Our consistent and dependable growth has been the direct result of extensive marketing planning and intensive implementation of those programmes, both at home and abroad.'

Starting with a Development Plan and Regulations in 1977 (Government of the Cayman Islands 1977), the Government of the Cayman Islands began to implement a linked set of tourism policies and plans. The 1977 plan recognised many environmental problems and led to a series of environmental laws that continues today. Also in 1977, the Government founded Cayman Airways, Ltd (CAL), recognising that growth in tourism was dependent upon reliable air transportation. The CAL's history has not been without financial problems, but the Government appears to have recognised these problems and decided that the only viable solution is a minimal service operation designed to plug the gaps in the route offerings of international carriers, yet maintaining a presence in the industry for reasons of national image.

In order to meet the tourism policy goals of the 1977 Development Plan, the Government commissioned a Ten-Year Tourism Development Plan in 1981 (Laventhol and Horwath 1981). The Government accepted a series of objectives proposed as future development criteria: controlled and planned growth of tourism; educational programmes; decreased dependency on expatriate labour; maximised land use to better accommodate and service foreign visitors; and quality superstructure and infrastructure with the fewest negative implications (Laventhol and Horwath 1981: V-4). Rather than expansion of the accommodation sector, the plan suggests that these objectives are best met through the enhancement and/or adaptive reuse of existing establishments, along with upgrading of the airports, ground transportation, road system, and cruise shuttle transport. It warns that further large-scale development of Grand Cayman could destroy its attractiveness and that Little Cayman and Cayman Brac should be kept small scale in order to protect their natural environments and their up-scale market niche (Laventhol and Horwath 1981: VI-32–5). This position was supported by a second Development Plan in 1987, in which the Government (Government of the Cayman Islands 1987: 27) defines its primary development goal as 'steady development in established directions and not one of radical change'. The Government (Government of the Cayman Islands 1987: 132) clearly reiterates its view that tourism development is to be a private sector concern, although it will be carefully controlled by government policy and planning.

The overall objective of the Government's (Government of the Cayman Islands 1992b) Ten Year Tourism Development Plan, the first tourism plan for the Cayman Islands, is

> to provide a clear set of policies, strategies and implementation guidelines to chart the way forward for tourism. . . . [W]hile tourism should continue to stimulate the economy for the benefit of the Caymanian people, it was imperative that this be achieved in the context of the preservation of the heritage, culture and environment of the Islands.
> (Government of the Cayman Islands 1992a: 165, 167)

This was followed by a Tourism Management Policy (Government of the Cayman Islands 1994) which continues the emphasis on seeking to attract 'up-market' stayover tourists, calling for continued discouragement of charter and other low-priced markets; however, charter flights continue, particularly in the summer slow season. There is also a major emphasis on promoting ecotourism.

In a statement that remains valid, Weaver (1990: 15) concludes that the Cayman Islands have developed a secure tourism sector, given the encouragement of a high degree of local control and participation and strong environmental controls. Similarly, Pratt (1993: 264) argues that the best explanation for the success of Caymanian tourism can be expressed in a single word: consistency. Focusing on the more up-scale and sophisticated traveller, marketing programmes have been aggressive, positioning the Islands as 'quiet, safe and friendly' – characteristics which these primary target vacationers want. In turn, the emphasis has been on providing a 'high quality' experience in a relatively expensive destination. The situation is not perfect, however, as growth in the tourism sector, even if somewhat restrained, has the potential to cause serious environmental problems.

Barbados

Although Barbados' tourism sector seemed to have stagnated in the early 1990s, it has picked up in recent years, with 544,696 stayover tourists, 533,279 cruise passengers, and US$711.3 million in tourism expenditures (33.0 per cent of GDP) in 2000 (CTO 2002: 125, 261). While the history of tourism in Barbados has been marked by a general lack of Government involvement in tourism policy and planning, nevertheless, various government activities have directly and indirectly affected tourism. Rather than having a tourism policy or a tourism plan *per se*, the Government of Barbados deals with tourism primarily through two mechanisms: national development plans and physical development plans. The former deal predominantly with social and economic policies, while the latter translate these into land-use policies. Both are guided by the ongoing development of government policy as presented in Throne Speeches and in the party manifesto of the governing party. The Government, therefore, is really only involved in tourism land-use planning (and marketing planning), and not tourism development policy or planning. One of the major results is a highly fragmented approach to tourism policy and planning with several major players (e.g. Board of Tourism, Barbados Hotel Association, Government) managing the tourism economy with, at least until recently, little evidence of coordination or cooperation.

This policy environment has been operationalised through a series of policies, plans and pieces of legislation. These begin with the first Physical Development Plan (Government of Barbados 1970) (which did not come into effect until 1976) which allocated lands for tourism development on the east, south, southeast, west and northwest coasts. It was followed by the 1979 National Development Plan (Government of Barbados 1979) which recognises the importance of tourism, not just for its direct and indirect economic impacts, but also for stimulating infrastructure development (e.g. airport, harbour). It expresses concern, however, that unplanned and unrestrained growth could weaken efforts to strengthen indigenous culture and adversely affect environmental quality of the country. The 1979 plan was replaced in 1983 by a second National Development Plan. It (Government of Barbados 1983: 78–92) recognises that the tourism sector performed below the expectation of the 1979 plan for a number of reasons: weak demand for travel due to worldwide recession; increased competition from other Caribbean destinations; and image problems in terms of price, standards and service. The plan recommends a shift in emphasis to quality rather than quantity.

With problems continuing into the late 1980s, the Government chose the path of outward-looking, export-led growth focusing on tourism, non-sugar agriculture, and manufacturing, as set out in the five-year Public Sector Investment Programme contained in the 1988 National Development Plan (Government of Barbados 1988a). This was supported by the 1988 Physical Development Plan (Government of Barbados 1988b: 2), the main objective of which is 'to indicate a national settlement development strategy and policy for the country ... to the year 2000'. It also incorporates physical development policies for land use, economic activities, housing, services, transportation, utilities, recreation and conservation. These policies include the concentration of tourism development on the west and south coasts, a sewerage project focused on the capital of Bridgetown, and tourism market diversification.

The Government has had a long-standing interest in promoting tourism as a major sector of the Barbadian economy. With exceptions, however, the Government seems to have adopted a laissez-faire attitude, preferring to encourage such policies as first growth, then quality control, and more recently market diversification, rather than becoming an active agent in planning for tourism development. The results, until recently, were a plateauing of tourist numbers and expenditure and an ageing accommodation stock, much in need of refurbishment and renovation, some major exceptions notwithstanding. Extensive marketing and government encouragement of deeply discounted air fares and hotel packages seems to have turned the picture around.

The Bahamas

On the surface, the tourism sector in the Bahamas seems to be a success story: 1,596,160 stayover tourists, 2,512,626 cruise passengers, and US$1,814 million in tourism expenditures (44.0 per cent of GDP) in 2000 (CTO 2002: 121, 261). An argument has been made, however, that, in real dollar terms, the economic benefits of tourism in the Bahamas actually peaked in the early 1970s (Wilkinson 1999). Unlike most other Caribbean islands, the Bahamas has a history of tourism and of government involvement – or perhaps, non-involvement – in tourism policy and planning that dates back to the mid-nineteenth century. The events are numerous and the documentation is voluminous; therefore, only the recent times will be highlighted here.

Following the failure of the 1955 Hawksbill Creek Agreement between the Government of the Commonwealth of the Bahamas (GOCB) and an MNC to create a deepwater harbour and industrial complex on Grand Bahama, a 1960 Supplemental Agreement was signed which encouraged rapid and massive private-sector involvement in resort development. The bubble soon burst, however, in the face of American recession, over-building and political instability. By 1970, new investment dried up, existing capital fled, land sales and tourism slumped, and unemployment soared.

This lesson that evolution might make better sense than revolution was not absorbed by the Government. It commissioned a tourism plan (Checchi 1969) for the entire country which called for expansion of tourism development on a massive scale, including a near-tripling of stayover tourists in five years. Just as the Government was beginning to think about implementing the plan, the problems in Grand Bahama became obvious, government-owned Bahamas Airways collapsed (to be replaced by government-owned Bahamasair in 1973), tourist numbers declined, and the government-owned Hotel Corporation of the Bahamas (HCB) purchased several hotels to save them from collapse (Holder 1993). Despite ongoing weakness in the tourism sector, the Government seemed determined to implement the Checchi theory that increased supply would lead to increased demand. Thus, the Government went on to develop the Cable Beach Hotel and Casino in New Providence which opened in 1984 at a cost of over US$100 million (GOCB 1988: 174). The Government still owns about 15 per cent of the total accommodation in the Bahamas – largely because the private sector is sceptical about the profitability of Bahamian tourism.

Despite tourism problems and its domination of the economy, the Government was without a real tourism plan until 1981. This ten-year Tourism Development Plan's (Dames and Moore 1981: i–2) overall goal was 'to maximise the country's competitive potential in the world tourism market while minimising the risks of economic loss'. The plan's objectives were to develop a comprehensive strategic plan, identify improvements required in tourism infrastructure, define priority projects for consideration by international lending agencies and perform pre-feasibility studies of several priority projects. Its growth policy was based on projections of massive increases in the number of tourists – growth which failed to materialise. Despite its argument that the proposed growth is economically viable, the plan ends on an unusual note which implies that the Bahamian financial climate must be made even more conducive in order to attract tourism investment (Dames and Moore 1981: 6–12). It points out that the Bahamas levies no income taxes, that most Government income is derived from import duties, and that the Hotels Encouragement Act already provides relief from property tax and import duties for items used in hotel construction. In the absence of any other important taxes, therefore, the Bahamas is unable to employ many of the forms of direct or indirect tax relief used by other governments as incentives. It then suggests other additional incentives: below-market loans, reduction of operating costs (service costs), and waiving the modest Business Licence Fee. It seems that even a virtually tax-free environment, therefore, is not sufficient to attract investment to the Bahamas. One has to wonder how this was factored into the consultants' financial analysis which 'demonstrated that all the developments proposed should be viable' (Dames and Moore 1981: 1–24).

Despite these concerns about the financial climate, the early 1980s was a time of large-scale expansion of the accommodation stock – but the tourists did not fill up the hotels. As a result, the Government in 1984 developed a Tourism Marketing Strategy and Development Plan, the main objective being to attract more stayover arrivals (GOCB 1984: 3.1). On the whole, however, this plan is basically a 'state of the problem' report and provides little concrete direction for either a marketing strategy or a development plan.

By the 1990s, a sense of desperation seems to have set in. Rather than take action, the result was the commissioning of series of plans and reports, all of which clearly outlined the seriousness of the tourism sector's problems – but which were shelved because of both the private sector and the Government's inability or unwillingness to act on their recommendations.

There is a temptation to end this analysis with a glib and probably overly dramatic conclusion: the Bahamian tourism sector is a house of cards. The Bahamas has a long and complex history of tourism with layers and nuances that may be hidden forever from view to all but a few, but definitely to the outsider. The stakes in this card game were high and it is not clear who won what and who continues to win. And it is not even clear who lost. But it all has occurred in a failed policy and planning context, despite high levels of government involvement.

Conclusion

In terms of these five case-studies, Tinsley's (1979: 310) contention that the state of the Caribbean tourism sector up to that point was 'the result of twenty-five years of non-planning' (and non-policy) does not hold true on the whole for the period before 1979 – at least in theory, if not in practice. Each of these states had explicit and/or implicit policies in place, even if they were never fully implemented through effective planning. For example, each had some form of legislation (variously termed hotel aid or encouragement ordinances or acts in four of the cases) to encourage tourism through investment in hotel development. All but one had some form of national development or structure plan, or physical development plan that addressed issues related to the tourism sector, although in many cases these plans were not formally approved by government; the Bahamas is the exception on a national level, but not so in terms of Grand Bahama. Except for Barbados, all had some form of report, plan or strategy setting a course of tourism development that in many cases remains very evident today.

Since 1979, the same pattern – explicit and implicit policy, but often ineffective plan implementation – holds true for Dominica, St Lucia and the Cayman Islands. Their current tourism sectors are the result of deliberate choices taken by the respective governments. That is not to say, however, that the routes taken are not without internal contradictions and inconsistencies (e.g. St Lucia's policy of encouraging local involvement in the sector, but continuing pattern of approving up-scale MNC developments; the recognition by the Cayman Islands of the dependence of tourism on the environment, but lack of control over growth). In contrast, Barbados and the Bahamas have essentially drifted in policy and planning vacuums into situations that could result in the continuing deterioration of virtually every aspect of their tourism sectors. Moreover, the only sign in these latter cases that a new approach to tourism policy and planning is possible is the fact that the problems in tourism are becoming very clear to almost everyone.

Not even those case-studies which do have some form of tourism policy and planning mechanism, however, are fully involved in all aspects of such policy and planning. The reasons are different in each case, but include a combination of factors (e.g. limited financial and human resources, institutional constraints, political will, economic imperatives) that have resulted in widely varying and incomplete patterns of involvement in tourism policy and planning.

The level of active involvement by governments in attempting to shape their tourism sectors was raised with Jenkins and Henry's (1982: 506) hypothesis that 'For each developing country, the degree of active involvement by government in the tourist sector will reflect the importance of tourism in the economy.' In terms of their measures of tourism's economic importance (contribution to GDP and national income, foreign exchange earnings, employment and income generated, and contribution to government revenues), the tourism sector in each of the case-studies is clearly of major economic significance. There is, however, a great deal of variation in its importance among the case-studies, both absolutely and relatively.

Nevertheless, if Jenkins and Henry's (1982: 506) hypothesis holds, 'one would expect government to intervene actively in the tourist sector'. Is this true for each of the case-studies? If the question were approached from a simplistic yes–no stance, the answer would be 'yes' for each of the case-studies, except perhaps Barbados. Each state has a tourism history which includes a variety of policies, plans, legislation, government-operated and/or financed infrastructure, etc., which could be described as representing policies reflecting active involvement and also, in most cases, passive involvement (Table 5.1). Moreover, each state has one or more of the five types of plans delineated by the World Tourism Organization (1980) as potentially being related to tourism: general national, national infrastructure, national tourism development, tourism infrastructure, and national promotion and marketing.

If, however, the question were approached from a more complex position that took a more evaluative judgement about the effectiveness of government involvement, then the answer varies for each state. In fact, if one sets aside such serious concerns as the obvious incompatibility between ever-growing numbers of visitors and environmental integrity, only the Cayman Islands can be described as being effective.

Moreover, if the question were approached from the even more demanding argument that 'It is probable that most governments would be

Table 5.1 Summary of government involvement in tourism policy and planning

	Dominica	St Lucia	Cayman Is.	Barbados	Bahamas
Govt Involvement in Tourism Policy					
Active: Managerial	X	X	X	X	X
Developmental		X	X	X	X
Passive: Mandatory	X		X	X	X
Supportive	X	X	X	X	X
Govt Involvement in Tourism Planning					
General national	n.a.	n.a.	X	X	
National infrastructure			X	X	
National tourism development	X	X	X		X
Tourism infrastructure	X	X	X		
National promotion and marketing	X	X	X	X	
Detailed Govt Involvement in Tourism Planning					
Foreign investment earnings: Active: Managerial	X	X	X		
Developmental					
Passive: Mandatory					
Supportive					
Foreign investment: Active: Managerial		X	X	X	X
Developmental				X	X
Passive: Mandatory					
Supportive					
Employment in tourism: Active: Managerial	X		X	X	X
Developmental		X			
Passive: Mandatory					
Supportive					
Land-use policies: Active: Managerial					
Developmental					
Passive: Mandatory	X		X	X	
Supportive					
Air transport: Active: Managerial					
Developmental					
Passive: Mandatory					
Supportive	X	X	X	X	X

X denotes involvement; n.a. = not approved.

involved in policy-making in each area [foreign exchange earnings, foreign investment, employment in tourism, land-use policies, and air transport], if only in a passive role' (Jenkins and Henry 1982: 506), then the answer is negative in all cases: none of them is involved in all of these areas, even in a passive role.

Despite this conclusion, some of the case-studies examined here do provide evidence that developing countries, even small ones, can go a long way in influencing – if not controlling – the shape and direction of their tourism economies. While more effective in terms of policy than planning, Dominica and the Cayman Islands provide the clearest examples of governments which chose a specific development path that has in fact been achieved. Despite its lack of control over the minor (as opposed to MNC) airlines that service its small airports, Dominica has the highest degree of control over its tourism economy because of the absence of MNC hotels and the high degree of local ownership of tourism-related businesses – although cruise ship MNCs are increasingly influential. The Cayman Islands, despite the influence of MNC airlines, cruise ships and hotels, has effectively created itself as a high-end, luxury destination. St Lucia has been less effective, with its contradiction of a policy of local involvement, but a practice of successfully attracting MNC all-inclusive and luxury resorts. In contrast, Barbados and the Bahamas have been notably ineffective in both policy and planning, drifting along under the influence of external forces; neither government seems to want to be involved in trying to direct their tourism future. The simple lesson is that, if a government has the political will to be strongly involved and is willing to take action to create a strong tourism policy and planning context, developing countries can heavily influence, if not totally control, the shape and direction of their tourism economies.

References

Acerenza, M.A. (1985) 'Planificación estratégica del turismo: esquema me todológico', *Estudios Turisticos* 85: 47–70.

Caribbean Tourism Organization (CTO) (2002) *Caribbean Tourism Statistical Report (2000–2001 Edition)*, St Michael, Barbados: Caribbean Tourism Organization.

Checchi and Co. (1969) 'A plan for managing the growth of tourism in the Commonwealth of the Bahamas', Report for Ministry of Tourism, Government of the Commonwealth of the Bahamas.

Clancy, M.J. (1999) 'Tourism and development: evidence from Mexico', *Annals of Tourism Research* 26(1): 1–20.

Dames and Moore Ltd (1981) 'Tourism development programme final report: Volume I', Report for Ministry of Tourism, Government of the Commonwealth of the Bahamas.

Dann, G. (1992) 'Socio-cultural impacts of tourism in Saint Lucia', *Studies in Tourism No. 3 (draft)*, Port-of-Spain, Trinidad: United Nations Economic Commission on Latin America and the Caribbean.

Getz, D. (1987) 'Tourism planning and research traditions, models and futures', Australian Travel Research Workshop, 5–6 November, Bunbury.

Giersch, W. (1987) 'Tourism Product and Marketing Concept for the Commonwealth of Dominica', Report for Commonwealth of Dominica.

Government of Barbados (1970) *Barbados Physical Development Plan*, Bridgetown: Government of Barbados.

—— (1979) *National Development Plan 1979–1983*, Bridgetown: Government of Barbados.

—— (1983) *National Development Plan 1983–1988*, Bridgetown: Government of Barbados.

—— (1988a) *Barbados National Development Plan 1988–1993*, Bridgetown: Government of Barbados.

—— (1988b) *Barbados Physical Development Plan (Amended 1986)*, Bridgetown: Town and Country Planning Office, Ministry of Finance, Government of Barbados.

Government of the Cayman Islands (1977) *The Development Plan 1977*, George Town: Central Planning Authority and Department of Planning, Government of the Cayman Islands.

—— (1987) *The Cayman Islands Development Plan 1988–1992*, George Town: Economic Development Unit.

—— (1992a) *Cayman Islands Annual Report 1991*, George Town: Government of the Cayman Islands.

—— (1992b) *A Ten Year Tourism Development Plan (1992–2002)*, George Town: Portfolio of Tourism, Aviation and Trade (Coopers and Lybrand Consulting, New York), Government of the Cayman Islands.

—— (1994) *Cayman Islands Tourism Management Policy 1995–1999*, George Town: Department of Tourism, Government of the Cayman Islands

Government of the Commonwealth of the Bahamas (GOCB) (1984) *Bahamas Tourism Marketing Strategy and Development Plan*, Nassau: Ministry of Tourism, Government of the Commonwealth of the Bahamas.

—— (1988) *Statistical Abstract 1968–1988*, Nassau: Ministry of Finance.

Government of the Commonwealth of Dominica (GOCD) (1976) *Dominica National Structure Plan 1976–1990*, Roseau: Tropical Printers, Government of the Commonwealth of Dominica.

—— (1985) *National Structure Plan*, Roseau: Government of the Commonwealth of Dominica

—— (1987) *Tourism Policy*, Roseau: Ministry of Agriculture, Trade, Industry and Tourism, Government of the Commonwealth of Dominica.

Government of St Lucia (1977) *St Lucia National Plan – Development Strategy*, Castries: Government of St Lucia.

—— (1989) 'National Tourism Policy.' Unpublished draft, Ministry of Industry, Trade and Tourism.

Holder, J.S. (1993) 'The Caribbean Tourism Organization in historical perspective', in D.J. Gayle and J.N. Goodrich (eds) *Tourism Marketing and Management in the Caribbean*, London: Routledge.

Inskeep, E. (1988) 'Tourism planning: an emerging specialization', *APA Journal*, Summer, 360–72.

Jenkins, C.L. and Henry, B.M. (1982) 'Government involvement in tourism in developing countries', *Annals of Tourism Research* 9(3): 499–521.

Kastarlak, B. (1975) *Tourism and its Development Potential in Dominica*, Castries, St Lucia: United Nations Development Programme.

Laventhol and Horwath Ltd (1981) 'Cayman Islands tourism industry: a 10 year development plan', Report for Government of the Cayman Islands.

Mathieson, A. and Wall, G. (1982) *Tourism: Economic, Physical and Social Impacts*, Harlow: Longman.

Pratt, M. (1993) 'Cayman Islands: successful tourism yesterday, today and tomorrow', in J.R.B. Ritchie, D.E. Hawkins, F. Go and D. Frechtling (eds) 'Special report: island tourism', *World Travel and Tourism Review: Indicators, Trends and Issues – Volume 3, 1993*, Wallingford, UK: CAB International.

Richter, L. (1989) *The Politics of Tourism in Asia*, Honolulu: University of Hawai'i Press.

Ritchie, J.R.B. and Crouch, G.I. (2000) 'The competitive destination: a sustainability perspective', *Tourism Management* 21(1): 1–7.

St Lucia Heritage Tourism Programme (2003) 'The St Lucia Heritage Tourism Programme', www.stluciaheritage.com.

Shankland Cox and Associates (1971) 'Dominica: a tourist development strategy – planning and policy document', Report for United Kingdom Overseas Development Agency.

Spinrad, B.K., Seward, S.B. and Bélisle, F.J. (1982) 'Introduction', in S.B. Seward and B.K. Spinrad (eds) *Tourism in the Caribbean: The Economic Impact*, Ottawa: International Development Research Centre.

Tinsley, J.F. (1979) 'Tourism in the Virgin Islands', in J.S. Holder (ed.) *Caribbean Tourism: Policies and Impacts*, Christ Church, Barbados: Caribbean Tourism Research Centre.

Tourism Planning and Research Associates (1991) *Dominica Tourism Sector Plan*, Report for Ministry of Trade, Industry and Tourism, Government of the Commonwealth of Dominica.

Weaver, D.B. (1990) 'Grand Cayman Island and the resort cycle concept', *Journal of Travel Research* 29(2): 9–15.

—— (1991) 'Alternative to mass tourism in Dominica', *Annals of Tourism Research* 18(3): 9–15.

—— (1993) 'Ecotourism in the small island Caribbean', *GeoJournal* 31(4): 457–65.

Wilkinson, P.F. (1997) *Tourism Policy and Planning: Case Studies from the Commonwealth Caribbean*, Elmsford, NY: Cognizant Communications Corporation.

—— (1999) 'Caribbean cruise tourism: delusion? illusion?', *Tourism Geographies* 1(3): 261–82.

—— (2000) 'Globalization', in J. Jafari (ed.) *Encyclopedia of Tourism*, London: Routledge.

World Tourism Organization (1980) *Physical Planning and Area Development for Tourism in the Six WTO Regions, 1980*, Madrid: World Tourism Organization.

6 Institutional arrangements for tourism in small twin-island states of the Caribbean

Leslie-Ann Jordan

Introduction

The purpose of this chapter is to address an existing gap between research on institutional arrangements for tourism in small island states and the internal core–periphery model. Using the twin-island states of Antigua and Barbuda, St Kitts and Nevis, and Trinidad and Tobago, it is argued that internal core–periphery relationships, as well as the earlier historical forces which have shaped political, economic and social structures throughout the region, have fostered specific institutional arrangements with respect to tourism development and management in each twin-island state. The chapter also discusses some of the implications of such arrangements in the context of institutional reform and makes some recommendations for suitable spatial and functional arrangements that are conducive to the development of effective tourism policy and planning.

Tourism policies in the Caribbean have largely centred on the facilitation and promotion of tourism for economic stability. More recently, however, some have sought to ensure that environmental and social–cultural concerns are incorporated into both policy statements and practice. Comparatively little research on institutional arrangements for tourism in small island states in the Caribbean has been undertaken, despite the fact that such arrangements facilitate the study of the interaction of several variables, such as the analysis of policy and related processes, economic and financial arrangements, organisational structures and political processes and the evolution of key events and ideas in history (Mitchell 1989).

Tourism and institutional arrangements

Institutional arrangements are one of the important components in the tourism public policy process. Brooks (1989: 131) suggests that they influence 'the process through which the policy agenda is shaped, problems are defined, alternatives are considered, and choices are ultimately made'. For Hall and Jenkins (1995: 18), 'understanding tourism public policy demands some understanding of and reference to the institutional arrangements in which tourism policy is made.' Consequently, the investigation of institutional arrangements needs to assess the interaction of several pervasive forces, including:

- legislation and regulations
- policies and guidelines
- administrative structures
- economic and financial arrangements
- political structures and processes
- historical and traditional customs and values
- key participants or actors
- inter-organisational relationships.

Undoubtedly, a greater understanding of the way in which these variables interact in the tourism public policy process in small island states will facilitate predictions about possible future policy-making as well as prescriptions for adjustments to existing institutional arrangements.

Both tourism policy and institutional arrangements are affected by a variety of conditions and circumstances (Taylor 1987; Pearce 1992), particularly:

- various economic, social, historical, cultural, political and geographic environments (which dictate and govern the extent to which tourism is developed)
- the type of tourism developed
- key origin markets
- the role or importance of tourism in the economy.

These conditions provide a social context within which relationships are negotiated. Accordingly, it is important to identify those features of the environment which have the most influence on tourism organisations and their associated arrangements (Pearce 1992). As such, it is argued here that no discourse on institutional arrangements for tourism in the Caribbean would be complete or accurate without a discussion of the role of internal core–periphery relationships.

Tourism and the internal core–periphery model

Institutional arrangements are also influenced by internal political dynamics, which can often mimic core–periphery relationships that are more commonly applied across, as opposed to within, nations or states. One of the defining features of the core–periphery relationship is the idea of domination of the periphery by the core, which was viewed as the place where administrative, cultural and economic power resided and was seen as the decision-making nucleus separated from the periphery (Paddison 1983; Knox and Agnew 1994; Dickenson *et al.* 1996).

Few tourism researchers have focused on internal core–periphery rela-tionships which are often present in small twin-island states and there has been little effort made to determine if those relationships have an impact on the business of tourism. For Weaver (1998: 293), 'internally induced core–periphery dynamics . . . have been neglected as a framework for the analy-sis of Third World tourism, as if domestic involvement in the national tourism sector were somehow assumed to be either implicitly benign, or neg-ligible.' While the core–periphery model has been used extensively in tourism studies from a dependency perspective at an international scale (see Turner and Ash 1975; Britton 1982; Blomgren and Sorensen 1998), the emergence of 'internal spatial dichotomies' are still 'fundamentally seen as corollaries of the international core periphery dynamic' (Weaver 1998: 293).

Institutional arrangements in twin-island Caribbean states

Antigua and Barbuda, St Kitts and Nevis, and Trinidad and Tobago all possess strong internal core–periphery relationships. Long before their union with the larger islands, Barbuda, Nevis and Tobago were accorded their own administrative machinery and functioned autonomously for much of their early history (Table 6.1). Consequently, they had evolved their own idiosyncratic customs, which they zealously guarded from infringe-ment (Ottley 1973; Lowenthal and Clarke 1979; Brereton 1981; Phillips 1985; Premdas 1998). They possessed a strong, long-standing sense of insular identity and viewed themselves as distinctive cultural, economic and political entities. Given these circumstances, it was not surprising that, as time went on, the partners in the union found themselves increasingly at loggerheads. Weaver (1998: 305) noted that, when these historical facts are considered along with the physical separation of the islands and the disparities in size and power, 'the emergence of a centrifugal internal core–periphery relationship is hardly surprising.'

The relationship between the central/federal governments and peripheral local government bodies within these twin-island states has been fraught with many problems. A common feature of the internal core–periphery relationship is small island dissatisfaction (Paddison 1983). For instance, Tobagonians have often complained of a legacy of neglect, domination and inequitable budgetary locations on the part of the Trinidad-based central government. Several factors have helped to reinforce Tobago's peripheral condition, including the need for Tobagonians to travel to Trinidad to access many essential government services, the absence for many years of Tobagonian representation in Parliament, and the lack of equitable treatment in terms of financial expenditures. Likewise, many Barbudans claim that they have been neglected and threatened by a domi-nant island partner that accounts for over 98 per cent of the population and most of the economic activity (Commonwealth Secretariat 2000; Constitutional Review Commission 2002). Barbudan apprehensions about

Table 6 1 Selected socio-economic and political characteristics of case-study
countries

	Antigua and Barbuda	St Kitts and Nevis	Trinidad and Tobago
Area (km²)	440	269	5,128
Population	Antigua 70,000 Barbuda 1,417	St Kitts 31,880 Nevis 10,080	Trinidad 1,300,000 Tobago 55,000
System of govt	Unitary	Federal	Unitary
Main organ of local govt	Barbuda Council	Nevis Island Administration (NIA)	Tobago House of Assembly (THA)
Date of independence	1 Nov. 1981	19 Sept. 1983	31 Aug. 1962
GDP (US$ million)[b]	459.43	207.2	5,927[a]
Stayover arrivals (000s)[c]	236.7	73.1	398.2
Cruise ship arrivals [c]	429.4	164.1	82.2
Visitor expenditure (US$ million)[c]	291.1	58.2	212.8
Tourism expenditure as % of GNP	72.3	31.1	–

Sources: Antigua and Barbuda Department of Tourism (2002); Caribbean Tourism Organization (1999, 2002); Government of Trinidad and Tobago (2002); St Kitts and Nevis Information Service (2001); St Kitts and Nevis Ministry of Tourism (2002).

Notes
a 1998.   b 2000.   c 2001.

the Central Government are largely based on fears that the main island's political interests were attempting to gain control over the communal lands for tourism purposes, in collusion with foreign interests (Weaver 1998). Similarly, Nevis has claimed domination and exploitation by St Kitts and has come to view St Kitts as the 'larger omnipresent looming partner' (Premdas 2000). Such mistreatment (whether real or perceived), combined with the subordinate island's distinctive cultural and historical identity, has fostered an ambivalent relationship between internal core and periphery. These accusations and counter-attacks have been entrenched in the countries' collective memory and have, to some degree, permeated many aspects of society.

In order to determine if the internal core–periphery relationship is reflected in, and influences, the coordination, planning and policy setting in the tourism sector in twin-island states of the Caribbean, qualitative interviews were conducted with employees from public sector tourism organisations on each island.[1] The results suggest that the internal core–periphery relationship has contributed to conflicts and tensions between key public sector organisations responsible for tourism policy-making in all three countries. Further, and as comments from the respondents reveal, those contemporary inter-organisational problems are the product of over one hundred years of inter-island conflict. The internal core–periphery relationship has been the framework within which the institutional arrangements for tourism have been formed and have continued to evolve in Antigua and Barbuda, St Kitts and Nevis, and Trinidad and Tobago. Overall, it seems that an internal core–periphery relationship has helped to create weak institutional arrangements for tourism, which has been characterised by:

- unclear distribution or location of the seat of power within the tourism policy-making process;
- ineffective constitutional and legislative framework to govern the activities of the public sector tourism organisations;
- centralisation of tourism policy and decision-making and limited decision-making power in the unitary states (Antigua and Barbuda and Trinidad and Tobago);
- lack of clearly defined roles, responsibilities and authority of public sector tourism organisations and their actors, which has led to duplication of efforts and resources, misunderstandings and inter-organisational conflicts;
- poor communication between and amongst public sector tourism organisations resulting in misunderstandings and conflict;
- reduced inter-organisational collaboration and cooperation, which hinders integrative planning and development of appropriate policies to address the changing environment within which tourism development occurs;
- lack of a common, shared sustained vision for tourism development;
- difficulty in maximising the benefits of private–public sector partnership and cooperation;
- inability of public sector organisations to develop and manage tourism in a controlled, integrated and sustainable manner.

Not surprisingly, many of these factors are inter-linked and may even act as causes or consequences for each other. The following sections will discuss these issues in greater detail with an emphasis on their impact on the institutional arrangements for tourism.

Constitutional and legislative framework: the political organisation of the state

The political organisation of the state is one of the most important factors that has influenced the design and effectiveness of the institutional arrangements for tourism in Antigua and Barbuda, St Kitts and Nevis, and Trinidad and Tobago. In the context of twin-island states, one of the common devices implemented as a means of managing claims and conflicts between core and periphery is decentralised internal autonomy (Premdas 2000). There are a number of states that have accommodated the representation of 'out islands' in their government, such as the Maldives, Mauritius, the Seychelles, Kiribati, Tuvalu and Vanuatu. They have done so in a variety of ways, such as nomination or specification of 'out islanders' filling elective posts or provision of local assemblies or any combination of these (Sutton 1987). Decentralised internal autonomy can also be manifested either in a federal arrangement, as is the case with St Kitts and Nevis, or constitutionally entrenched into the organisation of an unitary state, as is the case with Trinidad and Tobago and Antigua and Barbuda. Regardless of the arrangement chosen, the 'fundamental assumption is that the devolution of autonomous decision-making powers to a sub-state community creates in the diffusion of authority separate space and confers recognition in self-governing pride. It is pre-eminently a political arrangement' (Premdas 2000).

Antigua, St Kitts and Trinidad have attempted to accommodate Barbuda, Nevis and Tobago, respectively, in their government. They have done so by establishing the Barbuda Council, the Nevis Island Administration (NIA) and the Tobago House of Assembly (THA) respectively. However, there are some distinct differences in the outworking of these arrangements, owing to the type of government system operating in each country. One of the most important factors affecting a government's capacity to develop tourism policy is whether it has a federal or unitary form of government (Howlett and Ramesh 1995). For instance, federalism has been cited as a major reason for the weak policy capacity of governments in Australia, Canada and the USA (Howlett and Ramesh 1995; Jenkins 2000). It has constrained these states' capacity to develop consistent and coherent policies.

Whether tourism organisations and actors are operating within a federal system or a unitary form of government has an impact upon several key issues such as their autonomy, capacity and overall willingness to foster inter-organisational, inter-island cooperation. For example, St Kitts and Nevis has a federal system of government, which means that there are two autonomous levels of government within the country: the Federal Government and the NIA. These two organs of government are not bound together in a dominant/subordinate relationship but rather they enjoy more or less complete discretion in matters under their jurisdiction that were guaranteed by the constitution. The St Kitts and Nevis constitution clearly stipulates that the NIA is exclusively in charge of tourism development and tourism policy-making for Nevis. It has the power and resources to develop its own tourism plans and policies, collect its own taxes and to employ representatives for its overseas offices. This means that it is not heavily reliant or dependent on St Kitts for any resources or assistance and consequently, the tourism policy-makers in St Kitts are fully aware that they have no authority to dictate policy matters to Nevis. When Nevis's autonomy in the tourism policy process is added to the strained relationship between the Federal Government and the NIA, there tends to be less of a need or desire to communicate, cooperate and build inter-organisational relations with the tourism organisations in St Kitts.

On the other hand, in Antigua and Barbuda and Trinidad and Tobago the institutional arrangements for tourism have evolved differently due to the respective country's unitary system of government. Theoretically, in unitary systems, there is usually only one level of government and the local bodies owe their existence to the national government rather than to the constitution (Paddison 1983; Howlett and Ramesh 1995). However, although Antigua and Barbuda and Trinidad and Tobago are considered unitary states, the Barbuda Council and the THA, respectively, do not owe their existence to their central governments. Their existence has in fact been enshrined in the countries' respective constitutions. This divergence from conventional theory could be attributed to the dissatisfaction of these smaller islands with being forced to join with their larger neighbour and their insistence upon safeguarding their own independence.

However, despite the fact that the constitution guarantees the existence of the Barbuda Council and the THA, in the internal core–periphery relationship between Trinidad and Tobago and Antigua and Barbuda, Trinidad and Antigua, respectively, hold the advantage. They are the islands with the seat of central government and by extension the seat of ultimate power. They govern by virtue only of how the people vote in Trinidad and Antigua respectively; the vote in the subordinate islands does not have to matter. And, crucially, they can dictate tourism policy for Barbuda and Tobago respectively, as well as the size of budgetary allocations for the island, which is to say, the pace of tourism development. These are significant powers for one island to have in relation to another.

In the case of Trinidad and Tobago, there are contradictions between the constitution and the THA Act 1996 that have been a major source of confusion and frustration between tourism policy-makers in Trinidad and those in Tobago. The THA Act gives Tobago responsibility over specified policy areas, including tourism development, in a context of pre-existing, unreconciled laws, which gives the Central Government certain responsibilities in respect of the whole country. In other words, although the Act gives the THA responsibility for 'the formulation and implementation of policy' in tourism, the constitution then limits that responsibility by evoking Section 75(1), which gives the Cabinet (which is to say Trinidad) the power to intrude into an area of policy that is legally under the authority of the THA. In practice, this has been taken to mean that a minister in the Cabinet has the power to interfere with any functions for which the THA has been given responsibility. The minister, and not the THA, has the final word on the matter. Therefore, in effect, both the THA and the Central

Government have the authority to develop and implement tourism policy in Tobago and this has caused many conflicts and misunderstandings to occur between the Tourism and Industrial Development Company of Trinidad and Tobago (TIDCO), the Ministry of Tourism and the THA.

A similar scenario has unfolded in Antigua and Barbuda. The Barbuda Local Government Act 1976 gives the Barbuda Council authority to 'promote hotel and tourist development in accordance with and subject to any law relating to the allocation of land, foreign investment or tax incentives' and gives the Council the power to make by-laws relating to a wide range of matters including 'the provision of guides for tourists and the prevention of annoyance to tourists by beggars and idlers' and 'the control of and the imposition of fees, rates and taxes on restaurants and other eating places' (Barbuda Local Government Act 1976). However, the Central Government retains the power to give general or special direction to the Council as to the policy the Council should follow in the exercise of its powers with respect to the above subjects. In essence, Barbudans still do not have exclusive control over tourism policy or development on the island. The end result has been tension and conflict between the Central Government and the Barbuda Council.

The above discussion becomes more important in light of the fact that the central governments in both Antigua and Trinidad have contrasting perspectives on tourism development in Barbuda and Tobago respectively. In the case of Trinidad and Tobago, the policies of the Central Government have tended to favour large-scale developments, while the THA favours community/village-based tourism projects. Likewise, a major obstacle to smooth relations between the Barbuda Central Government and the Barbuda Council is the fact the Central Government is interested in accelerating the pace of development in Barbuda, while the Council and the people of Barbuda subscribe to a more measured pace of development and assert the right to be consulted on the nature of development to be pursued (Commonwealth Secretariat 2000).

In summary, in the unitary states of Trinidad and Tobago and Antigua and Barbuda, the constitutional and legislative frameworks give the periphery less autonomy and resources to develop and implement tourism policy. Consequently, they are more dependent and manipulated by the core. This situation has engendered resentment in the smaller islands towards the core, as they have to seek permission and resources from the core in order to manage their tourism industry.

Politics and power in the periphery

Politics and institutional arrangements for tourism are closely linked. The idea that power is geographically distributed, and that distribution in this case refers to the extent and manner of decentralisation within the state, has particular relevance to the internal core–periphery model. For twin-island states like Antigua and Barbuda, St Kitts and Nevis and Trinidad and Tobago, where a body of water separates the islands from each other, the issue of the distribution of power is crucial. Many of the national core–periphery issues centre around the question of who has power and where that power is located in a geographical sense. The problems of where power is located feeds into the problems of internal core–periphery relations and local autonomy, which in turn have helped to influence and shape institutional arrangements in the tourism industry.

All three sets of twin-island states profiled here have discovered that devising suitable institutional arrangements for geographically separate territories, in the context of the exercise of a common sovereignty within the framework of a unitary or federal state, is highly problematic. Given the particular history of the relations between the islands, such an effort presents a special challenge. Consequently, solutions must be sought taking into account the complex interplay between the psychological impact of that history and the insights derived from modern theories of governance (Commonwealth Secretariat 2000).

Notwithstanding the devolution of powers to peripheral local government bodies – the Barbuda Council, the NIA and the THA – there is still a lack of certainty and clarity in the division of powers between the two levels of government. Obviously, for institutional arrangements to be effective there needs to be a clear demarcation between the functions of the federal/central governments and those of the local governing bodies in the smaller islands. The specific roles and responsibilities of all the organisations responsible for tourism must be openly reviewed, extensively discussed and eventually, when revisions are made, there must be whole-hearted acceptance of those changes by all concerned. The mere existence of statutory and constitutional arrangements reflecting the distribution of powers and responsibilities between different levels of government does not in itself guarantee effective governance or effective inter-organisational tourism relationships. Given the history of disagreement, the long-standing suspicion and distrust and the conflict that characterises the internal core–periphery relationship of the countries under study, there needs to be a special mechanism for bringing parties together within a structured process of negotiations aimed at reconciling differences.

Absence of communication, cooperation, consultation and collaboration

One major similarity between all three case-studies is the lack of consistent communication, cooperation, consultation and collaboration between the public sector tourism organisations responsible for tourism policy and decision-making. To a large extent, this can be attributed to the nature of the internal core–periphery relationship. Figures 6.1, 6.2 and 6.3 show the extent of relationship(s) between various tourism organisations.

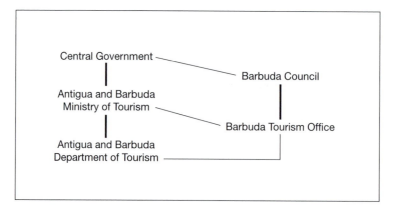

Figure 6.1 Organisational arrangements for tourism in Antigua and Barbuda
(thicker lines denote more consistent communication, cooperation and
consultation)

Table 6.2 shows a small sample of comments from individuals associated with specific organisations in each twin-island state. They illustrate
that there has been a conspicuous failure to ensure effective consultation
between tourism organisations located in the core and those located in the
periphery.

This absence of inter-organisational communication, cooperation,
consultation and collaboration is extremely important to the discourse on
institutional arrangements. If allowed to progress unchecked, it may have
serious implications for the ability of the public sector organisations under
study to develop and implement tourism policy.

Public sector organisations must be able and ready to respond to the
growing challenges facing the tourism industry, not only in their respective countries but also in the Caribbean region as a whole. When attempting
to address these challenges, it is not possible for individual organisations
to act in isolation, for each of them lacks the authority, resources and
ability required for concerted effort (Getz and Jamal 1994; Parker 1999;
Selin 1999). Instead, what is needed is some form of inter-organisational
collaboration. For this reason it is imperative that institutional arrangements be set up in such a way so as to facilitate the efficient administration
of tourism as well as the development of competitive tourism policy.

While a process of shared information and decision-making with all key
stakeholders involved allows for the generation of appropriate tourism
policy and subsequent planning endeavours with minimal negative impacts
(De Kadt 1979; Hall 1994, 2000; Tosun and Jenkins 1998; Bramwell and
Lane 1999; Timothy 1999; Ladkin and Bertramini 2002), weak institutional arrangements, fuelled by internal core–periphery conflict, threaten
the ability of public sector organisations to cooperate and collaborate in

Table 6.2 Respondents' comments on inter-organisational communication, cooperation, consultations and collaboration

Country	Comments
Antigua and Barbuda	'it has been difficult trying to get Barbuda to be incorporated into all the tourism plans'
	'it should be an automatic thing to include Barbuda . . . but it is not'
	'very often we [Barbuda Tourism Office] hear about the groups [travel agents and tour operators visiting the country] after they have left'
	'we [Barbuda Tourism Office] should feel as though we are part of the process and right now we don't feel as though we are a part of it'
	'Barbuda needs to advise the Ministry, and not just via the newspapers, what their vision is for the development of Barbuda's tourism'
St Kitts and Nevis	'Nevis thinks that we don't like them, so we don't bond . . . overall they feel that we take care of St Kitts first and then look after Nevis. So you find that we don't market together'
	'when we are selling the islands, we are just selling the one we belong to'
	'I don't have any idea how or what they do over in Nevis'
Trinidad and Tobago	'there is a lack of consistent sharing of information on projects and achievements, if any at all'
	'that is one of the main problems we have with TIDCO in that there is no consultation'
	'that's sad that a lot of things are being approved [in Trinidad] and we [THA] don't know anything about it'
	'the last time I [THA employee] had any major interaction with the Ministry was back in 1999'
	'we [THA] are not aware of what exactly goes on in TIDCO'
	'if you asked me [TIDCO employee] what THA has done, their Department of Tourism, within the last year or so, I couldn't tell you'

developing appropriate tourism plans and policies. This is an untenable situation for small island states. In order to respond to the fragmented nature of tourism development, there is a need for inter-organisational coordination, cooperation and collaboration in tourism planning and policy-making.

The collaborative approach is relevant to this discourse on institutional arrangements and the internal core–periphery model because it can specifically help destinations where fragmented and independent planning

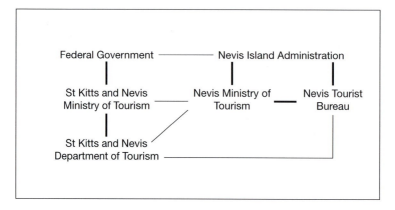

Figure 6.2 Organisational arrangements for tourism in St Kitts and Nevis (thicker lines denote more consistent communication, cooperation and consultation)

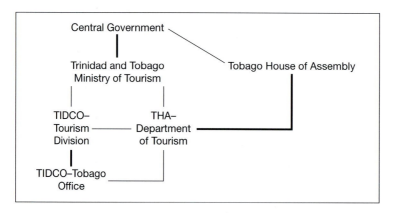

Figure 6.3 Organisational arrangements for tourism in Trinidad and Tobago (thicker lines denote more consistent communication, cooperation and consultation)

decisions by different tourism stakeholders give rise to power struggles over resources (Jamal and Getz 1999; Ladkin and Bertramini 2002). This has been the experience of the twin-island states discussed here, where the distribution of power in the tourism policy process is divided between the organisations in the core and in the periphery, which are competing for authority, autonomy and finances.

Cooperation is necessary both within and between various organisations and at different levels, especially between (1) government agencies, (2) different levels of government, (3) equally autonomous organisations at various administrative levels and (4) the public and private sector (Hall

1994; Timothy 1998). As part of the discussion on institutional arrangements for tourism, cooperation and coordination between public
sector organisations are essential tools for strengthening the institutional
framework. If adopted and employed, they can help to decrease the misunderstandings and conflicts related to overlapping and conflicting roles and
responsibilities that were mentioned by respondents.

A lack of inter-organisational cooperation and coordination usually
results in the inefficient use of scarce financial resources. For example, in
1998, in Trinidad and Tobago, TIDCO and the THA hired two different
consultancy firms to conduct national stakeholders' consultations, within
a month of each other. Currently, money and resources are being wasted
trying to synthesise these two documents to develop one national tourism
policy document. Another example is the situation in St Kitts and Nevis
where employees from the St Kitts Department of Tourism were contem-
plating developing a destination CD to be used as a marketing tool, not
realising that the Nevis Tourism Bureau had already designed one for its
own use. Even in Antigua and Barbuda, lack of cooperation and collaboration has resulted in foreign travel agents and writers visiting the island
unbeknownst to the Barbuda Tourism Office. In light of these problems,
perhaps we can agree with Timothy's (1998: 54) assertion that

> Efficiency could be improved if various agencies would cooperate by
> coordinating their efforts on development projects. Less money would
> have to be spent and more funds could be divided and allocated to
> other purposes. It might also eliminate some degree of redundancy that
> exists today in many countries.

In addition to administrative and policy coordination (see Hall 1994),
work coordination is also necessary to achieve inter-organisational equilibrium. Work coordination, according to Benson (1975), involves the
pattern of collaboration and cooperation between organisations, whereby
work is coordinated to ensure that the programmes and activities between
the organisations are geared into each other for maximum effectiveness
and efficiency. It therefore stands to reason that if tourism organisations
are not communicating, cooperating and consulting consistently with each
other, then their work coordination will be low, making it difficult for
stakeholders to grasp a holistic vision for tourism development for the
country, as they have no idea what they should be working together to
achieve. This lack of direction often leads to individuals taking their own
paths to tourism development, which may not be in the best interests of
the country overall. It is therefore not surprising that common solutions,
actions and budgets have not been developed to address critical problems
facing these countries' tourism efforts.

A final reason why the issue of inter-organisational communication,
cooperation, consultation and collaboration is so crucial to the discussion

on institutional arrangements, relates to the key role that the latter play in the tourism policy-making process. Brooks (1989: 131) stated that institutional arrangements are viewed as a filter that 'mediate(s) conflict by providing a set of rules and procedures that regulates how and where demands on public policy can be made, who has the authority to take certain decisions and actions, and how decisions and policies are implemented'. In order for this institutional framework to work effectively in practice the actors, who are responsible for shaping the policy agenda, defining the problems, considering and choosing between alternatives, implementing policy and adhering to the rules and regulations, must be in constant contact with each other to sort out misunderstandings and problems as quickly as they arise.

Undoubtedly, there is a need for small island states to adopt a cooperative collaborative approach to tourism policy-making. For twin-island states, the collaborative planning approach is necessary as it 'attempts to act as a tool to solve the many problems that arise when there is a lack of understanding and few shared common goals between the many parties often involved in tourism' (Ladkin and Bertramini 2002: 71). However, much of the discussion on cooperation and collaboration in the tourism literature often neglects the broader political context within which inter-organisational collaboration and cooperation is occurring (see Bramwell and Lane 1999). This is a grave oversight because pleas for better communication and greater cooperation are, more often than not, a mask of more fundamental issues, such as conflict of value systems between organisations, ambiguity of roles and responsibilities between individuals and institutions, and the need for legislative reform (Barrett and Fudge 1981). Given the history and politics of the internal core–periphery relationship existing between the islands discussed here, a collaborative planning approach needs to take into consideration constraining factors such as the distribution of power, resource flows, mistrust and suspicion amongst stakeholders, social, economic and political value systems, and personality and the legislative framework.

Revisiting the study of institutional arrangements for tourism

After an analysis of both the primary and the secondary data, five distinct characteristics of the role of the internal core–periphery model in institutional arrangements for tourism emerge. First, within the tourism public policy-making process, there is recognition of the multiplicity of relationships between public sector tourism organisations. Second, it emphasises the importance of the interaction between public sector officials. Third, these relationships are dynamic, continuous, day-to-day and informal. Fourth, the importance of the roles played by all public sector employees is reinforced. Finally, it emphasises the political nature of inter-organisational relationships and focuses on policies such as financial issues (who raises what amount and who spends it for whose benefit and with what results) and power issues (who gets what, where, how and why).

The internal core–periphery model can affect institutional arrangements in such a way so as to hinder an organisation's ability to act, while empowering others. Drawing on theories on the scope for action within the policy context (Barrett and Fudge 1981), it can be deduced that within the institutional framework, the internal core–periphery model can influence, shape and determine:

- the functions, responsibilities and statutory power conferred upon public sector tourism organisations from central/federal government;
- the political structure, accountability and reporting functions;
- access to resources (constitutional, legal, financial, technological and information) and amount of bargaining power;
- the formal limits of authority and control that organisations located in the internal core may have compared to organisations situated in the periphery;
- the structural relations and the interdependency between organisations and the machinery for communication, cooperation, consultation and collaboration; and
- the contexts for bargaining and negotiating.

These factors provide the setting within which groups of policy actors operate and interact with each other and therefore warrant further research to explore their linkages. It is important to note that there is no single best type of institutional arrangements for tourism, rather each country, depending on its history, political system, needs, infrastructure and resources, will have to design arrangements that best suit its environment. Nevertheless, there are several critical success factors for effective institutional arrangements for tourism. These include:

1 Adopting a collaborative planning and policy approach that identifies and legitimates the key public sector tourism policy-makers from both islands.

2 The need for all public sector tourism organisations and their actors to acknowledge, understand and appreciate the historical, social, economical and political environment, within which tourism policy is developed, implemented and monitored. For twin-island states, this includes being aware of the various ways in which the internal core–periphery relationship has influenced its inter-organisational relationships within the tourism sector.

3   The importance of having a balanced structure within the institutional framework with clear roles and responsibilities for all actors and organisations. It is also important for all actors and organisations involved in the tourism policy process to accept and respect each other's roles, responsibilities and authority. This would hopefully prevent overlapping roles, duplication of responsibilities, discord, communication barriers and ill management of scarce resources.

4   Within the twin-island system, there must be shared leadership of the tourism policy and decision-making process with well-defined, shared goals and vision, realistic expectations, as well as identification of benefits for each partner.

5   A flexible approach on the part of all tourism policy-makers must be characterised by a willingness to compromise and to understand each other's needs and to contribute to shared resources.

6   Good communication between all partners. This involves the establishment of formalised structures to facilitate the consistent exchange of information, ideas, innovation and resources. In addition to communication amongst key tourism policy-makers, there is also the need to communicate the nature and the workings of the institutional framework to all tourism stakeholders, in both the public and the private sector. The challenge will be to bring diverse and passionate views to the same table.

7   The development of an organisational culture, as well as a 'tourism policy-making culture', that seeks to foster cooperation and harmony between all organisations involved in the policy process, in a true spirit of trust and goodwill.

8   Periodic, systematic evaluation of the efficacy of each organisation's role and responsibilities.

9   The legislative and regulatory framework must be in harmony with the industry's vision and must be relevant and harmonious in dispensing the parameters of the respective responsibilities of organisations and actors. Additionally, this framework should encourage coordination among public organisations and promote the creation of joint bodies.

In addition to the above suggestions, a few recommendations can be made for improving inter-organisational relationships in twin-island states:

1   Establish a tourism clearing house – this could take the form of a website database designed to disseminate and share information between all tourism public sector organisations, as well as with the private sector organisations. The website could include information such as:

   a   a list of all the major tourism organisations involved in the tourism policy-making and planning process, clearly stating their main objectives and goals, as well as responsibilities;

b a list of all the employees with their main areas of responsibility stated;

c information on joint projects and programmes (past, present and forthcoming), including details on the project, implementing officers and organisations, and results/accomplishments;

d industry statistics and market trends.

2 Conduct joint nation-wide policy forums in order to address some of the weaknesses with the present institutional arrangements, such as overlapping and conflicting roles and responsibilities; poor inter-organisational communication, cooperation and consultation; inadequate legislative framework; the issue of inequitable sharing of financial resources. In addition to addressing these crucial issues, these forums should also be used to provide vision and leadership for the future growth of the industry.

3 Investigate the possibility of having employee exchange programmes, which was one of the unanimous suggestions that emerged from the study for improving inter-island, inter-organisational relationships.

These factors will all hopefully contribute to ensuring that tourism destinations maintain institutional arrangements that are efficient and effective.

Conclusion

The make-up of institutional arrangements for tourism in twin-island states is inextricably linked to their colonial heritage and by extension their present-day inter-island relationship. The internal core–periphery relationship has set the stage on which institutional structures are designed; organisational roles, responsibilities and authority are assigned; organisations and their actors interact and relate with each other; actors' values and interests are conditioned; distribution of power is determined; conflicts are played out; and how tourism plans and policies are debated, developed and implemented. It has contributed to present-day conflicts and tensions between the key public sector organisations responsible for tourism policy-making, relating to inter-organisational roles, responsibilities, authority, communication, cooperation and consultation.

Consequently, there needs to be an understanding and appreciation of the fact that the circumstances whereby these islands have come into their own as countries still influence their present-day systems and structures. Tourism practitioners and policy-makers should be aware of the way in which core–periphery relationships shape, influence and determine aspects of institutional arrangements for tourism. Such an understanding will undoubtedly assist small island states develop institutional arrangements for tourism that are best suited to their geographical, political, social and economic realities, as well as relevant to today's dynamic environment.

Note

1 Interviews were conducted with representatives from the Antigua and Barbuda Ministry of Tourism; the Antigua and Barbuda Department of Tourism and the Barbuda Tourism Office; the St Kitts and Nevis Ministry of Tourism, the St Kitts Department of Tourism, the Nevis Ministry of Tourism and the Nevis Tourism Bureau; and TIDCO, the Trinidad and Tobago Ministry of Tourism and the THA Department of Tourism (see Table 6.1). A total of fifty-four interviews were conducted: seven in Antigua and Barbuda; eighteen in St Kitts and Nevis; and twenty-nine in Trinidad and Tobago. Respondents were asked about a variety of issues including their roles and responsibilities; inter-organisational relationships focusing on coordination, communication, cooperation and conflict; power arrangements; and financial arrangements. Data were analysed using the framework method (Ritchie and Spencer 1994) at the single-case level and comparative analysis at the multiple-case level, which allowed easy identification of the salient issues and themes.

References

Antigua and Barbuda Department of Tourism (2002) *History and Culture*. Online: <http://www.antigua-barbuda.org/index.html> (accessed 9 September 2002).

Barrett, S. and Fudge, C. (1981) 'Examining the policy-action relationship', in S. Barrett and C. Fudge (eds) *Policy and Action*, London: Methuen.

Benson, J.K. (1975) 'The interorganizational network as a political economy', *Administrative Science Quarterly* 20: 229–49.

Blomgren, K.B. and Sorensen, A. (1998) 'Peripherality – factor or feature? Reflections on peripherality in tourism research', *Progress in Tourism and Hospitality Research* 4: 319–36.

Bramwell, B. and Lane, B. (1999) 'Collaboration and partnerships for sustainable tourism', *Journal of Sustainable Tourism* 7: 179–81.

Brereton, B. (1981) *A History of Modern Trinidad 1783–1962*, Jamaica: Heinemann Educational Books (Caribbean).

Britton, S. (1982) 'The political economy of tourism in the Third World', *Annals of Tourism Research* 9: 331–58.

Brooks, S. (1989) *Public Policy in Canada: An Introduction*, Toronto: McClelland & Stewart.

Caribbean Tourism Organization (CTO) (1999) *Caribbean Tourism Statistical Report*, Barbados: CTO.

—— (2002) Caribbean Tourism Statistical Report (2000–2001 Edition), St Michael, Barbados: CTO.

Commonwealth Secretariat (2000) *Review of the Operation of the Arrangements between the Government of Antigua and Barbuda and the Barbuda Local Council: Report of the Commonwealth Review Team*, London: Commonwealth Secretariat.

Constitutional Review Commission (2002) *Report of the Constitutional Review Commission*. Online: http://www.antigua-barbuda.com/busnss_politics/constitution_of_antigua_and_barbuda.pdf (accessed 29 October 2002).

De Kadt, E. (1979) *Tourism: Passport to Development?*, Oxford: Oxford University Press.

Dickenson, J., Gould, B., Clarke, C., Mather, S., Prothero, M., Siddle, D., Smith, C. and Thomas-Hope, E. (1996) *The Geography of the Third World*, 2nd edn, London: Routledge.

Getz, D. and Jamal, T. (1994) 'The environmental-community symbiosis: a case for collaborative tourism planning', *Journal of Sustainable Tourism* 2: 152–73.

Government of Trinidad and Tobago (2002) Official Government website. Online: <http://www.gov.tt/government> (accessed 25 August 2002).

Hall, C.M. (1994) *Tourism and Politics: Policy, Power and Place*, London: John Wiley and Sons.

—— (2000) *Tourism Planning Processes and Relationships*, Harlow: Prentice-Hall.

Hall, C.M. and Jenkins, J.M. (1995) *Tourism and Public Policy*, London: Routledge.

Howlett, M. and Ramesh, M. (1995) *Studying Public Policy: Policy Cycles and Policy Subsystems*, Toronto: Oxford University Press.

Jamal, T. and Getz, D. (1999) 'Community roundtables for tourism-related conflicts: the dialects of consensus and process structures', *Journal of Sustainable Tourism* 7: 290–313.

Jenkins, J. (2000) 'The dynamics of regional tourism organisations in New South Wales, Australia: history, structures and operations', *Current Issues in Tourism* 3: 175–203.

Knox, P. and Agnew, J. (1994) *The Geography of the World Economy*, 2nd edn, London: Arnold.

Ladkin, A. and Bertramini, A.M. (2002) 'Collaborative tourism planning: a case of Cusco, Peru', *Current Issues in Tourism* 5: 71–93.

Lowenthal, D. and Clarke, C.G. (1979) 'Common lands, common aims: the distinctive Barbudan community', in M. Cross and A. Marks (eds) *Peasants, Plantations and Rural Communities in the Caribbean*, Department of Sociology and Department of Caribbean studies, University of Surrey.

Mitchell, B. (1989) *Geography and Resource Analysis*, 2nd edn, Harlow: Longman.

Ottley, C.R. (1973) *The Story of Tobago*, Trinidad: Longman.

Paddison, R. (1983) *The Fragmented State: The Political Geography of Power*, Oxford: Basil Blackwell.

Parker, S. (1999) 'Collaboration on tourism policy making: environmental and commercial sustainability on Bonaire, NA', *Journal of Sustainable Tourism* 7: 240–59.

Pearce, D.G. (1992) *Tourist Organisations*, Harlow: Longman.

Phillips, F. (1985) *West Indian Constitutions: Post-Independence Reform*, New York: Oceana Publications.

Premdas, R.R. (1998) *Secession and Self-Determination in the Caribbean: Nevis and Tobago*, St Augustine: The University of the West Indies Press.

—— (2000) *Self-Determination and Decentralisation in the Caribbean: Tobago and Nevis*. Online: <http://uwichill.edu.bb/bnccde/sk&n/conference/papers/RRPremdas.html> (accessed 11 August 2001).

Ritchie, J. and Spencer, L. (1994) 'Qualitative data analysis for applied policy research', in A. Bryman and R. Burgess (eds) *Analyzing Qualitative Data*, London: Routledge.

Selin, S. (1999) 'Developing a typology of sustainable tourism partnerships', *Journal of Sustainable Tourism* 7: 260–73.

St Kitts and Nevis Information Service (2001) *St Kitts and Nevis Background Information*, Basseterre, St Kitts: Ministry of Tourism, Information, Telecommunications, Commerce and Consumer Affairs.

St Kitts and Nevis Ministry of Tourism (2002) *Visitor Arrivals by Category*, Basseterre, St Kitts: Internal publication, St Kitts and Nevis Ministry of Tourism.

Sutton, P. (1987) 'Political aspects', in C. Clarke and T. Payne (eds) *Politics, Security and Development in Small States*, London: Allen & Unwin.

Taylor, G.D. (1987) 'Research in national tourism organizations', in J.R.B. Ritchie and C.R. Goeldner (eds) *Travel, Tourism, and Hospitality Research: A Handbook for Managers and Researchers*, New York: John Wiley and Sons.

Timothy, D.J. (1998) 'Cooperative Tourism Planning in a Developing Destination', *Journal of Sustainable Tourism* 6: 52–68.

—— (1999) 'Cross-border partnership in tourism resource management: international parks along the US–Canada border', *Journal of Sustainable Tourism* 7: 182–205.

Tosun, C. and Jenkins, C.L. (1998) 'The evolution of tourism planning in Third-World countries: a critique', *Progress in Tourism & Hospitality Research* 4: 101–14.

Turner, L. and Ash, J. (1975) *The Golden Hordes: International Tourism and the Pleasure Periphery*, London: Constable & Company.

Weaver, D.B. (1998) 'Peripheries of the periphery: tourism in Tobago and Barbuda', *Annals of Tourism Research* 25: 292–313.

Legislations cited

Antigua and Barbuda

Antigua and Barbuda Constitution Order 1981.
Barbuda Local Government Act 1976.

St Kitts and Nevis

Nevis Tourism Promotion and Marketing Authority Ordinance 1999.
Saint Christopher and Nevis Constitution Order 1983 (Statutory Instruments 1983, No. 881), London: HMSO.
Saint Christopher Tourism Authority Act 1999.

Trinidad and Tobago

The Constitution of the Republic of Trinidad and Tobago 1962.
The Tobago House of Assembly Act No. 37 of 1980.
The Tobago House of Assembly Act No. 40 of 1996.
Tourism Development Act 2000.

7 Tourism and supranationalism in the Caribbean

Dallen J. Timothy

Introduction

Compared to many areas of the world, the islands of the Caribbean have enjoyed a rather peaceful existence with relatively few major power struggles and heated political altercations. However, the region has not been free of power conflicts altogether. Even as early as the Arawak habitation of the islands some 1,000–4,000 years ago and the subsequent invasion of the warring Carib tribes from South America, power struggles have existed in the Caribbean. At no time, however, was the struggle for power more clearly demonstrated than during the colonial period from the fifteenth century to the present day.

European powers spread throughout most of the world in an effort to conquer and claim. Many socio-political and cultural elements are evidence of this period of widespread exploration and exploitation in many parts of the world, and this is certainly the case in the Caribbean. The existence of Spanish, French, Dutch (nominally) and English as the primary lingua franca throughout the region and the contemporary racial mix of blacks, whites and South Asians on the islands testify to the once-powerful might of European metropoles in the Caribbean basin. During the past fifty years colonial territories began their status as independent nations, and these today, together with those who remain in colonial status, face assorted political and economic challenges and socio-cultural identity crises. Recent political events in, for example, Cuba (1950s), Grenada (1979–83), Haiti (1980s–90s), Puerto Rico (1950s–present), and St Kitts and Nevis (1990s–present) demonstrate power struggles and identity crises in the Caribbean stemming from, directly and/or indirectly, colonialism and political fragmentation.

Despite residual problems, colonialism and fragmentation in the Caribbean have generally created an enormous part of the tourist appeal of the islands. Aside from natural pleasures of climate, beaches and flora, the mix of African traditions with Curaçao's Dutch tang, the Dominican Republic's Latino pulse, the British Virgin Islands' English touch, and Martinique's typically French Caribbean flavour help create a cultural allure for millions of tourists. Additionally, remnants of the slave trade and urban colonial architecture provide the broadest category of heritage sites in the region (see also Found, Chapter 8).

While the Caribbean is fragmented politically, culturally and economically it is unified by a colonial past and a struggle for a common future. Moreover, nearly all of the islands depend heavily on tourism as a primary source of foreign exchange and employment. The region's islands and countries, therefore, have made efforts to unite in a joint cause to enhance their economic and political stability through multinational trade alliances, many of which deal directly or indirectly with tourism. The purpose of this chapter is to describe and examine various forms of supranational alliances, or economic trading blocs, which have shifted traditional power structures in the Caribbean and led to cooperative efforts in economic development, including tourism.

Tourism and regional cooperation in the Caribbean

Since the mid-1900s, countries have been examining the notion of cross-border regional (supranational) cooperation, particularly in areas of economic development, trade, human mobility and political stability. Many supranational alliances have been created in order to decrease the trade barrier effects of political boundaries. Many of these partnerships also involve other areas of human welfare and economic development, including cross-border migration, education, environmental conservation and tourism. Timothy and Teye (forthcoming) proposed four scales of supranational cooperation: global, regional, bilateral and inter-local. The scale of most concern in this chapter, however, is regional cooperation.

With the recent trend towards globalisation has come the growth of international trade alliances or economic communities as a way of assuring national economic survival (Jessop 1995). According to Balassa (1961), there are several stages, or types, of regional economic integration, including free trade areas, customs unions, common markets, economic unions, monetary unions and full economic unions. Regardless of the level of integration, the basic rationale for the growth of regional economic alliances is that individual countries can enlarge their trading hinterlands by uniting with other countries in a mutually beneficial relationship. In regions where nations are small and fragmented, small domestic markets are a major constraint to economic development and industrialisation, particularly when inadequate economies of scale fuel import substitutions. In these cases, supranational integration may improve access to larger markets and lessen the cost of imported items because they are purchased at a higher level of critical mass from outside the trading bloc (Campbell 1997; Müllerleile 1996; Timothy, forthcoming). According to Hussain (1999: 27), nations join regional alliances 'based on the realization, especially among smaller and less developed states, that national developmental objectives can be satisfactorily met only through collective regional effort'.

Cross-border cooperation is now beginning to be recognised as a tool for achieving sustainable tourism development goals, as collaboration has the potential to help form more equitable relationships, improve efficiency in operations and development, enhance ecological and cultural harmony, and create holistic management frameworks (Timothy 1999).

Most observers generally feel that globalisation and the rise of supranational trading blocs stimulate more growth in international tourism (Smith 1994; Taylor 1994; Teye 2000). They can potentially decrease the customary barrier effects of international borders and thus allow tourism to benefit from the more dynamic international relations that follow (Timothy, forthcoming). With a decline in the barrier effects of political boundaries through supranationalism, larger 'functional nations' are created in both demographic and geographic terms. Visitors from outside the region can travel more freely between nations, which may attract larger numbers of tourists (Timothy 1995). In addition to creating larger destinations, larger origins are also created. Most economic communities attempt to raise standards of living among all member nations to a more equal level and to improve economies of member states in general. According to Smith (1994) and Taylor (1994), this will result in higher numbers of business and pleasure trips inside supranational communities and further afield.

Many regional alliances have developed since World War II in nearly all parts of the world (e.g. European Union (EU), North American Free Trade Agreement (NAFTA), Association of Southeast Asian Nations (ASEAN)). In the Caribbean, economic communities have existed since the mid-1900s, when most of the islands were still under colonial rule. These alliances have evolved considerably through the years, however, as new-found sovereignty and autonomy among the region's nations, changing political regimes, and shifting economic structures have brought about a need to re-evaluate the purposes, extents and types of multilateral relationships among neighbours. Today there are several supranational alliances in the Caribbean, many of which overlap in their membership and geographical coverage.

Müllerleile (1996: 31–2) highlighted six reasons the islands of the Caribbean have been eager to enact multilateral trade agreements:

- to prevent complete economic collapse or significant decline with each microstate acting on its own;
- the fear that globalisation and technological development might be detrimental to Caribbean identity;
- unification could strengthen the region politically, at the risk of some loss of individual national sovereignty;
- by collaborating in the form of regional alliances, a critical mass could be created to enable the various nations to become self-reliant;
- simple survival, particularly in terms of foreign trade policy and security interests;

- the quality of leadership and experience, perhaps more effective through larger political units.

Owing to the economic importance of tourism in the region, nearly all of the Caribbean's trade alliances deal directly or indirectly with various aspects of tourism. On a global scale this is quite uncommon. Few economic community treaties and agreements deal directly with tourism, the most notable exceptions being ASEAN and the EU. In most other instances tourism is implied under arrangements that deal primarily with human movement, trade in goods and services, and transportation. Timothy (forthcoming, 2002, 2001) and Timothy and Teye (forthcoming) have suggested several issues most trading blocs consider that deal directly with tourism, even when tourism is not mentioned explicitly in the negotiations. These include transportation, environmental and community sustainability, flow of people, trade in goods and services, unified economics and banking, and image and promotion.

Some of the most critical areas of cooperation in transportation include railway and road development, airport construction and public transportation. Air alliances are an important part of these negotiations and are becoming more commonplace in regional economic blocs worldwide. In theory, regional cooperation in transportation can enhance accessibility and reduce unnecessary duplication of services and facilities. This may result in more efficient production and standardised quality. Transportation is one of the most commonly cited problems of tourism in the broader Caribbean (see Duval, Chapter 17). Wilkinson (1997: 40–1) noted several issues that have affected air and sea travel in the Caribbean and that have spurred regional alliances in the area to act on transportation. First, there has traditionally been an overdependence on foreign carriers, primarily owing to their control over computer reservation systems and their ability to form service partnerships. These external hegemonic relationships benefit from economies of scale that are far greater than those that could be realised by small regional carriers. Second, airline deregulation in the United States in the early 1980s resulted in bankruptcies among some airlines that serviced the Caribbean, with only some of the routes being picked up later by other airlines. Third, since most of the region's airlines are small, almost all state-owned, underequipped and at the mercy of foreign carriers, they must be more aggressive in their endeavours to profit from tourism.

Environmental issues have also been an important part of most supranational agreements, but they have been all but ignored in the context of tourism. Regional agreements on environmental issues are particularly important when specific resources extend throughout an entire region, for cross-border partnerships can encourage more balanced use of resources and more consistent systems of conservation, and sound regional policies may be developed to assure more sustainable uses of the environment (Timothy 1999). Cultural conservation has rarely been acknowledged in supranational treaties, although this is beginning to change in relation to arts and built cultural heritage.

One of the main goals of most supranational coalitions is to ease the barrier effects of borders for people and goods. The three groups of people most affected by this are tourists from within the alliance, tourists from outside the alliance and workers from within the alliance (Timothy 2001). Economic communities aim to relieve immigration procedures for citizens of member states and sometimes even for outsiders. Visa requirements are sometimes simplified or eliminated altogether. In some alliances, member citizens do not require visas to travel to other member nations, and some have instituted a no-passport policy, wherein travellers do not even require a passport if they are carrying national identity cards. Visa waivers for migrant workers are also becoming more common elements of supranational alliances, which have significant implications for tourism-based human resource demands. Politically and economically, many alliances aim to create single customs unions, combined tariff barriers and, in some rare cases, a common currency. With the exception of the European Union, few economic communities have considered implementing a common currency among all member nations, although several have begun to consider issuing joint community passports.

Many of the issues described so far have considerable implications for regional marketing and promotion. Combined advertising campaigns, common promotional budgets, equal exposure in promotion outlets and joint marketing research may have the effect of saving funds when entire regions are marketed as large destinations wherein people can visit multiple countries (Timothy and Teye, forthcoming). Several alliances (e.g. The South African Development Community (SADC) and ASEAN) have recently established tourism units specifically to handle pan-regional promotional efforts and market-related data collection, which have achieved only marginal levels of success. The following sections examine these issues from the standpoint of various multinational alliances in the Caribbean.

The Caribbean Community (CARICOM)

In 1973, CARICOM was created as the successor body to a loose grouping of countries known until that time as the Caribbean Free Trade Area (CARIFTA) (Erisman 1992; Grugel 1995). Its original members included Barbados, Guyana, Jamaica and Trinidad and Tobago. In 1995, Suriname became the first non-English-speaking country to join the Community. A total of fifteen island states comprise CARICOM today (Table 7.1).

At inception, the objectives of CARICOM were: (1) to integrate member states economically through the establishment of a common market; (2) to develop and coordinate joint foreign policies; and (3) to promote functional cooperation (CARICOM Secretariat 1997a; Gill 1997). Many of the

Table 7.1 Members of CARICOM, 2003

Antigua and Barbuda	Grenada	St Kitts and Nevis
Bahamas	Guyana	St Lucia
Barbados	Haiti	St Vincent and the Grenadines
Belize	Jamaica	Suriname
Dominica	Montserrat	Trinidad and Tobago

association's most important efforts to date have dealt with communication and information, science and technology, health, drug abuse, education, meteorology, disaster preparedness and legal issues (Gill 1997).

Within CARICOM, air transportation has been hotly contested for years, particularly, as Wilkinson (1997) mentions, in response to foreign monopolies in air transportation to the islands. CARICOM is strongly urging its member states to form coalitions in air transport to combat some of the problems arising from outside interests in the airline industry, especially in terms of travel to and between the islands and finding ways of attracting more tourists through reasonably priced direct flights from the major markets (Müllerleile 1996). CARICOM's efforts began to pay off in 1994, when LIAT, the Caribbean's largest air carrier, was privatised with shareholders, including eleven Caribbean governments, making it a common regional carrier (Blacklock 1997; Müllerleile 1996). Caribbean Star and Caribbean Sun airlines have also developed in the past few years from CARICOM's efforts at air transportation cooperation.

CARICOM has been the most active Caribbean alliance in terms of the flow of people and goods. It has made many efforts to unite its members in eliminating travel restrictions within the region for member citizens. The CARICOM Single Market aims to promote the hassle-free and unrestricted movement of its citizens as tourists, job-seekers, entrepreneurs and traders. Its efforts also include eradicating restrictions on trade in goods, services and capital that originate within CARICOM (CARICOM Secretariat 1997b). The alliance is also working to eliminate all internal tariffs and quota restrictions for the movement of goods, as well as harmonise regional standards for goods and services. So far, these intentions have encountered significant limitations (discussed below), with the most notable result being the elimination of the need for passports for CARICOM citizens from several member states, thus expediting immigration procedures for people with CARICOM identifications and passports. Free movement of workers has also been extended to citizens of CARICOM nations. Work permits have been eliminated between some member nations, and university graduates are permitted to move throughout the region without work permits in at least seven countries. Freedom of movement is presently being extended to athletes, artists, musicians and media personnel as well (CARICOM Secretariat 1997b).

While CARICOM is certainly interested in sustainable development generally, there has been relatively little discussion or advancement regarding how tourism fits within the broader goals of sustainability. However, during CARICOM's 1992 Summit on Tourism, sustainability was an important consideration. The primary issue of concern was cruise ships because so many CARICOM states are highly dependent on this form of tourism. The key theme was large numbers of ship passengers being unloaded for short periods of time on relatively small and fragile islands, particularly in light of the fact that they spend very little money once in port compared to what they spend on the ships. The second concern relates to the ecological impacts caused by the ships themselves and by the construction of major port facilities (Wilkinson 1997: 41; Wood, Chapter 9). The primary action taken so far in response to this summit is the charging of a standard port fee per person disembarking from the ship, which is meant to help island governments offset some of the ecological, social and economic disadvantages of cruise tourism – an action which has been met with considerable debate on its level of efficacy.

Since its inception, CARICOM has worked diligently towards becoming a single market and economy and has had some, albeit limited, success in creating a common external tariff instead of several relatively high tariffs levied by each member state (Campbell 1997; CARICOM Secretariat 1997a). In recent years, proposals have been made to establish a CARICOM common currency (favouring the Caribbean dollar (CARICOM Secretariat 1997b)), although this recommendation has yet to be implemented or even justified.

As would be expected, political integration has been much slower than economic integration. Discussions frequently surface in CARICOM meetings to address the issue of standardising foreign policy among member states (CARICOM Secretariat 1997b; Gill 1997). This is particularly so in the case of diplomatic representation abroad. Another movement has been the desire within CARICOM for common passports, which could raise problems regarding overlap in membership between alliances.

Although marketing and pan-regional promotion at a global level were lacking in the early 1990s (Wilkinson 1997), several collaborative efforts to promote and market the entire region abroad have since been enacted. Recent efforts, under the auspices of CARICOM and the Association of Caribbean States (ACS), have sought to create a 'Caribbean brand' marketing campaign (CARICOM Secretariat 2002). Such efforts in recent years have intensified in recognition of growing competition from similar destinations.

Organisation of Eastern Caribbean States (OECS)

As the British colonies of the eastern Caribbean gained independence from Great Britain through the 1960s and 1970s, it became clear that there was a need to create a formal alliance that would assist each country in its development efforts and support the other new countries of the former British Empire. As a result, the colonial West Indies Associate States Council of Ministers (established in 1966) and the Eastern Caribbean Common Market (begun in 1968) were superseded by the Organisation of Eastern Caribbean States in June 1981 when seven eastern Caribbean countries signed the Treaty of Basseterre, thus agreeing to cooperate and promote solidarity among member states (OECS 2002). The OECS membership today includes Anguilla, Antigua and Barbuda, the British Virgin Islands, Dominica, Grenada, Montserrat, St Kitts and Nevis, St Lucia, and St Vincent and the Grenadines. While not all members are fully independent and sovereign, they are all part of the British Commonwealth or still under British rule.

The primary aim of the OECS is to contribute to the sustainable development of all member countries. Article 3 of the Treaty lists the tasks of the OECS as:

- cooperation between and among member states; promotion of unity and solidarity;
- defence of sovereignty, territorial integrity and independence of member states;
- common foreign policy and joint representation abroad; and
- the promotion of economic integration (Müllerleile 1996).

In July 2001, all OECS heads of government met and agreed to deepen economic integration by creating an economic union that would assist those member countries facing the challenges of globalisation and trade liberalisation. The objectives of the new treaty include facilitating the free movement of people, goods, services and capital between member nations and to improve economic diversification and human resource development (OECS 2002). While CARICOM and the ACS are most heavily involved in issues related to air transportation, the OECS also plays a role. Its Directorate of Civil Aviation is one of the most established common institutions in the region and deals with all OECS governments in matters pertaining to civil aviation, including airport development, implementation of international air travel conventions, and quality and adequacy of air services (Müllerleile 1996).

In economic terms, the founding of the East Caribbean Currency Authority in 1965 was rather ground-breaking, for it established a common currency (East Caribbean Dollar – EC$) long before most of the present alliances were created. Today, all members of the OECS, with the exception of the British Virgin Islands, are members of the monetary union under the Eastern Caribbean Central Bank and use the EC dollar as their monetary unit. The monetary union of the OECS has been a benefit for OECS states because it has ensured currency stability (Gill 1997; Müllerleile

1996). A common OECS passport was planned to be issued in January 2003, but as of April 2003 this goal had not yet been met.

Association of Caribbean States (ACS)

The ACS is the newest supranational regional alliance in the Caribbean. Founded in 1994, it was formed as a result of efforts among the broader Caribbean region to widen the integration process beyond CARICOM and the OECS. Present membership includes all CARICOM members (except Montserrat), several Latin American countries, and Cuba and the Dominican Republic (see Table 7.2). Additionally, France, Aruba and the Netherlands Antilles are associate members, and all non-independent islands of the wider Caribbean are technically eligible for associate status (Insanally 1997: 54).

The ACS was established as an association for consultation, collaboration and concerted action in economic integration. In common with OECS and CARICOM, the main goals of the ACS are to strengthen regional cooperation and economic integration, but its goals also include preserving the environmental integrity of the Caribbean Sea and promoting the sustainable development of the entire region. To accomplish these goals, several subcommittees have been set up to carry out stated objectives. These committees oversee natural resources; protection and conservation of the environment; science, technology, health, education and culture; tourism; trade development and external economic relations; and transportation (Insanally 1997: 56). Since its inception, the ACS has been highly concerned with tourism and has pursued its development and sustainability vigorously (ACS 2002). The ACS is the primary supranational alliance dealing directly with tourism in the Caribbean region (not including the Caribbean Tourism Organization, discussed below). As a result of the region-wide importance of tourism, the Ministerial Council (the governing body of the ACS) met often throughout the 1990s and developed a Plan of Action on Tourism, Trade and Transportation. Such efforts are based on three premises: that tourism and transportation are activities common

Table 7.2 Members of ACS, 2003

Antigua and Barbuda	El Salvador	Panama
Bahamas	Grenada	St Kitts and Nevis
Belize	Guatemala	St Lucia
Barbados	Guyana	St Vincent and the Grenadines
Colombia	Haiti	Suriname
Costa Rica	Honduras	Trinidad and Tobago
Cuba	Jamaica	Venezuela
Dominica	Mexico	
Dominican Republic	Nicaragua	

to all member countries, that there was a need for a regional strategy to assist the Caribbean in competing in the global marketplace, and that tourism is an industry which relies on a sustainable and high-quality natural and cultural environment.

While all of the Caribbean multinational alliances deal, to some degree, with transportation, the ACS is particularly involved in this facet of tourism. One of the ACS's strategies for improving tourism is the development of air and sea transport services and the strengthening of interconnections in the Caribbean. The objectives in this endeavour aim to guarantee air access to ACS countries and promote socially and ecologically responsible cruise tourism (Hatton 1997). One programme currently being developed by the ACS is the 'Uniting the Caribbean by Air and Sea' campaign through its transportation committee. This means that airport infrastructures in ACS member states are being evaluated and modernised. Regulatory and operation frameworks for inter-island air transport are also being established, particularly those concerned with air safety and an open skies policy. In terms of sea travel, evaluations and modernisation of port infrastructures are also being discussed and gradually operationalised. Issues including security, routing, transport of toxic waste, information exchange, energy efficiency and provision of port services are all being considered by the ACS (Insanally 1997).

Under the direction of the ACS the Caribbean Airline Alliance Committee (CAAC) acts as a liaison between regional airlines in operational matters. Its work includes arranging the use of the same gates and check-in areas by Caribbean airlines and jointly contracting a single ground handler for all the regional carriers. Its efforts also include negotiating partnerships in the areas of training, maintenance, insurance, revenue accounting, e-ticketing, reservation services and fuel purchases (Blacklock 1997: 160).

In the area of sustainability in tourism, the ACS has been quite active. During the 1990s, the alliance established a Special Committee on Sustainable Tourism, which developed a sustainable tourism plan, encompassing many elements of the industry, including transportation, environment and human resources. The plan included several strategies for achieving sustainable tourism. Many of the strategies were designed to protect the environment and the communities within which the industry functions. As part of its first strategy, the ACS noted the need to protect cultural integrity on the islands by identifying, promoting and encouraging the preservation of cultural values of the Caribbean; regulating the region's image; broadening the definition of what is considered the heritage of humankind in the Caribbean area and acquiring funds to protect and salvage the region's cultural heritage (Hatton 1997).

Another strategy adopted by the ACS for achieving sustainable development in tourism is to seek out national experiences in the region upon which models of preservation, conservation and use of the environment can be based. Issues of particular interest at the supranational level are the negative ecological and economic results of tourist concentrations in specific areas, patterns of interaction between nature and society, and water, drainage and waste management and control (Hatton 1997). In fact, the ACS sustainable tourism plan goes beyond what most supranational alliances cover in sustainability terms by including a strategy to educe community member and other stakeholder participation in tourism-planning and decision-making, encourage residents to profit from the benefits of tourism by stimulating more entrepreneurial activity, and promote cooperation between government agencies and between the private and public sectors.

Establishing a region-wide 'zone of sustainable tourism' was a significant ambition for the Committee on Sustainable Tourism throughout the 1990s, where the principles and goals it established could be better carried out on a regional level. In the mid-1990s, the Committee began drafting an agreement on the designation of the zone of sustainable tourism. Finally, in 2001, the ambition of the ACS was realised when the Convention on the Sustainable Tourism Zone of the Caribbean (STZC) was established as a tool for enforcing the aim of the plan and the committee: to ensure that Caribbean destinations can attract visitors, but at the same time, do so in a way that does not harm the physical environment or the communities where tourism takes place (ACS 2002).

Human resources have been cursorily addressed by CARICOM and OECS, but it has been primarily the ACS that has addressed this issue specifically in the realm of tourism. Education and training to prepare people to work in tourism and to understand the issues of sustainable tourism are among the most important goals of the ACS sustainability programme. Specific areas currently targeted include professional training programmes, preparation of community members to service the needs of visitors, sensitisation programmes for local authorities and community leaders, and even training programmes for tourists who will visit the unique environments of the Caribbean (Hatton 1997).

From a marketing perspective, there have been attempts to promote the region holistically, and particularly popular are new movements to formulate multi-island vacation packages. The ACS is in the process of developing strategies that will encourage multi-destination tourism and achieve complementary relations rather than competitiveness in the industry between islands. In so doing, the unique characteristics of each destination are being highlighted in regional marketing (Hatton 1997). The building of a Caribbean information centre is also among the ACS's plans for sustainable tourism (ACS 2002).

Caribbean Tourism Organization (CTO)

CARICOM, OECS and ACS are supranational governance bodies that deal with economic and political issues between member states, and they are endowed with authority through formal treaty to legislate economic, ecological, social and political actions. There is, however, a new development in the Caribbean that resembles supranationalism, although owing to its lack of the above characteristics, is not supranationalism in the strictest sense. The Caribbean Tourism Organization (CTO) was established in January 1989 from a merger of the Caribbean Tourism Association (CTA) and the Caribbean Tourism Research and Development Centre (CTRDC). The CTO is an international development agency with offices in the Caribbean and North America. Its primary duties include marketing, human resource development, tourism research and information management, product development, statistical collection and presentation, technical assistance and consulting services (Holder 1997).

As part of its marketing endeavours, the CTO provides tour operators and travel agents with information about the entire region, yet what is particularly important is its efforts to give special support to its member countries that are unable to represent themselves (Holder 1997). There is cooperation between the CTO and the Caribbean regional trading blocs, but it is not sponsored by them. Since its establishment in 1989 the CTO has taken the responsibility of many of the tourism concerns of CARICOM, OECS and ACS.

Weaknesses in Caribbean supranational cooperation

Despite its potential to assist in achieving sustainability in tourism, supranationalism and cross-border cooperation in general face many challenges. Several of these are addressed below.

Small size and population

Perhaps the most glaring challenge in the Caribbean is small size in both area and population (Gill 1997). The population of OECS member countries, for example, is just over 500,000, and the land area is some 2,900 square kilometres, which makes achieving economies of scale difficult, even within cooperative frameworks such as the OECS. Small domestic markets make domestic production unfeasible in most islands, and production for export is perhaps even more difficult. The small physical size of many islands translates simply into limited natural resources that can be used for economic development. As a result, many Caribbean islands are overly dependent on tourism, which has many dangerous repercussions. Limited resources also restrict the range of cultural and natural resources that can be used for tourism. As a result, individual islands are excessively dependent on one specific type of tourism, which is also hazardous owing to various changes in product demand. Additionally, the small population creates shortages of human resources and of skilled leaders in tourism and policy-making who could undertake far-reaching regional leadership commitments.

Geographical fragmentation and underdevelopment

Unlike the European Union and other supranationalist movements, the Caribbean is comprised of geographically diverse and scattered islands, which typically makes travelling in the region more difficult and costly. Another result of small size and fragmentation is the considerable differences in socio-economic development and population sizes between member nations. Hatton (1997) notes that, for the purposes of the ACS's Zone of Sustainable Tourism, different levels of development can create significant obstacles to implementing such a plan and prolong the process. In the least developed members of CARICOM, ACS and OECS, it is often difficult to justify being involved in regional communities when citizens of their own countries are impoverished, undernourished and undereducated. Differences in levels of development between member states can be problematic, because trade among 'countries at different levels of development, more often than not, favours the more developed countries to the detriment of the less developed ones' (Bista 1991: 18).

In ecological terms, balanced partnerships are difficult to achieve when significant socio-economic differences prevail (Williams *et al.* 2001). Issues that receive a great deal of urgent attention in more affluent societies (e.g. environmental conservation) are less important in poorer nations, because their primary goal is to provide food, jobs and education for their populations. Imbalances therefore occur when one partner has the knowledge and monetary means for tourism development and conservation but other members do not (Parent 1990). Uneven standards of development also typically mean there will be inconsistent standards of environmental protection (Timothy and Teye, forthcoming).

Homogenising the product

From a marketing perspective, there is also a common fear that collaborative networks in the Caribbean may result in the market being unable to distinguish between the islands (Wilkinson 1997: 208). In other words, it reflects a potential danger of creating homogeneity to the extreme rather than each island remaining individualised with its own competitive advantages. Müllerleile (1996) noted that, to some degree, this problem already exists since almost all Commonwealth Caribbean states currently offer similar attractions to the same clienteles.

Policy restrictions and lack of political will

Political or policy-based limitations also exist. Joining a supranational alliance necessarily means giving up some degree of sovereignty, and because sovereignty is generally viewed as pure control over national space and national issues, there traditionally has been a lack of commitment on the part of national governments to ratify all supranational agreements. This reluctance stems from the fact that national concerns nearly always prevail over broader regional concerns, and countries are generally unwilling to concede any degree of power. This is particularly so when a multilateral union has authority to legislate actions and policies in the area of economic development, industry and trade, including tourism. Gill (1997: 102) illustrates this problem from a currency perspective: 'Since the OECS countries share a common currency ... which is managed autonomously by the Eastern Caribbean Central Bank, individual countries cannot employ exchange rate mechanisms as a lever for increasing competitiveness, as other CARICOM countries do.' Another example relates to states' willingness to reduce or eliminate intra-regional import and export tariffs. Although most intra-regional trade is currently free of illegal restrictions, some member states in CARICOM continue to apply illegal and discriminatory tariff measures (duties and other customs taxes) and illegal licensing requirements on imported goods and services (Gill 1997). This is common throughout the world, where there is a reluctance on the part of member nations to eliminate tariff and non-tariff barriers owing to a desire to protect individual national economies and to continue raising revenue (Bista 1991).

Another policy-oriented drawback relates to the 'unanimity rule'. Most Caribbean alliances, in particular CARICOM and the OECS, require unanimity among member states in policy- and decision-making. This rule, according to some observers, 'places the entire membership at the mercy of a single dissenter' (Gill 1997: 100), which frequently results in inaction and non-implementation of decisions. This problem has led to breached contracts and lost credibility within the region and throughout the world as individual states abandon the goals set in CARICOM, OECS, ACS decisions.

Political tensions

As mentioned earlier, power struggles in the Caribbean do not compare with those in many other parts of the world, but they do constitute a major concern for the Caribbean's supranational alliances and tourism. CARICOM is particularly concerned with heated border disputes currently going on between Suriname and Guyana. Dissent in the region regarding Nevis's efforts to declare its independence from St Kitts has generated tension among several neighbouring islands. The coups and military conflicts in Haiti in recent years, and the ban on trade with, and travel to, Cuba by the United States government are significant concerns for all of the Caribbean's trading blocs. There are also several nascent independence movements in the region. These types of political tensions have clear implications for tourism, which is a highly conflict-sensitive service industry.

As such, the CTO and the economic alliances in the region are concerned how these issues will affect the continued success of tourism.

Political tensions also include the fact that supranationalism is difficult to manage. It often creates harmful competition, is value-laden and time-consuming, and may tend to favour elites through political opportunism (Blatter 1997; Church and Reid 1996, 1999; Timothy 1999), thereby marginalising further the positions of less-affluent community members and communities.

Conclusions

The idea of supranational associations at the regional level has become quite popular in the Caribbean region over the past thirty years as a way of enhancing regional economies. Tourism plays an important, albeit small, role in these negotiations. Given the prominence of tourism in the region, it is surprising that it has not received attention in economic bloc treaties commensurate with its importance. Despite their limited direct coverage of tourism, Caribbean trading blocs have dealt with the industry through negotiations regarding transportation, environmental and community sustainability, flow of people, unified economies and banking, and image and promotion. These are fairly consistent with issues of supranational coalitions throughout the world. However, there are several barriers to regional cooperation that are unique to the Caribbean. These include small size, geographical and political fragmentation, and varying levels of development between member states. These and other political and socio-economic limitations have contributed significantly to delays in shaping common regional policies related to tourism and its ancillary areas. Thus, because of its limitations, of which most countries are aware, supranationalism in the Caribbean has little chance of achieving all of its goals adequately. That so many countries have memberships in overlapping alliances is perhaps evidence of individual economic communities not meeting the needs of Caribbean nations.

Despite these difficulties, it is clear that supranational cooperation will continue in the Caribbean far into the future. In fact, many Caribbean islands are members of other trading blocs that were not considered in this chapter, including SELA (Latin American Economic System) and LAFTA (Latin American Free Trade Association). Since 1990, there has been talk of a Free Trade Area of the Americas (FTAA), and several summits have endorsed the idea. This new proposal has its share of critics, who suggest such an alliance will be of most benefit to the most powerful partners (i.e. the United States, Canada and Mexico) at the expense of the smaller and weaker partners. Nonetheless, most of the Caribbean islands (with the exception of Cuba) are enthusiastic about the prospect of forming such an expansive partnership with all countries in the Americas (Campbell 1997; Gill 1997).

References

Association of Caribbean States (ACS) (2002) About the Association of Caribbean States. Online: <www.acs-aec.org>.

Balassa, B. (1961) *The Theory of Economic Integration*, Homewood: Richard D. Irwin.

Bista, N.K. (1991) *PTA in Intra-ASEAN Trade: Issues of Relevance to SAARC*, Singapore: Institute of Southeast Asian Studies.

Blacklock, M. (1997) 'Cooperating to compete in a global environment – airlines in the ACS', in ACS (ed.) *Association of Caribbean States: Trade, Transport and Tourism*, Port of Spain: Association of Caribbean States.

Blatter, J. (1997) 'Explaining crossborder cooperation: a border-focused and border-external approach', *Journal of Borderlands Studies* 12: 151–74.

Campbell, F.A. (1997) 'Trade, investment and regional cooperation: the Caribbean experience', in ACS (ed.) *Association of Caribbean States: Trade, Transport and Tourism*, Port of Spain: Association of Caribbean States.

CARICOM Secretariat (1997a) 'CARICOM works to become a single market and economy', in ACS (ed.) *Association of Caribbean States: Trade, Transport and Tourism*, Port of Spain: Association of Caribbean States.

—— (1997b) *Removing the Barriers: Facts on CARICOM Single Market and Economy*, Georgetown: CARICOM Secretariat.

—— (2002) Caribbean Community. Online: <www.caricom.org>.

Church, A. and Reid, P. (1996) 'Urban power, international networks and competition: the example of cross-border cooperation', *Urban Studies* 33: 1297–1318.

—— (1999) 'Cross-border co-operation, institutionalization and political space across the English Channel', *Regional Studies* 33: 643–56.

Erisman, M (1992) *Pursuing Postdependency Politics: South-South Relations in the Caribbean*, Boulder, Colo.: Lynne Rienner Publishers.

Gill, H.S. (1997) 'CARICOM and hemispheric trade liberalization', in A.J. Jatar and S. Weintraub (eds) *Integrating the Hemisphere: Perspectives from Latin America and the Caribbean*, Washington, DC: Inter-American Dialogue.

Grugel, J. (1995) *Politics and Development in the Caribbean Basin: Central America in the New World Order*, Bloomington: Indiana University Press.

Hatton, M.C. (1997) 'Towards a sustainable tourism zone in the wider Caribbean', in ACS (ed.) *Association of Caribbean States: Trade, Transport and Tourism*, Port of Spain: Association of Caribbean States.

Holder, J. (1997) 'The Caribbean Tourism Organization – A study in regional unity', in ACS (ed.) *Association of Caribbean States: Trade, Transport and Tourism*, Port of Spain: Association of Caribbean States.

Hussain, R.M. (1999) 'SAARC 1985–1995: a review and analysis of progress', in E. Gonsalves and N. Jetly (eds) *The Dynamics of South Asia: Regional Cooperation and SAARC*, New Delhi: Sage.

Insanally, R. (1997) 'The ACS and the regional integration process', in ACS (ed.) *Association of Caribbean States: Trade, Transport and Tourism*, Port of Spain: Association of Caribbean States.

Jessop, B. (1995) 'Regional economic blocs, cross-border cooperation, and local economic strategies in postcolonialism', *American Behavioral Scientist* 38: 674–715.

Müllerleile, C. (1996) *CARICOM Integration: Progress and Hurdles, A European View*, Kingston: Kingston Publishers Limited.

OECS (2002) Welcome to the Organisation of Eastern Caribbean States. Online. Available HTTP: <www.oecs.org>.

Parent, L. (1990) 'Tex-Mex Park: making Mexico's Sierra del Carmen and sister park to Big Bend', *National Parks* 64: 30–6.

Smith, G. (1994) 'Implications for the North American Free Trade Agreement for the US tourism industry', *Tourism Management* 15: 323–6.

Taylor, G.D. (1994) 'The implications of free trade agreements for tourism in Canada', *Tourism Management* 15: 315–56.

Teye, V.B. (2000) 'Regional cooperation and tourism development in Africa', in P.U.C. Dieke (ed.) *The Political Economy of Tourism Development in Africa*, Elmsford, NY: Cognizant Communications Corporation.

Timothy, D.J. (1995) 'Political boundaries and tourism: Borders as tourist attractions', *Tourism Management* 16: 525–32.

—— (1999) 'Cross-border partnership in tourism resource management: International parks along the US–Canada border', *Journal of Sustainable Tourism* 7: 182–205.

—— (2001) *Tourism and Political Boundaries*, London: Routledge.

—— (2002) 'Tourism in borderlands: Competition, complementarity, and cross-frontier cooperation', in S. Krakover and Y. Gradus (eds) *Tourism in Frontier Areas*, Lanham, MD: Lexington Books.

—— (forthcoming) 'Supranationalist alliances and tourism: insights from ASEAN and SAARC', *Current Issues in Tourism*.

Timothy, D.J. and Teye, V.B. (forthcoming) 'Political boundaries and regional cooperation in tourism', in A. Lew, C.M. Hall and A. Williams (eds) *Geography Companion to Tourism*, Oxford: Blackwell.

Wilkinson, P.F. (1997) *Tourism Policy and Planning: Case Studies from the Commonwealth Caribbean*, Elmsford, NY: Cognizant Communications Corporation.

Williams, A.M., Baláž, V. and Bodnárová, B. (2001) 'Border regions and trans-border mobility: Slovakia in economic transition', *Regional Studies* 35: 831–46.

8 Historic sites, material culture and tourism in the Caribbean islands

William C. Found

Introduction

While the tourist's view of the Caribbean region is dominated by images of sand, sun and sea, the various islands possess a wealth of historic sites and material culture that are becoming increasingly popular with visitors to the region (e.g. Slinger 2000; Weaver 1995, 2001a). This provides a basis for the Caribbean to share in the global trend towards heritage tourism. Important steps in the identification, documentation, preservation, restoration and exposure of the Caribbean's precious historic resources have been made – necessary preconditions for the continued development of the related tourism industry.

This chapter first outlines the prehistoric and historic basis for the creation of Caribbean heritage sites and artifacts. The review includes three particularly important observations: (1) the region has played host to a substantial volume of migration and intra-regional population movements; (2) the context for alternative tourism experiences, and their increasing role in the overall development of Caribbean tourism, is grounded in the identification and presentation of sites and artifacts associated with the differing population groups; and (3) the need to address key anthropological, historical and archaeological interpretations of artifacts as hallmarks of past lifeways is critical. The chapter then demonstrates how such artifacts have become central to a growing segment of the tourism market in the Caribbean region. The chapter concludes with suggestions for enhancing the growth of this part of the tourist economy.

The emergence of artifacts in the Caribbean islands

The prehistoric period

As outlined elsewhere (Duval, Chapter 1), human occupation in the Caribbean has been synonymous with mobility and population movements. Archaeological evidence in the islands of the Caribbean indicates a date of first human occupancy as early as 4000 BC (Rouse 1992: 32; Allaire 1997: 21). While the migration routes and numbers of migrations into the chain of islands in this early period are still far from certain, chipped stone tools from the early Lithic period have been located in a number of different sites in the Greater Antilles and the Bahamas. Prehistoric sites in Trinidad dating from at least as early as 4000 BC have yielded stone-ground, bone and shell tools, types of artifact not found in most other Caribbean sites until about 2000 BC during the Archaic period. The hunter-gatherers who lived during the first few thousand years of settlement survived primarily on the rich resources of the land–sea margins, with diets including a range of fish, shell-fish, mammals (e.g. hutia), reptiles (e.g. turtle), herbs and fruit. Artifacts from the early period of human habitation include only tools of stone, bone and shell, located in the middens that mark the locations of former settlements.

By 200 BC new migrations of agriculturalists from South America (many by way of the mouth of the Orinoco River and Trinidad) had led to their rapid resettlement of the chain of Caribbean islands. The new settlers, identified by archaeologists as Saladoid people, planted crops (including cassava and sweet potatoes) in semi-permanent locations, and made pottery. The shapes, decorations and colouring of the ceramic pots and griddles have become the primary means for distinguishing different prehistoric groups, and for tracking their likely paths of migration from South America through the island chain. Despite great advances in archaeological research during the past few decades, the number of migration episodes, communication back-and-forth among Caribbean islands, and the extent to which cultures in particular Caribbean locations developed *in situ* rather than through geographical diffusion is still uncertain. Archaeologists have identified dozens of different cultural sub-groups, many of which are named after locations in or near South America with similar ceramic artifacts.

By the time of European contact in 1492, the Caribbean islands were settled by groups who varied in population density, political organisation, the production of artifacts and language. The Taino of Hispaniola occupied the area of greatest population density, political organisation and village-type urbanisation; and early European descriptions of the indigenous population are derived largely from contact with these people (Las Casas 1971; Martyr 1970; Oviedo 1959; Sauer 1966: 37–9). Columbus and subsequent European authorities were initially attracted to the areas of dense Taino population, as this geographic area seemed to offer the greatest prospects for gold, human labour, food production and cattle ranching. Islands of the eastern Caribbean were much less attractive for permanent settlement. They were much smaller than the Greater Antilles, further removed from Central America (which became the primary focus of Spanish attention after the Cortes conquest of 1519), and were the home of the Kalinas (Island Carib) Indians, who proved to be quite resistant to any who attempted permanent settlement.

Artifacts from the prehistoric ceramic period are plentiful, and include a rich variety of pottery; stone, shell and bone tools; a small quantity of organic objects; artifacts for artistic, ornamental and spiritual use; petro-glyphs and pictographs; and the remains of buildings and settlements (e.g. discoloured soil, indicating the location of former wooden posts) (Olazagasti 1997; Pendergast 1998; Righter 1997). The remnants of plazas, lined with earth or decorated stones, and ball courts have survived in Hispaniola and Puerto Rico. While conucos (small agricultural clearings) and rural land-scapes have not survived, the legacy of the ceramic period is continued through the contemporary use of traditional food crops and a limited use of fire in clearing fields for planting. Similarly, contemporary canoes and fish traps are modified versions of prehistoric creations for fishing and travel.

Early European/African settlement

The period of early European settlement (which involved the first importa-tion of slaves from Africa) in the Caribbean islands yielded a number of highly visible artifacts which survive today. Most prominent are the cores of the fortified urban settlements, where the Spanish constructed very perma-nent stone, concrete and brick buildings, located on road grids built around central squares or plazas. Fortresses, religious buildings, administrative headquarters, houses and places of business were built to last; and they have survived hurricanes, fires and military conflict remarkably well. Prominent examples are the colonial cores of Santo Domingo, San Juan, Santiago de Cuba, Trinidad (Cuba) and Havana. Remnants of the Spanish occupation of Jamaica are much less visible. New Seville, the original Spanish capital, is rubble above ground level; and details of the settlement here and at the subsequent capital, Spanish Town, are only traceable through excavation.

Historical artifacts from the early Spanish period are less obvious in rural areas, although the impact of early Spanish settlement in the coun-tryside was great. Large tracts of land were granted to individuals who, after unsuccessful efforts to 'enslave' the native population, established ranches for raising cattle. Perhaps the greatest impact on the landscape resulted from the population explosions of cattle, pigs, horses, goats, sheep and dogs, which helped lead to the destruction of the traditional Amerindian conucos, aided in deforestation, and introduced the first large mammals to the terrestrial landscape (Watts 1987: 78–121). Feral and semi-domesticated descendants of these animals still roam parts of the rural landscapes of Cuba and Hispaniola. Early landscapes were also changed through the introduction of new cultivated plants (e.g. citrus, sugar and bananas), through the creation of gold mines (particularly in Hispaniola) and through the construction of military fortifications at strategic coastal locations.

Settlement in the eastern Caribbean

A major transformation of the Caribbean began in 1624, with the estab-lishment of permanent settlements in the eastern islands, first by the British, and then followed quickly by the Dutch and the French. The towns established by the British and French were smaller and more fragile than towns in the Greater Antilles, partly because much construction was of wood, and partly because the new arrivals tended to emphasise development of the countryside. After an initial, largely unsuccessful experiment with small farming based on the production of tobacco, cotton and indigo, the settlements in St Kitts, Nevis, Montserrat, Antigua, Guadeloupe, Martinique and Barbados switched to the large-scale production of cane sugar, using the 'Brazil' system imported to the region by the Dutch. The British replicated the same system in Jamaica, when they took the island from the Spanish in 1655. Thus developed the large plantation system, with its large acreages, large sugar mills (powered by animals, water or wind), complexes of estate buildings and heavy reliance on slaves imported from Africa (Richardson 1992). The landscape artifact was part agricultural and part industrial, given the large capital investment in buildings and equipment necessary for producing sugar. It has proven to be a highly persistent artifact, as landscapes in those islands that were transformed by cane-sugar production in the seventeenth century still reflect the patterns established 350 years ago.

While the Dutch were the primary instigators of the new Brazil system of cane production – helping to develop the system in British and French locations through capital investment, technological know-how and transportation of the product – they tended to develop their own small island possessions in the eastern Caribbean as centres for trade (particularly in Curaçao and St Eustatius) and as sources of salt. The historical remnants of particular interest in these centres, consequently, are urban.

Given the almost constant state of warfare among the European powers that had acquired islands in the Caribbean by 1700, all island landscapes included prominent military installations at strategic coastal locations – installations whose development and enlargement continued well into the nineteenth century (and most of which survive today). The larger establishments also included hospitals, given the high rates of illness and death of military personnel (tropical diseases killed far more people than warfare). Other important institutions included churches and cemeteries, characteristically distributed throughout towns and the countryside in the British and French islands; and Catholic monasteries and convents in the Spanish and French islands. The monasteries in the French islands often served as industrial as well as religious sites (e.g. for sugar processing or brick manufacturing).

The late eighteenth and the nineteenth centuries

The late 1700s brought three important transformations to the Caribbean region – the permanent settlement by European powers of islands in the eastern Caribbean where previous efforts had been successfully resisted by remnants of the indigenous population (Dominica, St Lucia, St Vincent, Grenada); acquisition and early settlement of Trinidad by Britain; and the successful slave revolt in St Domingue to form the first independent Caribbean state, Haiti. The first of these transformations led to the establishment of plantations in the coastal peripheries and floodplains of Dominica, etc. While sugar was produced, environmental conditions were not ideal; and a variety of plantation crops were grown. The mountainous interiors of these islands remained in forest. Trinidad, which was far removed from other areas of intense Spanish interest, had only scattered settlements (many operated as small farms by French settlers) when the British acquired the island in 1798. This quickly changed as large-scale cane-sugar production was introduced. St Domingue (the western third of Hispaniola) was the world's leading producer of cane sugar, and the region's most technically advanced plantation system before France lost the colony with the revolution that began in 1791. With the revolt came the destruction of many plantations, and the development of a largely rural economy based on small subsistence farms operated by a broadly dispersed population. A particularly significant artifact created in the early days of the new nation was the huge citadel constructed by Henri Christophe, who ruled northern Haiti.

Some planters who fled from St Domingue with their slaves settled in nearby Cuba, and they quickly established the same technologically advanced kinds of sugar cultivation left behind in Haiti. Climatic and soil conditions in Cuba were ideal for growing sugar cane. Production rose steadily throughout the nineteenth century until Cuba was the world's largest producer. Technical advances improved efficiency and profitability, and plantations grew ever larger. Huge sugar factories (centrales) were developed, particularly following the introduction of the first steam trains in the Caribbean islands. The trains increased the size of the areas that could be served by a single sugar factory, thus allowing operators to take advantage of optimum economies of scale. Cuba's economy thrived as sugar, tobacco and coffee products gained worldwide popularity. Havana became a large, wealthy 'world' city, particularly after the defeat of the Spanish in the Spanish–American war (1898) paved the way for immense American investment. During prohibition in the USA (1919–33) Havana developed even further as an entertainment 'playground' for nearby Americans. As the city continued to grow, architects from around the world experimented with the latest in styles and innovations (e.g. Art Deco).

The other Caribbean island to undergo rapid development in the nineteenth century was Trinidad. Again, cane sugar was the primary source of wealth. Development was aided by the importation of indentured labourers from India, following the emancipation of slaves in 1834. Early in the twentieth century Trinidad added an important industrial base to the economy with the development of its large petroleum reserves.

Abolition of the slave trade and emancipation affected every Caribbean island in the nineteenth century, although the dates and impacts varied significantly from place to place. In some of the traditional centres of cane cultivation (e.g. Barbados), sugar production continued unabated, as freed slaves and their descendants worked as paid labourers on plantations that appeared to change little through time. In other islands, where freed slaves could find alternative ways of earning a living (e.g. on peasant farms in the interior of Jamaica), the landscape changed significantly as sugar production declined rapidly. In general, the freed population tended to occupy land that had been considered marginal for plantations, particularly where churches or (eventually) governments helped to make land available. Many also migrated to towns, seeking paid employment and contributing to steadily increasing urbanisation.

Steam power was introduced to most Caribbean islands by the middle of the nineteenth century, which changed the appearance of the sugar factories, and led to the establishment of railways on most islands. Electricity arrived in towns and cities towards the end of the century. This, and the advent of motorised vehicles, dramatically changed town and countryside alike.

The twentieth century

The twentieth century saw many great changes in the artifacts of the Caribbean. Landscapes changed immensely as several of the smaller islands abandoned the cultivation of cane sugar, leaving behind only the decaying remnants of former plantations.

Industrialisation of many kinds and the development of financial services transformed one island after another, particularly after several former colonies achieved political independence and sought to diversify their economies. The growth of towns, and dense settlement of the countryside surrounding towns, transformed the urban and peri-urban landscapes. Supermarkets, rooftop satellite dishes and fast-food outlets reflected creeping globalisation. Railways disappeared, except in parts of Cuba and St Kitts. The Cuban revolution transformed the appearance of Havana, with many decaying architectural gems from the past taking on new functions. Tourism, which had only faint beginnings prior to the twentieth century, developed rapidly, particularly with the advent of mass tourism after World War II. The end of the century also saw growth in ecotourism and heritage tourism, in keeping with international trends.

Identification, protection and restoration of historic sites and material culture

One finds occasional reference in the more distant past to efforts to identify and protect Caribbean artifacts (e.g. items connected with Columbus, or specific grand buildings), but the period of intense interest and activity began only in the last half of the twentieth century. Fortunately, a large number of energetic people and organisations have preserved many of the items collected privately prior to 1950, and they have done their best to maintain and enhance collections in the small number of museums established in earlier times. At the same time, efforts, both local and international, have been focused on new initiatives to document items of prehistoric and historic interest, and to protect or restore a range of historic sites and material culture. Much research has been undertaken, many new museums have been established, a multitude of buildings have been restored, and printed and electronic materials have been made available to the growing number of people interested in authentic explanations of the past.

Prehistoric material culture

Much prehistoric research and documentation is initiated and supervised by local professional archaeologists. At the same time, many projects, particularly in the smaller islands, involve collaborators from outside of the region. Collaboration with North American and European universities and museums is particularly common. At one time many excavated artifacts were taken to institutions outside of the Caribbean, as researchers sought to study the materials at their leisure in environments where the artifacts could be well protected and preserved. A number of excellent facilities for storing and displaying prehistoric materials have been constructed in the Caribbean over the past few decades, however, so that newly discovered artifacts normally remain in the region. Some of the museums are large and quite comprehensive (e.g. Cuba, Puerto Rico, Martinique), while others are small but very effective in their presentations (e.g. Antigua, St Martin, Dominica, Aruba).

As Timothy and Boyd (2003) note, museums of heritage attractions now incorporate public viewing in addition to conservation services. The Caribbean pre-Columbian museums are designed with the general public in mind. Signage is sometimes multilingual, and some displays are organised for both young people and adults. Pamphlets are aimed at an audience of interested but largely uninformed clients. Books available for purchase tend to be quite technical, and beyond the easy understanding of the general public. A few museums (e.g. 'On the Trail of the Arawaks' in St Martin, and the Barbados Museum and Historical Association) provide opportunities for outsiders to participate in supervised archaeological digs.

Several museums include displays of reconstructed or simulated prehistoric villages, either in miniature or life-size. Huts, complete with hammocks, ceramic objects, woven fabrics and artifacts (wood, stone and thatch), provide highly informative displays. Some museums also include photographs of native people from South America whose appearance and culture are similar to some prehistoric Caribbean groups. An outstanding 'reconstruction' of a prehistoric settlement is located at the Tibes Indian

Ceremonial Park, near Ponce in Puerto Rico. Built on an excavated Igneri/ Taíno site, this reconstruction includes a small village of huts, a selection of native plants, conucos and seven bateys/ballfields – all just a short walk from an outstanding pre-Columbian museum.

Artifacts from the historic period

Most of the attention on artifacts from the Caribbean past has been directed at the historic period, and a vast amount of energy has been applied to the identification, protection and restoration of the finest remaining items. Buildings, including houses, military establishments, churches, sugar mills and government headquarters, have been the primary foci for action; and there has been particular concern for structures from the long colonial era. Sites and artifacts associated with the former landowning elite have received much attention, with a corresponding lack of attention to the poor, slaves and 'ordinary' people of the colonial past.

One of the most significant artifacts from the past, and the largest in scale, has not received the recognition that it deserves – the rural landscape. The primary Caribbean economic activity throughout most of the post-1492 period has been agriculture, focusing on the production of tropical crops for export. The emergence of landscapes reflecting this focus, largely unplanned aside from the property surveys that designated initial ownership units, has been the dominant geographic feature of most Caribbean islands (see e.g. Higman 1988). It continues to be a most distinctive attribute, yet it seldom receives the kind of attention directed to individual buildings, or even to urban streetscapes. Governments are concerned with contemporary economic and social development, with facilitating sensible urbanisation, with transportation facilities and with the designation and protection of 'natural' sites for conservation. At the same time rural/farming landscapes tend to be either unappreciated or taken for granted. The cane-sugar landscape of St Kitts, little changed in many ways over 350 years, represents an artifact of immense historical significance. Localities throughout the Caribbean where collections of small peasant farms emerged after emancipation (sometimes the sites of 'free villages') may be difficult for the uninitiated to identify, but they are no less significant or aesthetically pleasing than landscapes dominated by large plantations. The texture of rural landscapes varies greatly from island to island, and they form the often subconscious backdrop for those enjoying a journey through the countryside, or just commuting between different destinations. It is noteworthy that two of the 'cultural' locations on the UNESCO list of World Heritage Sites are rural landscapes in Cuba – the Archaeological Landscape of the First Coffee Plantations in the South-East of Cuba, and the Viñales Valley, an isolated area of traditional farming and historic vernacular architecture in the extreme west end of the country. The identification and at least partial preservation of other historic rural landscapes throughout the Caribbean would stimulate appreciation for these precious artifacts.

Some of the earliest and most effective attention to historic sites has focused on the original cores of the major Spanish colonial cities. 'Old San Juan' in Puerto Rico, complete with fortified city wall, represents the most extensive preservation and restoration. A combination of government action, enlightened town planning, public funding and much financing from individual owners of historic buildings has produced what many consider to be the finest historic site in the Caribbean. The old city and the Fortaleza defensive structures form a UNESCO World Heritage Site; and the adjacent fortress, El Castillo de San Felipe del Morro, is now a US National Park. The old-town portion of Santo Domingo, Dominican Republic – the earliest Spanish capital that still survives – is also a carefully restored UNESCO World Heritage Site. Public buildings, including the Diego Columbus house (1510–14), the Fortaleza Ozama, many religious structures and museums created from former institutional structures, form the core of this popular and important location. Three colonial centres in Cuba – Havana, Santiago de Cuba and Trinidad – complete the list of urban UNESCO World Heritage Sites in the Caribbean islands. Each has been attentively restored, supported by international funding and a heavily committed Cuban Government. Cuba has produced some of the region's best experts in Spanish colonial restoration, and the Government maintains a research centre where original structures and art objects are analysed and restored.

Streetscapes in a number of smaller Caribbean towns have been subject to varying degrees of restoration. In Nassau in the Bahamas, a backwater for at least a century after migrating British Loyalists constructed Georgian homes after the American Revolution, a large number of handsome buildings have been restored with generous funding from the financial institutions that now occupy them. Proximity to the USA was a major factor prompting Nassau to become a centre for offshore banking in the middle of the twentieth century. This locational advantage also stimulated an early tourist industry, and led to its prominence as a major port for cruise ships. Careful planning and private financing have assured the preservation of the historic town despite the construction of many huge hotels and thriving mass tourism. The arrival of cruise ships has stimulated preservation and restoration of streetscapes, or at least a number of individual buildings, in several Caribbean locations. Willemstad, Curaçao, developed as an early cruise-ship destination and offshore banking centre, and by the 1950s a large hotel had been constructed partially in an old fort that guarded the entrance to the harbour. Early eighteenth-century merchants' houses/ warehouses on Handelskade (the street facing the waterfront in the Punda district) form one of the most colourful historic streets in the Caribbean. Nearby Oranjestad, Aruba, has no historic buildings to rival those in Willemstad; but newly constructed streetscapes attempt to emulate the early Dutch architecture of Curaçao. In St John, Antigua, recently refurbished historic buildings line the streets leading from the cruise-ship dock, providing bars, restaurants and shops for luxury goods. Even the newly constructed shopping area is named Heritage Quay. In Roseau, Dominica, the recent construction of cruise-ship facilities has created a ripple effect whereby Victorian buildings on nearby streets have been repaired, painted and converted to businesses serving the tourist public. The list of locations where the construction of cruise-ship facilities has helped to stimulate concern for historic buildings goes on (Bridgetown, Castries, Fort de France, etc.).

Some Caribbean ports that cater to yachts rather than larger ships have also undergone a 'heritage' restoration. The best example is English Harbour, Antigua, where the Nelson's Dockyard area has been wonderfully and authentically restored. Gustavia (St Barthelemy) and Marigot (St Martin) are other yachting havens where modern businesses have located in well-restored historic buildings, and planners have sought to preserve a strong sense of historic streetscapes. Few yachts venture to St Eustatius, but citizens of the old town of Oranjestad (particularly members of the local Historical Association) stand ready to provide walking tours of one of the most fascinating historical towns in the Caribbean. Fort Oranje is completely restored, but the Lower Town has remained largely untouched since it was destroyed in the 1780s. This adds to the sense of mystery and charm (Found 2002).

Local historical associations, National Trusts, governments, many concerned individuals and international agencies have made individual historic buildings the major foci of preservation and restoration throughout the Caribbean. Fortresses have received much attention, and two – Henri Christophe's Citadel and adjacent Sans Souci Palace at Ramier, in northern Haiti; and Brimstone Hill, St Kitts – are designated as UNESCO World Heritage Sites. Many others have been well preserved, and even small fortifications are gradually being identified and protected. Other public buildings receiving attention include historic administrative headquarters, churches, monasteries and convents. Many former industrial sites have been protected, particularly sugar mills, boiling houses and distilleries. Some are featured in public parks or museums (e.g. the Annaberg Estate in Virgin Islands National Park, St John), and many others have been converted into hotels, restaurants, stores or the centrepieces of golf courses. A few have been restored (e.g. the Morgan Lewis windmill in Barbados, which can still be used to grind cane). At Betty's Hope in Antigua even the underground cisterns have been reconstructed.

By far the greatest number of protected or restored buildings are houses – usually large, rural, great houses or the urban residences of members of the former aristocracy. Some have been converted into hotels, restaurants or museums, but many serve as private residences. A number are divided into a private portion and a part which is open to the public. Preserving the housing of the elite from colonial times has been a priority throughout the region.

Museums serve as the major places for observing smaller artifacts from the past. The number and quality of museums, both public and private, has been increasing steadily. The larger public museums tend to publish brochures, monographs or journals regarding collections or local historical research (e.g. the Barbados Museum and Historical Society, the Institute of Jamaica). Even small private museums tend to include a gift shop where some documentation on local history can be purchased.

Access to historic sites and material culture by tourists

The culturally rich historic sites and artifacts that have been documented and preserved in the Caribbean islands have become major attractions for tourists, particularly over the past fifty years; yet tourism in the Caribbean is still unbalanced in the direction of the three Ss of sea, sand and sun (Ashworth and Tunbridge 2000: 195–7; Weaver 2001b: 288–9, but cf. Weaver 1995 and Chapter 10; Duval 1998). It would be a momentous task to obtain precise data concerning the extent to which tourists visit historic sites, as such data are not centrally assembled, even within individual islands. Indeed, requests for such information from this author are often met with statements of regret that numbers are not kept, or puzzlement that anyone would care. Even allowing for the need of some to keep information confidential, it is clear that reliable counts of visits to historic sites and material culture have not been made. At the same time, casual observations and discussions with officials reveal that a number of excellent sites attract far fewer tourists than they could accommodate.

Informal interviews conducted with several heritage site operators in Barbados, Jamaica and Trinidad by the author and research assistants in December 2002 and January 2003 suggest a number of important trends. Some major institutions, such as the national museums, host very small numbers of visitors, including tourists (one national museum had under 14,000 visitors in each of 2001 and 2002 – less than fifty per day). The annual numbers of visitors to historic sites, such as those administered by National Trusts, vary immensely from one site to another (the numbers range from a few hundred to over 100,000, depending on the site). A major factor accounting for this variation is whether or not excursions from cruise ships choose the site as a destination. This observation indicates, of course, that tourists are the major visitors to the most popular sites. The number of visitors per month varies with the total number of national tourist arrivals, which corroborates the vital importance of tourists as clients. Operators of private historic sites (e.g. colonial houses) report that the major factor determining the number of visitors is the number of organised tours from cruise ships – the same pattern as observed in National Trust sites. It is clear that marketing arrangements between the operators of historic sites and the cruise-ship companies, or the agencies which provide tours for those companies, are of utmost importance in the development of heritage tourism. The perceived quality of historic sites from the perspective of large numbers of tourists is important in determining which sites will be visited by cruise-ship passengers, but it is not the only factor. Accessibility of sites, both in terms of distance from the cruise ship or with respect to parking arrangements for buses, is a very significant factor. The attractiveness of a site is enhanced if it is located on a route that can include a number of interesting locations, strung together in an efficient tour.

Public safety is significant in determining which sites are popular, as tourists (and tour companies) are likely to avoid locations where visitors feel threatened or harassed. Ideal locations are well-presented buildings within easy walking distance of cruise ships (the Dominica Museum in Roseau being an excellent example). Operators of historic sites in Barbados, Jamaica and Trinidad indicate that cruise-ship tourists are far from homogeneous, and that the specific type of visitor has an important bearing on their likelihood of visiting heritage sites. For example, operators feel that the lower prices available for some cruise vacations over the past few years are responsible for declines in numbers of cruise-ship visitors. To them, tourists who take inexpensive vacations have little disposable income to spend on visiting historic sites.

While cruise-ship tourists provide the greatest single source of visitors to historic sites in several islands, stayover tourists also create a significant demand. Normally these tourists include visits to historic sites as a complement to their beach-based vacations rather than as a primary focus. Marketing arrangements with tour companies are important in providing historic sites with access to stayover tourists, while cruise-ship passengers are more dependent on pre-packaged tours of relatively short duration. Stayover tourists can arrange for more leisurely visits to historic sites, either as part of packaged tours or as individual trips by taxi or rented car.

Information available to potential visitors to historic sites across the Caribbean is uneven overall, but generally adequate. Many commercial international guidebooks include useful but brief descriptions of relevant sites. A few excellent inventories of historic architecture throughout the region have been published during the past few years (Crain 1994; Douglas 1996; Gravette 2000). Books by local authors outlining historic landscapes and buildings or about museum collections are available in individual islands. Some are colourful coffee-table books, others learned journals, and several are local histories. They are often unavailable beyond the island or region, however, as they are not marketed by international publishers. Distribution within individual islands is unpredictable; quite often excellent volumes cannot be obtained in resorts or in small bookstores. CD-ROMs and videos are becoming more common, but only a few (usually produced in Europe) have good content about historic sites or authentic material culture. Websites sponsored by island governments or by international tour companies are often disappointing when it comes to historic content. Their major concern seems to be to sell the sun/sand/sea image, tempered by graphic or textual references to parodies of pirates, cannons and old maps. Some historic organisations, including several National Trusts, have produced informative websites; but most islands could benefit from comprehensive websites describing a full range of historic sites. Good websites for museums, perhaps with virtual tours, would be invaluable.

Promoting historic sites and material culture: issues and challenges

A first step in improving tourist participation in 'historic' and 'prehistoric' tourism is to recognise the factors that will make it possible. First and foremost, the Caribbean islands have a rich legacy of interesting historic/prehistoric artifacts, ranging from landscapes, to buildings, to museum pieces. The substantive underpinnings for history-based tourism are authentic, strong and increasingly well documented and presented. In crass commercial terms, the product is sound. It is worthwhile to mention, however, that the recent Caribbean Strategic Plan as outlined by the Caribbean Tourism Organization in the Bahamas in October 2003 (CTO 2002) makes little mention of the role of heritage sites in the region for future tourism development.

Second, world tourism trends favour increased emphases on historic, cultural or heritage sites (Timothy 1997). While Europe has been the leader in tourism related to historic sites or museums, North Americans, who form the largest single group of Caribbean tourists, are becoming increasingly interested as well (Dickinson 1996; Pitman 1999). The growth of ecotourism is highly significant, as that movement includes an appreciation for authentic culture and involves the movement of tourists well beyond the hotel gate or cruise-ship gangplank into the rich context of local life and exposure to historic landscapes and material culture.

The Caribbean region is off to a good start. The identification and documentation of historic/prehistoric artifacts has increased dramatically in the past fifty years, and many preservation and restoration projects have been instigated. The region can be proud of its many dedicated and skilled individuals who have painstakingly raised the consciousness of others, and who have built museums, written books, formed organisations, arranged archaeological digs and protected buildings and streetscapes. Governments have shown important leadership, perhaps most in Cuba, the French islands and Puerto Rico, but increasingly in others as well (e.g. St Lucia has recently named heritage tourism as a national priority). International organisations have provided important assistance, a pattern that is likely to continue. Caribbean governments and non-profit organisations have learned how to articulate their needs, and to work with the many international donors who can provide funding and other forms of support.

Still, a number of challenges remain. The Caribbean is not seen internationally as a major area of historical interest, given its predominant sun/sand/sea image. Change will take time. To diversify the Caribbean image effectively will require a greater degree of regional coordination and commitment to common policy than has been the case so far (Bryan 2001: 20). Substantial investment will be required – investment to restore buildings and landscapes, upgrade museums, train interpreters, provide more printed documentation and add substantially to Internet resources (see e.g. Champion 1997). The comparatively wealthy states of the Caribbean have an advantage, from both the perspective of government and private financial investment. For other states finance will be a challenge, given the many pressing problems related to basic needs, but inventive solutions involving local and external financing can be found (Organization of American States 2002). Progress in promoting heritage tourism will be greatly enhanced if national governments declare this as a policy objective, and if the governments ensure that planning among different agencies (e.g. transportation, tourism, parks, education, advertising) is well coordinated.

Many have argued for the 'democratization of cultural attractions' (Boniface 1995: 23), which would encourage a broad range of local residents and tourists alike to participate in history-based tourism. Much documentation and preservation of historical sites in the Caribbean has focused on elite societies of the past, and there is room for more attention to finding or re-creating artifacts from former societies of slaves, the poor or others who may be forgotten. Historic sites that reflect a broad range of economic and social circumstances can also be highly marketable (Fallis 1998). Further, local projects that are community-based, involving people from the grass-roots through to the national levels, can have great power to generate interest, education, pride and sustainable activity. And projects that provide tourists with an active role (e.g. in terrestrial or underwater archaeology) can do much to raise their interest and participation.

The marketing of tourism based on historic/prehistoric artifacts is a challenge which could be addressed aggressively. Marketing strategy would have to involve organisations at all levels – from the region to the village – and would need to address the key agencies (e.g. the cruise-ship lines) involved in determining the tourist's selection of sites to visit. Such efforts could significantly enhance the heritage tourism product in the region, and would serve to support and encourage those local individuals and groups who have already brought important sites and artifacts to international attention.

Acknowledgements

I am indebted to the following for their assistance in the preparation of this chapter: Professor Karl Watson, Head of the M.A. Programme in Heritage Studies at the University of the West Indies, Cave Hill Campus;

Andrea Alleyne, graduate student at the Cave Hill Campus (and visiting graduate student at York University); Jodi McGaw and Alexandra Tieman, graduate students at York University; and Dr David Duval, for his astute comments and suggestions, in both university and field settings.

References

Allaire, L. (1997) 'The Lesser Antilles before Columbus', in S. Wilson (ed.) *The Indigenous People of the Caribbean*, Gainesville: University Press of Florida.

Ashworth, G. and Tunbridge, J. (2000) *The Tourist-Historic City: Retrospect and Prospect of Managing the Heritage City*, Kidlington: Elsevier Science.

Boniface, P. (1995) *Managing Quality Cultural Tourism*, London: Routledge.

Bryan, A. (2001) 'Caribbean tourism: igniting the engines of sustainable growth', *The North–South Agenda*, Paper 52.

Caribbean Tourism Organization (2002) 'Regional Strategic Plan'. Online: <http://www.onecaribbean.org> (accessed 15 March 2003).

Champion, S. (1997) 'Archaeology on the World Wide Web: a User's Field-Guide', *Antiquity* 71: 274. Online: <http://intarch.ac.uk/antiquity/electronics/champion.html>.

Crain, E. (1994) *Historic Architecture in the Caribbean Islands*, Gainesville: University Press of Florida.

Dickinson, R. (1996) 'Heritage tourism is hot', *American Demographics* 18(9): 13–14.

Douglas, R. (1996) *Caribbean Heritage: Architecture of the Islands*, Port of Spain: Darkstream Publications.

Duval, D.T. (1998) 'Alternative tourism in St Vincent', *Caribbean Geography* 9(1): 44–57.

Fallis, M. (1998) 'Heritage tourism explosion (growth of tourist destinations that market to African Americans)', *American Visions* 13(5): 4.

Found, W. (2002) *St Eustatius: The Golden Rock* (documentary film), Toronto: CNS Faculty Support Services, York University (http://www.yorku.ca/geograph/video).

Gravette, A. (2000) *Architectural Heritage of the Caribbean: An A–Z of Historical Buildings*, Kingston: Ian Randle Publishers.

Higman, B. (1988) *Jamaica Surveyed: Plantation Maps and Plans of the Eighteenth and Nineteenth Centuries*, Kingston: Institute of Jamaica Publications.

Las Casas, B. (1971) *Bartolomé de Las Casas: History of the Indies*, trans. A. Collard, New York: Harper & Row.

Martyr, P. (1970) *De Orbe Novo: The Eight Decades of Peter Martyr D'Anghera*, trans. F. MacNutt, New York: Burt Franklin.

Olazagasti, I. (1997) 'The material culture of the Taino Indians', in S. Wilson (ed.) *The Indigenous People of the Caribbean*, Gainesville: University Press of Florida.

Organization of American States (2002) *The Financing Requirements of Nature and Heritage Tourism in the Caribbean*, Washington: O.A.S. (Department of Regional Development and Environment).

Oviedo y Valdés, G. (1959) *Natural History of the West Indies*, trans. S. Stoudemire, Chapel Hill: University of North Carolina Press.

Pendergast, D. (1998) *The Royal Ontario Museum in Cuba*, Toronto: Royal Ontario Museum.

Pitman, B (1999) 'Muses, museums, and memories', *Daedalus* 128(3): 1–31.

Richardson, B. (1992) *The Caribbean in the Wider World, 1492–1992: A Regional Geography*, Cambridge: Cambridge University Press.

Righter, E. (1997) 'The ceramics, art, and material culture of the early ceramic period in the Caribbean islands', in S. Wilson (ed.) *The Indigenous People of the Caribbean*, Gainesville: University Press of Florida.

Rouse, I. (1992) *The Tainos: Rise and Fall of the People who Greeted Columbus*, New Haven: Yale University Press.

Sauer, C. (1966) *The Early Spanish Main*, Berkeley: University of California Press.

Slinger, V. (2000) 'Ecotourism in the last indigenous Caribbean community', *Annals of Tourism Research* 27(2): 520–3.

Timothy, D.J. (1997) 'Tourism and the personal heritage experience', *Annals of Tourism Research* 34: 751–4.

Timothy, D.J. and Boyd, S.J. (2003) *Heritage Tourism*, Harlow: Personal Education.

Watts, D. (1987) *The West Indies: Patterns of Development, Culture and Environmental Change since 1492*, Cambridge: Cambridge University Press.

Weaver, D.B. (1995) 'Alternative tourism in Montserrat', *Tourism Management* 16: 593–604.

—— (2001a) 'Mass tourism and alternative tourism in the Caribbean', in D. Harrison (ed.) *Tourism and the Less Developed World: Issues and Case Studies*, Wallingford: CAB International.

—— (2001b) *Ecotourism*, Sydney: John Wiley & Sons.

9 Global currents

Cruise ships in the Caribbean Sea

Robert E. Wood

Introduction

For five hundred years, no region has been more exposed to global currents of change than the Caribbean. Today, the imperatives of neoliberal globalisation are at once dismantling the remaining props of the islands' agricultural export industries and at the same time increasing their reliance on that most global of industries, tourism. As the Caribbean Tourism Organization (CTO) secretary-general, Jean Holder (2001), has observed:

> The tourism consumer is scattered across the entire globe and every country on the globe is competing for his business. Everyone is competing on quality or price or on both, and it's an entirely free market. There are no preferences and no protection. Imagine trying to sell one's sugar or bananas or some of the products of our infant manufacturing industries on this basis! There is probably no other product in the world which finds itself in similar circumstances to that of tourism.

The intimate connections between tourism and globalisation have been noted for some time (Waters 1995; Bauman 1998; Wood 2000; Hannam 2002), but the most rapidly growing type of tourism in the Caribbean region is almost certainly the most globalised form of all: cruise tourism. With its massive but mobile ships flying flags of convenience from around the world, owned by companies incorporated in nations which their ships may never visit, exempted from most types of labour, health, safety and environmental regulations imposed on other industries (including the rest of the tourism industry), searching out both customers and destinations around the globe, the cruise industry is uniquely detached from any particular location and yet almost everywhere at the same time. Out of its uniquely global and *deterritorialised* nature grow a complex mix of potentialities and challenges for Caribbean tourism and for Caribbean societies generally.

This chapter assesses several of these potentialities and challenges. The first section provides a brief historical overview of cruise tourism in the

Caribbean. The second section explores the uniquely deterritorialised nature of the cruise industry and its implications. The third section explores cruising and sustainability in terms of environmental and economic impact. A concluding section assesses the uneasy relationship between the cruise industry and Caribbean efforts to bring it under greater regional control.[1]

Historical background and contemporary status

From the 1880s on, early tourists to the region came on 'banana boats' – steamships transporting cargoes of bananas and other tropical products and also offering first-class passenger accommodations. The United Fruit Company, with a fleet that grew to over sixty ships, was by far the largest of the companies offering such cruises, which generally lasted three to four weeks. Regularly scheduled banana-boat sailings were supplemented by an assortment of winter cruises from a variety of other shipping and cruise companies (Lawton and Butler 1987: 331). *The Caribbean Cruise*, a guidebook for cruise passengers, listed thirty-seven companies with cruises to the region (Foster 1928: 11–16).

The Great White Fleet of the United Fruit Company was the dominant cruise presence throughout this period (Plates 9.1, 9.2 and 9.3). So-named because of its white hulls, yet recalling the Great White Fleet of Theodore Roosevelt's Navy that was sent around the world to assert US sea power in 1907–9, it was inevitable that the fleet name would conjure up racial and neo-colonial associations as well. In addition to being a symbol of US economic power, United Fruit's Great White Fleet became a symbol of US-style racial discrimination. The ships engaged in a broad range of racially discriminatory practices, such as making black Caribbean passengers occupy inferior cabins and eat early before the white passengers – and often be denied bookings at all. Such policies confirmed the widespread fear in the Caribbean, as articulated by the *Jamaica Times* in 1912, 'that with the entrance of wealthy American tourists here we must look for attempts to introduce into Jamaica the colour discrimination on ships, in hotels, and even in churches that disgraces the United States' (Taylor 1993: 36). As for United Fruit's Jamaican Titchfield Hotel, to which cruise tourists repaired for a day or two:

> Not only was it patronized by Americans, but it was owned by Americans, run by Americans, and staffed (from chef to waitresses) by Americans – in sum, it was totally symbolic of the new economic imperialism ushered into the Caribbean by Americans. . . . The hotel industry served to resuscitate the dying master–servant culture of the Great House era in Jamaica. Little wonder, then, that the hotels became an arena of subtle black–white confrontation in which the specter of past struggles came to life in a new guise.
>
> (Taylor 1993: 87, 89)

Plate 9.1 1914 Great White Fleet advertisement (courtesy of the New Orleans Public Library)

WEST INDIES

CRUISES

An even keel, warm summer skies, the air soft and fragrant with the verdure and blossoms of tropical islands; flying fish and sporting dolphins; palm-fringed shores; and, tucked away behind moss-covered forts, strange cities and strange people. Pages of pirates and buccaneers from your best loved storybook; a moving, living picture of adventure and romance, with imagination riding free.

No holiday to compare with these cruises; no trips more carefully planned or more adequately conducted than the tours of the American Express Travel Department this winter to the American Tropics.

Luxurious passenger steamers of the Great White Fleet will be your hotel. Every comfort and convenience on sea, every facility to see and enjoy the most interesting places ashore; all under the personal direction of men most experienced in West India travel. Our record of past service and the international reputation of the American Express Travel Department is your guarantee.

Cruises sailing in January and February. Duration, twenty-four days. Itinerary:—New York; Havana and Santiago, Cuba; Port Antonio and Kingston, Jamaica; Panama, Canal Zone; Port Limon and San Jose, Costa Rica; Nassau; New York.

Bookings should be made at once. Write for illustrated, descriptive booklet of these cruises, diagram of steamers, rates, etc.

AMERICAN EXPRESS
TRAVEL DEPARTMENT
65 Broadway **New York**

Plate 9.2 1919 Great White Fleet advertisement

'These are the ghosts which must be exorcized if a modern tourism sector is to take hold in the Region and gain popular acceptance', CTO's Holder (2001) has said, partly in reference to the 'stories of the resentment felt by locals at the beginning of the 20th century meted out to them by visitors'. Taylor's historical account makes clear that aggressive begging and other forms of harassment, along with overt expressions of animosity towards cruise tourists, were all commonplace early in the twentieth century. With not a single one Caribbean-owned or managed, the roughly seventy (vastly larger) cruise ships that ply the Caribbean Sea today are seen by many in the region as a new Great White Fleet.

In the post-World War II period, relatively inexpensive air fares ushered in the era of mass tourism in the Caribbean. Eighty-one per cent of tourists to Jamaica in 1952, for example, were stayover tourists arriving by air; only 19 per cent were cruise tourists. For the Caribbean as a whole, the proportion of stayover tourists arriving by air has been gradually declining ever since, and the proportion of cruise tourists rising. By 1980, 35 per cent of tourists to the Caribbean region were cruise tourists, and by 2001, almost 44 per cent.

As Table 9.1 shows, cruise passenger arrivals in the Caribbean more than doubled from 3.8 to 7.8 million between 1980 and 1990. The figure doubled again between 1990 and 2001, when cruise ships disgorged an estimated 15.2 million cruise passengers into Caribbean ports. While stayover tourist arrivals increased at an average rate of 4.3 per cent between 1990 and 2000, the rate of increase for cruise passengers was 6.5 per cent (CT0 2002: 21). Cruise arrivals in the Caribbean went up 9.9 per cent in 2001, even registering gains in the final four months of the year; stayover arrivals declined. Cruise arrivals grew while stayover arrivals fell again in 2002.

For a number of countries, cruise-passenger arrivals have outnumbered stayover-tourist arrivals for some time. As Table 9.2 shows, of twenty-six Caribbean destinations reporting both cruise and stayover visitors in 2000, cruise passenger arrivals were greater than stayover arrivals in half. Cruise arrivals as a percentage of total tourist arrivals ranged from 9 per cent in Belize to 73 per cent in Dominica. Passenger arrivals were quite

Table 9.1 Stayover and cruise tourist arrivals, 1980–2001

	Stayover tourist arrivals (millions)	Cruise tourist arrivals (millions)	Cruise tourists as per cent of stayover tourists	Cruise tourists as per cent of total tourists
1980	6.9	3.8	55.1	35.5
1990	12.8	7.8	60.9	37.9
2000	20.3	14.5	71.4	41.7
2001	19.5	15.2	77.9	43.8

Source: Caribbean Tourism Organization (2002).

Jamaica, loveliest of the West Indies

HERE you will find two delightful resort hotels —The Myrtle Bank at Kingston and The Titchfield at Port Antonio, both owned and operated by the United Fruit Company.

On your Great White Fleet Cruise you may motor across Jamaica, stopping for a brief space at these hotels or remain aboard ship—whichever suits your individual taste.

Twenty-three days of complete change and relaxation visiting Cuba, Jamaica, Panama, Costa Rica, Colombia and Guatemala—depending on the Cruise selected. No passports or sailing permits required from American citizens.

Great White Fleet ships are specially constructed for Cruise Service. The newest and finest vessels sailing to the Caribbean. Unexcelled in appointments and cuisine.

From New York and New Orleans. Only one class—first class.

Write for free illustrated booklet "Following the Conquerors through the Caribbean," cabin plans and fare information.

Address Passenger Department
UNITED FRUIT COMPANY
Room 1630, 17 Battery Place, New York
General Offices: 131 State St., Boston, Mass.

On Your
GREAT WHITE FLEET
CARIBBEAN CRUISE

Plate 9.3 1921 Great White Fleet advertisement

concentrated, with the US Virgin Islands, Puerto Rico and the Bahamas accounting for 48 per cent of the total (CTO 2002: 94). However, as *Cruise Industry News* (2002: 277) notes, 'the chief trend is toward the west. It is estimated that the ports of the Yucatan Peninsula will significantly exceed the passenger count of one-time leader St Thomas, and will also surpass those of the Bahamas.' The trade publication attributes this to the greater

Table 9.2 Cruise ship and stayover arrivals and cruise passenger taxes and expenditures, 2000

	Cruise ship calls	Cruise arrivals (thousands)	Stayover arrivals (thousands)	Cruise arrivals as per cent of total arrivals	Cruise passenger taxes (US$)	Total cruise passenger expenditures (US$ millions)	Average cruise passenger expenditure (US$)	
							CTO	FCCA
Antigua and Barbuda	331	429.4	236.7	64.5	6.00	11.9	27.7	86.81
Aruba	330	490.1	721.2	40.5	3.50	n.a.	n.a.	82.02
Bahamas	1,892	2,512.6	1,596.2	61.2	15.00	148.0	58.9	77.90
Barbados	485	533.3	544.7	49.5	6.00	54.9	102.9	81.12
Belize	70	58.1	195.6	22.9	5.00	n.a.	n.a.	n.a.
Bermuda	165	209.7	328.3	39.0	60.00	44.0	209.8	n.a.
Bonaire	71	43.5	51.3	45.9	10.00	3.7	85.1	n.a.
Br. Virgin Islands	230	188.5	281.1	40.1	7.00	6.0	32.0	n.a.
Cayman Islands	605	1,030.9	354.1	74.4	10.00	112.4	109.0	79.42
Cozumel	885	1,504.6	230.0	86.7	3.00	n.a.	n.a.	131.40
Curaçao	213	309.4	191.2	61.8	3.50	24.5	79.2	n.a.
Dominica	287	239.8	69.6	77.5	5.00	6.5	27.1	n.a.
Dominican Rep.	n.a.	182.4	2,972.6	5.8	1.00	n.a.	n.a.	n.a.
Grenada	360	180.3	128.9	58.3	3.00	3.1	17.2	n.a.
Guadeloupe	249	392.3	807	32.7	1.50	n.a.	n.a.	n.a.
Haiti	n.a.	304.5	1,322.7	40.7	6.00	n.a.	n.a.	n.a.
Jamaica	504	907.6	140.5	68.4	15.00	73.0	80.4	73.15
Martinique	n.a.	286.2	562.3	33.7	none	7.9	27.6	n.a.
Puerto Rico	698	1,301.9	3,341.4	28.0	10.30	n.a.	n.a.	53.84
St Kitts and Nevis	343	164.1	73.1	69.2	1.50	4.7	28.6	56.22
St Lucia	389	443.6	269.9	62.2	6.50	23.0	51.8	n.a.
St Maarten	492	868.3	432.3	66.8	5.00	n.a.	n.a.	n.a.
St Vincent and G'ines	n.a.	86.2	72.9	54.2	10.00	1.4	16.2	n.a.
Trinidad and Tobago	n.a.	82.2	398.2	17.1	5.00	n.a.	n.a.	n.a.
US Virgin Islands	1,014	1,768.4	607.2	74.4	4.00	459.5	259.8	173.24

Source: Derived from Caribbean Tourism Organization (2002) and Florida–Caribbean Cruise Association (2001).

cultural diversity offered on western Caribbean cruises in comparison to the 'relatively homogeneous Eastern Caribbean fare' and to the repositioning of ships to Gulf Coast ports in the aftermath of 11 September.

While the CTO welcomes increases in the numbers of cruise tourists and has been committed to arresting the long-term decline in the Caribbean market share of worldwide cruise tourists, it has at the same time been troubled by the declining proportion of stayover tourists. The reason for this is that cruise tourists spend only a small fraction of what stayover tourists spend. While cruise tourists constituted about 42 per cent of all tourists to the Caribbean in 2000, they accounted for only 12 per cent of tourist expenditures. Table 9.2 provides varying estimates from the CTO and the Florida–Caribbean Cruise Association (FCCA).

In a controversial calculation, Hall and Braithwaite (1990: 342–4) argue that cruise-passenger spending has a higher multiplier effect than spending by stayover tourists. Their reasoning, based on a dichotomous model of leakage effects and types of expenditures, is purely hypothetical and is not supported by the available empirical research. The FCCA and other studies find a high level of spending (generally over half) on watches, jewellery and clothing, all of which tend to be imported. Given that 'shopping' is a relatively small category of stayover expenditures, which focus much more on accommodation, food, entertainment, local guides and transportation, it seems implausible to believe that cruise-passenger expenditures are more beneficial. A study commissioned by the Caribbean Development Bank found 51 per cent of such expenditures going to duty-free imports in Barbados and concluded that its findings 'do not find any especially large impact from cruise tourism' and found cruise expenditures as low as 6.3 per cent of stopover tourist expenditures in Dominica (Caribbean Development Bank 1996: 3, 50, 76).

A uniquely deterritorialised industry

Not a single ship that cruises the Caribbean Sea is Caribbean-owned. But while most of these ships sail from companies whose headquarters are located in the United States, neither are most US-owned. Carnival Corporation, whose acquisition of Princess Cruise Line in 2003 raised its share of Caribbean-cruise capacity to 43.3 per cent, and Royal Caribbean International, with a capacity share of 35.4 per cent, together account for 78.7 per cent of passenger capacity in the Caribbean. Both have their corporate headquarters in south Florida. Yet Carnival is incorporated in Panama and Royal Caribbean in Liberia. The next largest Caribbean cruise company, Norwegian Cruise Lines (NCL), is owned by Malaysian-based Star Cruises, which recently changed its incorporation from Isle of Man to Bermuda. Together these three companies account for 85.7 per cent of Caribbean cruise capacity (*Cruise Industry News* 2002: 282), making the Caribbean-cruise industry one of the most concentrated industries in the world.

Cruising's detachment from place does not end there. Carnival's ships fly the flags of Liberia and Panama. Royal Caribbean ships mainly sail under Norwegian and Liberian flags. Star Cruises' NCL ships are mainly registered in the Bahamas and Panama. Celebrity's ships mainly fly the Liberian flag. And so it goes for most cruise ships operating in the Caribbean.[2]

The significance of this is that under international maritime law, enforcement of labour, environmental, health and safety laws is left to the 'flag' or 'registry' state. Countries that attract ships owned in other nations by making it be known that enforcement as well as fees will be minimal, are known as 'flag of convenience' states. If a cruise ship violates international pollution laws, for example, the arbiter of any claims is generally the flag state. A US General Accounting Office Study found that of 111 cases of marine pollution referred by the US to foreign flag states in the 1990s, no penalties were imposed in 109 of them and only minor fines were imposed in the other two (General Accounting Office (GAO) 2000: 9, 20). The flag of convenience regime, enshrined in the Law of the Sea and other international treaties, effectively insulates cruise companies from territorially based state and regional regulation in the Caribbean.

The Caribbean Hotel Association has complained for years about the unfair advantages that the cruise industry enjoys relative to the land-based hotel industry. While CTO's Holder (2001) has noted the lack of 'hard research data on the facts of the case', it is clear that the flag of convenience system undergirds the competitiveness and profitability of the cruise sector. Its ability to recruit its labour force literally from anywhere in the globe and to pay wages only a small fraction of prevailing wages in the Caribbean gives the cruise industry an enormous advantage, quite apart from the many other tax and regulatory freedoms it also enjoys. Indeed, one critical analysis of the negative effects of the flag of convenience system rejects the idea of trying to eliminate flags of convenience precisely because such an action 'would be financially devastating to the cruise industry' (Schulkin 2002: 125).

As mobile floating chunks of multinational capital, cruise ships have no permanent home, whatever their current 'homeport' may be. Some Caribbean countries benefited from this in the aftermath of 11 September 2001, when cruise companies on a moment's notice shifted between 15,000 and 18,000 berths from Europe to the Caribbean. But by the same token, cruise companies have no loyalty to the Caribbean in the event that more profitable opportunities beckon elsewhere.

Cruise ships' detachment from the physical territory of the Caribbean is perhaps most evident on the cruise ships themselves. Employees from south and south-east Asia, and from eastern Europe, vastly outnumber employees of Caribbean origin.[3] Ship motifs and décor have nothing to do with the Caribbean – the most striking perhaps being the ships of Holland America Line, whose ships almost exclusively highlight Dutch and Indonesian motifs and employ almost exclusively Indonesian and Filipino workers (Plate 9.4).

Plate 9.4 Officer and south-east Asian crew on Holland America's *Zaandam, 2001*

Caribbean motifs, cuisine and music – to say nothing of Caribbean history and society – are entirely absent from the cruise experience on most cruise ships in the region. On Carnival Cruise's 'fun ships', most of the deck chairs are set up by the crew to face inward towards the pool and activity areas, not outwards towards the sea and the islands (Wood 2000, 2002).

Cruise ships and sustainability

Sustainability became the buzz-word for both tourism and development in the 1990s. From the perspective of the CTO (2002: 248):

> As the most tourism dependent region in the world, Caribbean tourism stakeholders are faced with immense challenges: maintaining the

tourist flows necessary to guarantee economical stability; ensuring the proper use of resources for the benefit of both visitors and locals; guaranteeing that the resources that now attract visitors to our region, will continue to exist and attract visitors.

The relationship of cruise tourism to sustainable development is a contentious one. While the industry's self-portrayal (see WTTC *et al.* 2002) is one-sided at best, the possibility of 'Cruise-tourism: a chance of sustainability' has been mooted by two geographers, Ritter and Schafer (1998), who argue that cruise tourism compares favourably in terms of its environmental impact with other forms of tourism and has the potential to get even better. In contrast, a recent article in the *Georgetown International Environmental Law Review* concludes: 'Cruise ship pollution represents a significant and growing threat to human health and environment throughout the world' (Schulkin 2002: 113). From the vantage point of ecotourism, as articulated by Honey (1999: 39):

> Perhaps more than any other sector of the mass tourism industry, Caribbean cruise ships are anathema to the concepts and practices of ecotourism. These high-volume, prepaid, packaged holidays, with their celebration of sun-and-fun overconsumption and self-indulgence and their brief ports of call to allow tourists to buy souvenirs or duty-free First World luxuries, are the mirror opposite of the small-scale, locally owned, culturally sensitive, environmentally low-impact, and educational precepts of ecotourism.

Assessing sustainability claims about the industry is made difficult by the breadth and variability of the concept, inadequate data and the difficulty of assessing future trends. The following sections briefly address the cruise industry and sustainability in terms of the marine environment and in terms of its economic contribution.

Environmental sustainability

Cruise ships are only a tiny fraction of the world's shipping fleet, but because of their large size, their large number of passengers (over 3,000 on the larger ships), and the extravagant lifestyle they sustain, they have been estimated to produce 77 per cent of all marine pollution worldwide. On a single voyage, a large cruise ship on average produces 210,000 gallons of sewage, 1,000,000 gallons of graywater, 125 gallons of toxic chemicals and hazardous waste, eight tons of garbage and 25,000 gallons of oily bilge water (Schmidt 2000; see also Klein 2002: ch. 4; Ocean Conservancy 2002). Until very recently, virtually all of this was dumped at sea, and much of it still is. In addition, cruise ships are major sources of air pollution in ports (Bluewater Network 2000).

Relatively little data exist on cruise-ship pollution in the Caribbean. Few Caribbean countries have either the resources or the incentive (given their dependence on tourism) to collect and share information on this subject. Most of what is known about cruise-ship pollution involves illegal discharges in US coastal waters. Since most cruise ships in US coastal waters sail to the Caribbean, these data give some sense of the problem in Caribbean waters as well.

It is important to recognise at the outset that most ocean pollution is completely legal under international law. The failure of key countries like the United States to support strong regulatory regimes – the US has not even signed the United Nations Convention on the Law of the Sea (UNCLOS) – along with the power given to flag of convenience states both to block new laws and to enforce existing ones – has resulted in an extremely weak regulatory environment for the cruise-ship industry. The International Convention for the Prevention of Pollution from Ships (MARPOL) makes it illegal to dispose of plastic and oil anywhere at sea. It is legal to dump everything else – from raw sewage to toxic chemicals – subject only to limited national restrictions (generally not extending beyond twelve miles from shore and often much less). Under MARPOL Annex 5, the Caribbean is due to become a 'special area' subject to stricter regulations, but the failure to obtain the requisite treaty signatures to date has so far prevented it from coming into being.

The International Council of Cruise Lines (ICCL), of which most Caribbean cruise operators are members, adopted in 2001 a set of waste-management standards to which its members have to adhere. The standards are mostly based on MARPOL, which as noted above are extremely minimal. They include no mechanisms for monitoring or enforcement.

Public attention was focused on cruise-ship pollution in the case of Royal Caribbean, which after a lengthy legal battle admitted that it had engaged in a criminal fleet-wide conspiracy to disable anti-pollution devices and to dump oily bilge into waters off Puerto Rico, the US Virgin Islands and elsewhere. In 1999 the company agreed to an $18 million fine. Successful US prosecutions of several other major cruise companies quickly followed.

Cruise-industry organisations and companies have scrambled to portray themselves as environmentally friendly. They insist that they recognise that the health of their industry depends on the health of the seas that they sail in and that they are taking the necessary steps to achieve this. Undoubtedly there is some truth in this, but in a context of minimal regulation and enforcement, the highly competitive nature of the industry (involving cut-throat competition both with other cruise companies and land-based resorts), along with seemingly obsessive profit maximisation, has produced an organisational incentive structure in which polluting but cost-saving short-cuts have brought financial rewards to those engaging in them. Illegal polluting, covered up by the falsification of ship logs, continues to this day, as evidenced by convictions of Carnival and NCL in 2002.[4]

The alarming degradation of the Caribbean sea and coastal areas clearly stems from many sources. Land-based sources contribute a higher proportion of marine pollution than do ships (Schumacher *et al.* 1996; Eco-Exchange 2000). However, particular cases of environmental destruction can be traced to cruise ships. Three hundred acres of coral reef have been destroyed by cruise-ship anchors around Georgetown, Grand Cayman, and 80 per cent of a coral reef in a marine park off Cancun, Mexico, was destroyed when a cruise ship ran aground. Given the centrality of a healthy sea to both cruise and resort tourism, the bottom line is that sea-based and land-based tourism interests in the Caribbean have common and interrelated environmental concerns.

In this connection, it is interesting to examine the conflict that arose and still simmers over taxing cruise ships for land-based pollution facilities. In 1995, a $50.5 million project was instituted in six Organization of Eastern Caribbean States (OECS) countries to address both land-based and cruise ship solid-waste management in the islands. Operation of such solid-waste facilities is required for countries to ratify the MARPOL agreement making the Caribbean a special area, and thus constitutes an important step towards this goal. As part of the agreement with the World Bank, the six OECS countries were required to collect a $1.50 levy from each tourist, whether they arrived by air or by cruise ship, to finance part of the project.

The Florida–Caribbean Cruise Association (FCCA), which represents the major cruise companies operating in the Caribbean, immediately opposed the levy and threatened boycotts – claiming that its member ships had 'zero discharge'. This was hardly an idle threat since the cruise companies had implemented boycotts and played OECS countries off against each other in 1992, successfully rolling back an attempt by OECS countries to raise their port fees from $3.00 to $9.25 per visitor. It took intervention by the World Bank to force the cruise companies to back down. In a section of a World Bank policy research paper entitled 'Pollution: facing down the cruise lines', Schiff and Winters (2002: 19) write:

> The program had to withstand a last-ditch stand by the cruise lines, which tried again in 1997 to split the islands with threats of boycotts. The donors mediated the dispute – with what one participant described as 'a bit of arm twisting' – and a $1.50 charge per cruise ship passenger was charged from May 1998.

As Klein (2002: 113–14) points out, the cruise industry had a perverse incentive to oppose the levy, because the project was a necessary precondition for bringing MARPOL Annex V into effect, which will make the Wider Caribbean (encompassing both the Caribbean Sea and the Gulf of Mexico) a special area subject to a ban on almost all forms of dumping anywhere. But the industry's opposition also reflected the mindset behind the various schemes to reduce anti-pollution costs noted earlier. Klein (2002: 114) quotes a Carnival executive:

> The reason that Carnival Corp. makes the kind of money we do is because we pay great, great attention to controlling our costs. Sure, it's just $1.50. But it's $1.50 here, then $1.50 there, then $1.50 over there. When you allow people to unfairly charge for things, then you open a Pandora's box.

In a broader sense, the position of the cruise industry in this case reflects the kind of deterritorialisation discussed in the previous section. Claiming to be champions of clean seas, the companies sought to reject participation in a programme that not only would handle ship waste but also address the dominant, land-based, sources of marine pollution.

In sum, the impact of cruise tourism in the Caribbean on environmental sustainability is decidedly mixed. The industry is investing in more efficient and anti-pollution technologies in its new ships (many of them cost-saving in the long run), although it has been slow to retrofit its older ones. Its overall record is improving, but the increasing size of its ships makes the potential impact of any accident or disaster that much greater. It is important to remember, however, that even a total elimination of illegal dumping would still leave intact the right, and the practice, of dumping most polluting substances at sea. What is legal under MARPOL and other agreements is clearly unacceptable in terms of long-term environmental sustainability. On land, the influx of thousands of short-term cruise visitors, with only a few hours at their disposal, is putting severe environmental pressure on popular natural sites where tour buses go (Pattullo 1996: 125–8; Caribbean Development Bank 1996: 72–3). Meanwhile, 75 per cent of Caribbean hotels and resorts have wastewater systems that do not comply with basic standards (Bryan 2001: 9). What is most clear is that environmental issues of land and sea, and of stayover tourism and cruise tourism, are deeply intertwined, and require a regionally based system of governance that cuts across both.

Sustainable tourism and sustainable development

A once-thriving Great Lakes cruise market virtually disappeared after a terrible fire in 1949 that took 118 lives (Oceans Blue Foundation 2002: 12). The contemporary Caribbean cruise industry in its early days had to survive fire disasters on the *Yarmouth Castle* and on the *Viking Princess* in 1965 and 1966 that together took ninety-two lives (Cudahy 2001: 6–15). Since September 2001, there has been much concern about a more deadly replay of the terrorist hijacking of the *Achille Lauro* in 1985. Barring such unknowable eventualities, however, the sustainability of Caribbean-cruise tourism might seem assured.

With only 12–13 per cent of Americans having cruised, and yet a good 50 per cent expressing interest in doing so, the cruise industry is confident about its future (AAPA 2000). The companies are investing in new and bigger ships, and as we have seen, there is every reason to believe that cruise passengers will soon surpass stayover tourists in the Caribbean region.

The FCCA estimates direct passenger and crew spending throughout the Caribbean to have come to $1.4 billion in the 1999/2000 cruise year, with an additional $1.2 billion generated by the 'multiplier effect' as these expenditures induce further spending, for a total of $2.6 billion. The study estimates that cruise-related expenditures created 60,136 jobs throughout the Caribbean paying $285 million to Caribbean residents (FCCA 2001: ix, 9–15).

FCCA figures in previous reports have been critiqued as overestimates (Pattullo 1996: 165–7; Caribbean Development Bank 1996: 5). Whatever the exact facts of the case, the more interesting questions involve the composition and effects of cruise-tourist expenditures and the effect on stayover tourism. The two are related in that the limited contributions of cruise tourism are often accepted on the grounds that cruise tourists may return as stayover tourists.

As noted above, cruise-passenger expenditures are far less than stayover tourists and tend to focus on duty-free goods with high leakage effects. Caribbean provisioning of cruise ships has increased, but remains negligible overall. Since the proportion of cruise-ship workers who are Caribbean nationals is small (published estimates range from 7 to 26 per cent), the direct employment benefits remain very limited. In addition, the introduction of mega-cruise ships has necessitated major and expensive port upgrading.

Despite periodic efforts to raise them, port passenger fees remain extremely low in the Caribbean, testimony to the ability of the cruise lines to play destinations off against each other. Bermuda to the north charges $60 per visitor, but as Table 9.2 shows, no Caribbean port comes close to this. (Bermuda also requires cruise ships to stay two and a half days, whereas almost no Caribbean port rates even twenty-four hours.) The great majority remain well below $10. Half receive $5 a head or less. Cruise-passenger fees are almost everywhere in the Caribbean a small proportion of departure fees levied at airports on stayover tourists, which themselves have been characterised as 'minimal revenue generators' by a World Bank Environment Department (2000) report. They have been further reduced by the use of private islands and by additional 'days at sea' in some Caribbean cruise itineraries (Showalter 1994; Wood 2000).[5]

Efforts to form a united front to negotiate collectively with the cruise lines have not in the past been successful. As noted earlier, a coordinated effort of OECS states collapsed in the face of divide and conquer strategies of the cruise lines. CARICOM leaders shortly thereafter mandated a study of the possibility of 'a regional regulatory body and licensing system to oversee the operations of cruise ships in the Caribbean', which was duly commissioned by the CTO, with financing from the European Union's Caribbean Tourism Development Programme. The report (CTO 1994), which proposed an ambitious Caribbean Cruise Shipping Authority, promptly sank into oblivion in the face of regional disunity and cruise-industry opposition.

Although it hardly represents an official World Bank position at this point, the idea of imposing an environmental levy on cruise passengers (and other tourists) was given intellectual support in a report of its Environment Department in collaboration with the European Commission in 2000. The report defines the excess return to tourism assets that results from the 'distinctive and attractive environmental assets of the Caribbean' as *rents*, and argues that a larger proportion of these rents should be captured by Caribbean governments. Among other policy options, the report proposes environmental user fees for all arriving cruise passengers (World Bank Environment Department 2000: viii–x, 30).

While such ideas enjoy broad support among Caribbean leaders, their political viability remains much in doubt. Both CARICOM and the CTO formed cruise committees in 2001, but the new efforts seemed informed mainly by a weary sense that Caribbean countries have to come to terms with the cruise industry, which has become and will remain a central fact of life.

'Cruise conversion' – the recruitment of cruise tourists to return as stay-over tourists – is a central component of the CTO's new initiative. FCCA studies over the past decade have consistently found a little more than half of cruise passengers saying that they might return as stayover tourists to Caribbean destinations. The FCCA President estimates that 25 per cent actually do so (Perry 1999). The fact is that there are virtually no data on this important question. A study in the Bahamas found 6.2 per cent of tourists to the Bahamas in 1996 had been influenced by a previous cruise visit (Wilkinson 1999: 278). Since cruise tourists greatly outnumber stay-over tourists in the Bahamas, this would seem to suggest a quite low conversion rate. One comparative study of cruise and resort tourists concludes that cruise tourists may be a quite specific travel group not prone to be lured back to resorts (Morrison *et al.* 1996). Certainly cruise companies go to great lengths to keep their passengers coming back onboard, both by offering discounts and upgrades and by maintaining a steady flow of advertising that compares land vacations unfavourably to cruise vacations. More generally, there is virtually nothing about the onboard cruise ship experience that 'sells' the Caribbean *per se*. As examples of what sociologist George Ritzer (1999) has called 'cathedrals of consumption', ever competing to create more artificial and simulated environments of enchantment and extravagance, cruise ships seemed designed to shape expectations difficult to achieve in anything but a floating theme park.

Conclusion: the Great White Fleet revisited?

Cruise ships are the Great White Fleet of today, but what they represent is the coming of age of neoliberal globalisation. No other industry is more globalised than the cruise industry, and as the largest regional cruise industry, the Caribbean is therefore particularly important for assessing the relationship between global currents and tourism.[6]

Globalisation in the Caribbean combines with the historical legacies of colonialism and slavery to produce potentially explosive situations. For example, the decline of the banana and sugar industries, as a result of World Trade Organization rulings against traditional preferences, has increased unemployment that in turn has translated into increased harassment and aggressive solicitation of cruise tourists in Jamaica (McDowell 1998). Cruise ships in response have been eliminating from their itineraries ports where such problems are prevalent. Caribbean destinations find themselves competing with each other on a new terrain, while deep-rooted local resentments continue to grow, threatening the land-based tourism industry as well.

As I have argued previously (Wood 2000), in this and other ways Caribbean cruise tourism also represents 'globalisation at sea' in the sense of a voyage into unchartered waters without a known destination – environmentally, socially or economically. The allegedly beneficial outcomes of a system built on neoliberal principles are only now being subject to test. The results so far do not warrant optimism in terms of the broader health of Caribbean tourism and societies.

Cruise tourism has been increasing its market share steadily and is likely to surpass land-based stayover tourism relatively soon. The industry's uniquely deterritorialised nature has given it competitive advantages while freeing it from any significant regulation in terms of its environmental and social practices. The challenge posed by the CTO's 1994 study remains salient:

> [T]here is little doubt at all that cruise tourism in the Caribbean is here to stay as an integral part of the overall tourism sector. The region is not so much required to choose one sector to the exclusion of the other, as to strike an appropriate balance between the two in a complementary and synergistic fashion, which enhances the aggregate benefits to the destination countries concerned. The task of establishing and maintaining such a balance is surely one of the biggest challenges facing regional governments . . .
>
> (CTO 1994: C4–5)

Caribbean countries are not helpless in this struggle, but as in much of the region's history, broader global currents will shape their prospects. The assistance Caribbean intergovernmental associations have received from the World Bank and European Union in dealing with the cruise industry represents a promising development, as does an increasingly broad set of NGO initiatives, such as the Cruise Ship Stewardship Initiative of the Oceans Blue Foundation (2002), Bluewater Network's (2003) cruise ship campaign, and War on Want's and the International Transport Workers Federation's (2002) 'Sweatship' campaign. The success of efforts by Caribbean nations to exercise greater control over the uniquely globalised and deterritorialised cruise industry, and to fit it into a strategy of sustainable development, will partly be shaped by the outcome of such efforts to bring global currents under more effective and humane governance.

Notes

1   Reflecting current political realities connected with the North American-based cruise industry, Cuba will not be discussed significantly in this chapter, although it does receive some European cruise ships and could have a huge impact on Caribbean itineraries if it were incorporated into the dominant Caribbean cruise market.
2   A few national registries have successfully lured back their fleets by loosening their regulations in what are often called 'second registries'. Norway has partly done this with Royal Caribbean, and Holland America reflagged its fleet in the Netherlands after the latter rewrote its legislation. Princess Cruise Line recently reflagged most of its fleet to Great Britain.
3   Indeed, on one voyage on Holland America's *Zaandam* that the author sailed on, there was exactly one employee of African or Caribbean descent among the entire crew of 650 or so.
4   Since polluting in non-coastal waters is technically handled by the flag states, who routinely do nothing, the United States has resorted to prosecuting companies for presenting falsified logs at US ports, a violation of US law.
5   A 1988 joint report by the Organization of American States and the Caribbean Tourism Research and Development Center expresses alarm about 'the threat of the non-destination', but acknowledges that the problem stems in part from disappointing or unpleasant experiences in real destinations.
6   Several commentators (e.g. Bloor *et al.* 2000) have noted the uniquely global nature of the shipping industry as a whole, but cruise ships are unique in the global diversity of both passengers and crew and are also specifically exempted from some regulations that apply to cargo ships. It is therefore arguable that the cruise-shipping sector is the most globalised of all.

References

American Association of Port Authorities (AAPA) (2000) 'Cruise Industry Information'. Online: <http://www.aapa-ports.org/industryinfo/cruise_industry. html> (accessed 2 October 2002).

Bauman, Z. (1998) *Globalisation: The Human Consequences*, New York: Columbia University Press.

Bloor, M., Thomas, M. and Lane, T. (2000) 'Health risks in the global shipping industry: an overview', *Health, Risk and Society* 2(3): 329–40.

Bluewater Network (17 July 2000) 'A stacked deck: air pollution from large ships'. Online: <http://www.earthisland.org/bw/stacked.pdf> (accessed 26 June 2001).

—— (2003) 'Safeguarding the seas'. Online: <http://www.bluewaternetwork.org/campaign_ss_cruises.shtml> (accessed 30 January 2003).

Bryan, A.T. (2001) 'Caribbean tourism: igniting the engines of sustainable growth', Miami, FL: North-South Center, University of Miami. Online: <http://www.ciaonet.org/wps/bra03/bra03.html>.

Caribbean Development Bank (CDB) (1996) 'A study to assess the economic impact of tourism on selected CDB borrowing member countries', prepared by ARA Consulting Group Inc., Systems Caribbean Ltd., Ione Marshall, KMPG Peat Marwick.

Caribbean Tourism Organization (CTO) (1994) 'Report on the requirements for establishing a regional regulatory body and licensing system to oversee the operations of cruise ships in the Caribbean sea', prepared by A. Ralph Carnegie, St Michael, Barbados: Caribbean Tourism Organization.

—— (2002) *Caribbean Tourism Statistical Report (2000–2001 Edition)*, St Michael, Barbados: Caribbean Tourism Organization.

Cruise Industry News (2002) *Cruise Industry News Annual 2002 (Fifteenth edition)*, New York.

Cudahy, B.J. (2001) *The Cruise Ship Phenomenon in North America*, Centerville, Maryland: Cornell Maritime Press.

Eco-Exchange (2000) 'Caribbean treaty looks for sea change'. Online: <http://www.rainforest-alliance.org/programs/cmc/newsletter/jun00–1.html> (accessed 11 January 2003).

Florida–Caribbean Cruise Association (FCCA) (2001) 'Cruise industry's economic impact on the Caribbean. Prepared by PriceWaterhouseCoopers and Business Research and Economic Advisors', Pembroke Pines, FL.

Foster, H.L. (1928) *The Caribbean Cruise*, New York: Dodd, Mead and Company.

General Accounting Office (GAO) (2000) *Marine Pollution: Progress Made to Reduce Marine Pollution by Cruise Ships but Important Issues Remain*, Washington, DC.

Hall, J.A. and Braithwaite, R. (1990) 'Caribbean cruise tourism: a business of transnational partnerships', *Tourism Management* 11(4): 339–47.

Hannam, K. (2002) 'Tourism and development I: globalisation and power', *Progress in Development Studies* 2(3): 227–34.

Holder, J. (2001) 'Meeting the challenge of change: address delivered by the Secretary General of the Caribbean Tourism Organization at the second Caribbean tourism summit, Nassau, the Bahamas', 8–9 December 2001 (10 Dec.). Online: <http://www.caricom.org/pressreleases/pres149_01.htm> (accessed 30 December 2001).

Honey, M. (1999) *Ecotourism and Sustainable Development: Who Owns Paradise?*, Washington, DC and Covelo, CA: Island Press.

Klein, R.A. (2002) *Cruise Ship Blues: The Underside of the Cruise Industry*, Gabriola Island, BC: New Society Publishers.

Lawton, L.J. and Butler, R.W. (1987) 'Cruise ship industry – patterns in the Caribbean, 1880–1986', *Tourism Management* 8: 329–43.

McDowell, E. (1998) 'Jamaica sweeps off its welcome mat', *New York Times*, 21 June.

Morrison, A., Yang, C., O'Leary, J.T. and Nadkarni, N. (1996) 'Comparative profiles of travellers on cruises and land-based resort vacations', *Journal of Tourism Studies* 7(2): 15–27.

Ocean Conservancy (May 2002) 'Cruise control: a report on how cruise ships affect the marine environment', Washington, DC. Online: <http://www.oceanconservancy.org/dynamic/aboutUs/publications/cruiseControl.pdf>.

Oceans Blue Foundation (2002) ' "Blowing the whistle" and the case for cruise certification: a matter of environmental and social justice under international law', Vancouver, BC. Online: <http://www.oceansblue.org/bluetourism/chartacourse/cruiseship/cruisereport.html>.

Pattullo, P. (1996) *Last Resorts: The Cost of Tourism in the Caribbean*, London: Cassell.

Perry, D. (1999) 'Cruises seen as mixed bag in Virgin Islands' (7 Feb.). Online: <http://www.s-t.com/daily/02–99/02–07–99/e09he189.htm> (accessed 25 January 2003).

Ritter, W. and Schafer, C. (1998) 'Cruise-tourism: a chance of sustainability', *Tourism Recreation Research* 23(1): 65–71.

Ritzer, G. (1999) *Enchanting a Disenchanting World: Revolutionizing the Means of Consumption*, Thousand Oaks, CA: Pine Forge Press.

Schiff, M. and Winters, L.A. (2002) 'Regional cooperation and the role of international organizations and regional integration', Washington, DC: World Bank Research Working Paper 2872.

Schmidt, K. (2000) *Cruising for Trouble: Stemming the Tide of Cruise Ship Pollution*, San Francisco: Bluewater Network. Online: <http://www.earthisland.org/bw/stacked.pdf>.

Schulkin, A. (2002) 'Safe harbors: crafting an international solution to cruise ship pollution', *Georgetown International Environmental Law Review* 15(1): 105–32.

Schumacher, M., Hoagland, P. and Gaines, A. (1996) 'Land-based marine pollution in the Caribbean: incentives and prospects for an effective regional protocol', *Marine Policy* 20(2): 99–121.

Showalter, G.R. (1994) 'Cruise ships and private islands in the Caribbean', *Journal of Travel and Tourism Marketing* 3(4): 107–18.

Taylor, F.F. (1993) *To Hell With Paradise: A History of the Jamaican Tourist Industry*, Pittsburgh: University of Pittsburgh Press.

War on Want and International Transport Workers Federation (2002) 'Sweatships'. Online: <http://www.waronwant.org/download.php?id=71>.

Waters, M. (1995) *Globalisation*, London: Routledge.

Wilkinson, P.F. (1999) 'Caribbean cruise tourism: delusion? illusion?', *Tourism Geographies* 1(3): 261–82.

Wood, R.E. (2000) 'Caribbean cruise tourism: globalisation at sea', *Annals of Tourism Research* 27(2): 345–70.

—— (2002) 'Caribbean of the east? Global interconnections and the south-east Asian cruise industry', *Asian Journal of Social Science* 30(2): 420–40.

World Bank Environment Department and in collaboration with the European Commission (2000) 'Tourism and the environment in the Caribbean: an economic framework (Discussion Draft 20453-LAC)', Washington, DC.

World Travel and Tourism Council (WTTC), International Federation of Tour Operators (IFTO), International Hotel and Restaurant Association (IH&RA) and International Council of Cruise Lines (ICCL) (2002) 'Industry as a partner for sustainable development: tourism', London. Online: <http://www.uneptie.org/outreach/wssd/docs/sectors/final/tourism.pdf>.

10 Manifestations of ecotourism in the Caribbean

David B. Weaver

Introduction

Since the term first appeared in the tourism literature in 1985 (Romeril 1985), ecotourism has emerged as a major focus within the field of tourism studies (see for example Fennell 1999; Wearing and Neil 1999; Weaver 2001a, 2001b; Page and Dowling 2002), and across an increasingly broad array of destinations, including most Caribbean states and dependencies. Evidence of its importance and formalisation at the global level includes the United Nations declaration of 2002 as the International Year of Ecotourism. The purpose of this chapter is to describe the scope of the ecotourism sector within the insular Caribbean and to raise issues that are pertinent to its future development. The first section considers the definition of ecotourism in terms of three core criteria, and this is followed by the presentation of a generic ecotourism model that takes into account the structural and strategic dimensions of the sector. The contemporary status of ecotourism in the region is then described, and relevant issues are raised.

Definition of ecotourism

A major problem associated with the evolution of ecotourism has been definitional confusion, which in turn has given rise to extensive misuse of the term, both deliberately and inadvertently (Wheeller 1994). This confusion is also largely responsible for the widely disparate estimates that have been proffered as to the magnitude and growth of the sector. For example, the World Tourism Organization (WTO) in early 1997 estimated that ecotourism accounted for between 10 and 15 per cent of global tourism, but subsequently raised this figure to 20 per cent later in the same year (Wight 2001). However, in conjunction with the formalisation process and related attempts to specify the parameters of the sector, it is now widely recognised that ecotourism involves three core criteria. These are (1) a primary focus on natural attractions, (2) an educational or learning element, and (3) management that strives for environmental, economic and socio-cultural sustainability (Blamey 1997, 2001; Clarke 2002).

Focus on natural attractions

Virtually all ecotourism definitions assert that the sector is focused primarily on natural attractions (Page and Dowling 2002). It follows logically that relatively undisturbed natural habitats, and particularly those associated with higher order protected areas (e.g. IUCN category II units, or 'National Parks'), constitute the optimal and most popular ecotourism venue (Lawton 2001). However, two factors qualify the 'natural attraction –natural habitat' nexus. First, most definitions also concede that the associated cultural features of a relatively undisturbed natural area constitute a secondary category of ecotourism attraction. This is a logical and welcome corollary given that no area, no matter how ostensibly undisturbed, is completely free from direct or indirect human influences. More specifically, the line between nature and culture in areas inhabited by indigenous people, as in parts of the Australian Outback, is often blurred to the point where the two are almost indistinguishable (Butler and Hinch 1996; Hinch 2001). The inclusion of these cultural influences in product interpretation (see below), therefore, provides a more complete understanding of the dynamics associated with 'natural' ecotourism attractions. Fennell (1999) has coined the acronym 'ACE' tourism to describe products where the *a*dventure tourism, *c*ultural tourism and *e*cotourism elements are amalgamated to such an extent that it is impossible to clearly distinguish that product as belonging to any one of those individual components.

The second factor considers the scope of the natural attraction resource base. At the holistic end of this spectrum, a destination would emphasise the entire ecosystem (e.g. the 'desert' or 'tropical rainforest') without according special status to particular sub-components. In this case, the availability of a representative natural habitat is both appropriate and necessary. At the other end of the spectrum, many destinations focus only on certain high-demand, charismatic mega-fauna or mega-flora such as giant pandas or redwoods. Some of these species periodically take advantage of food and shelter opportunities in highly modified environments, as for example in the case of migrating whooping cranes encountered in the grain fields of Saskatchewan, or storks in the chimneys of urban housing in Europe (Lawton and Weaver 2001). The implication of both these factors is that while high-order protected areas under most circumstances offer the choicest ecotourism venues, virtually all places offer at least some ecotourism potential. Modified spaces and lower-order protected areas, moreover, offer these opportunities without raising the environmental sustainability issues that are encountered when considering the impacts of ecotourism in relatively undisturbed and hence more vulnerable settings.

Provision of educational or learning opportunities

The emphasis on nature/culture-focused learning opportunities differentiates ecotourism from other forms of nature-based tourism such as adventure tourism and 3S (sea, sand, sun) tourism, where the natural environment provides a convenient setting to fulfil non-nature-focused motivations such as hedonism (in the case of 3S tourism) or thrill-seeking (in the case of adventure tourism) (Weaver 2001a). Learning opportunities are facilitated by the on-site provision of interpretive signage, visitors' centres and/or tour guides, reflecting Weiler and Ham's (2001: 549) view of interpretation 'as an indispensable tool for achieving the goals of ecotourism'. However, learning can also be facilitated by guidebooks and external tour guides, while many definitions allow for 'appreciative' or 'spiritual' learning opportunities that are highly personal, unmediated and inchoate in terms of the outcomes that are provided. Moreover, the provision of learning opportunities in whatever form does not ensure that visitors receive or assimilate the desired information, which often combines factual information and themes about the resource with minimal impact messages, preferably delivered in a way that maximises visitor satisfaction (McArthur 1998; Wearing and Neil 1999). The common denominator is that the managers of the site should provide a setting conducive to learning, whether through active on-site interpretation, personal reflection or something in between.

Sustainability

Environmental sustainability is a crucial component of all ecotourism definitions, while most also advocate sustainable socio-cultural and economic outcomes as expressed through support for 'community' development and other benefits to local residents (Fennell 1999). Sustainability, therefore, is almost universally supported as a principle, yet paradoxically it is the most contentious and elusive of the three core criteria. One factor is the malleability of related terms such as 'sustainable development' and 'sustainable tourism'. Hunter (1997) regards this flexibility as an asset because it takes into account the diverse settings in which tourism occurs; what is acceptable for an urban setting, for example, is likely not appropriate for a wilderness environment. However, flexibility can also lead to appropriation, wherein these terms are widely supported simply because they can be used to affirm and reinforce ideologies that emphasise contradictory impulses, such as continuous economic growth or environmental preservation (Weaver, forthcoming: a).

Weaver (forthcoming: a) also regards the complexity of tourism systems as problematic to the implementation of sustainability, in that the contribution of tourism activity to sectors such as transportation is extremely difficult to isolate, as are the myriad indirect and induced effects of tourism. How much responsibility should tourism therefore take for the atmospheric pollution generated by automobiles, or for the environmental problems caused by housing construction induced by tourism employees? In addition, complex systems such as tourism are not conducive to the linear extrapolation of trends. Emission levels (e.g. from a hotel) may appear to be sustainable over a long period of time, but may suddenly give way to an 'avalanche effect' when a hitherto unknown critical mass of contamination is attained and generates a red tide. A similar phenomenon exists in social structures when an apparently trivial action triggers widespread rioting in a destination that had appeared stable prior to that catalyst. Spatial, temporal and sectoral discontinuities between cause and effect are another problem associated with complex tourism systems. For example, measures taken in one community to become more sustainable (e.g. a ban on development) may serve to divert development at a later stage to more vulnerable locations, while the opening of an upstream open-pit mining operation could in turn undermine those same sustainability initiatives. Added to the fact that all destinations are unique, the above factors cumulatively make the selection of appropriate sustainability indicators and their critical threshold values an extremely difficult if not impossible process.

All this suggests the futility of asserting that any destination or product is sustainable *beyond doubt*, and that ecotourism *must* be associated with sustainable outcomes. It appears more sensible to focus on the *intent* of sustainability, wherein ostensibly positive impacts are deliberate, negative impacts are inadvertent, and timely action is taken to rectify negative impacts once they come to light (Weaver 2002). A final comment should be made about financial sustainability. While no definitions of ecotourism to the author's knowledge include this 'fourth sustainability', it is clear that no tourism, however sustainable in the other respects, will survive if it is not financially viable. Hence, sustainability stratagems must at least implicitly recognise this truism.

Structural and strategic dimensions of ecotourism

The above section has elaborated on and raised issues associated with each of the three core criteria of ecotourism, towards achieving a broader understanding of the sector's possibilities and limitations. Within these parameters, it is now appropriate to present a classification model that will assist in the description and analysis of the Caribbean ecotourism sector. According to this model (Figure 10.1), ecotourism products range along both a structural and strategic spectrum, as described below.

Structural spectrum: hard and soft ecotourism

Many definitions of ecotourism, and especially those proffered during the 1980s, are restrictive in that they associate the sector with some or all of the following traits: (1) a high level of environmental commitment among all participants, (2) a product and conduct emphasis on 'enhancement' sustainability (or practices that enhance the environmental and/or socio-cultural status of the destination), (3) participants as specialist ecotourists

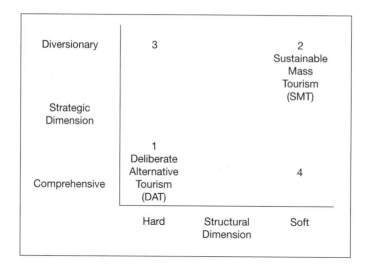

Figure 10.1 Strategic and structural dimensions of ecotourism

participating in specialised ecotourism experiences, (4) longer trips involving smaller groups of tourists engaging in physically and mentally challenging activities, (5) minimal reliance on services, (6) deep interactions with nature, and (7) an emphasis on FIT-oriented (i.e. free and independent travel) experiences. This amalgam of characteristics indicates a 'hard' ecotourism ideal type (i.e. an undistorted hard ecotourism model against which actual situations can be compared) (Weaver 2001a). The emphasis on the hard dimension in early definitions is associated with the origins of ecotourism in the 1980s as a form of nature-based, small-scale 'alternative tourism' that emerged as a reaction to the perceived environmental, economic and socio-cultural excesses of large-scale tourism development.

At the opposite end of the structural axis in Figure 10.1 is the 'soft' ecotourism ideal type, which amalgamates the following characteristics: (1) a moderate or 'veneer' commitment to environmental issues, (2) a product and conduct emphasis on 'steady state' sustainability (i.e. maintaining the status quo), (3) participants are conventional tourists experiencing ecotourism as one element of a diversified tourism experience, (4) shorter trips involving larger groups of tourists engaging in activities that are not physically or mentally challenging, (5) extensive reliance on services, (6) shallow interactions with nature, and (7) an emphasis on package travel mediated through travel agencies and tour operators. Some ecotourism purists oppose the inclusion of the soft or 'mass' dimension as a legitimate manifestation of ecotourism on grounds that the scale and client characteristics are not conducive to sustainability. However, it can also be argued that soft ecotourism can not only adhere to the three core criteria described above, but can do so under some circumstances in a way that is even more conducive to site sustainability than hard ecotourism. For example, it is only through the critical mass of visitation offered by soft ecotourism that this sector can fulfil its potential to offer financial incentives for the preservation and enhancement of the natural environment. This critical mass also helps to position ecotourism as a resource stakeholder that can hold its own with competing lobbies such as the forestry and mining industries. Moreover, soft ecotourism can be channelled into site-hardened intensive-use zones that occupy only a very small portion of a given high order protected area (Weaver 2002). It is through the inclusion of the soft or 'mass ecotourism' dimension into the ecotourism equation that the higher estimates of this sector's contribution are given credence.

Strategic spectrum: comprehensive, regional and diversionary dimensions

Simultaneous with the structural spectrum, ecotourism products can also be categorised along a strategic spectrum (Figure 10.1). Comprehensive ecotourism occurs when an entire jurisdiction at the state or dependency level is focused on ecotourism and allied activities (Weaver 1993). This is an uncommon option usually found in microstates. In contrast, the diversionary type occurs where ecotourism products are provided in close proximity to non-ecotourism tourism products as a supplement to the latter. Control over the ecotourism product is often invested in the conventional tourism business. For example, a cruise-ship line might offer tours to a small nature preserve on one of its privately owned islands, a hotel might maintain a small botanical garden of native flora within its grounds, or the same hotel might encourage snorkelling above an offshore coral reef. Urban parks and cemeteries are potential public venues for diversionary ecotourism (Lawton and Weaver 2001). Between these two extremes, there is the option of regional ecotourism, which entails a focus on ecotourism within a particular area or region of a state or dependency, while other parts of the latter continue to emphasise more conventional varieties of tourism. Potential settings include mountainous interiors, peripheral islands, less developed sections of coastline (including mangroves, sand dunes and estuaries), modified rural spaces, extensive interior wetlands and coral reefs.

Ecotourism as deliberate alternative tourism and sustainable mass tourism

The structural options combine with the strategic options to create a matrix consisting of several possible ecotourism combinations (Figure 10.1). Where hard ecotourism (and activities leaning towards hard ecotourism) combine with the comprehensive or regional strategic option (quadrant 1), deliberate alternative tourism is evident. The qualifier 'deliberate' is added to recognise the presence of a regulatory environment that ensures adherence to the principles of alternative tourism. 'Circumstantial' alternative tourism, in contrast, occurs when a destination resembles alternative tourism (i.e. it is small-scale, locally controlled, etc.), but the resemblance is superficial since it merely indicates that the destination is in the exploration or involvement stage of the resort cycle. No regulations, therefore, exist that attempt to keep the destination in this condition (Weaver 1993). The contrasting possibility is the combination of soft and diversionary ecotourism (quadrant 2), which produces a form of sustainable mass tourism. With regard to the remaining options, hard ecotourism is not likely to combine with diversionary ecotourism (quadrant 3) given the nature of the mass tourist market that participates in the latter, while soft ecotourism could possibly combine with comprehensive ecotourism (quadrant 4). This matrix, along with the earlier material on core ecotourism criteria, will inform the following discussion of the Caribbean ecotourism sector.

Ecotourism in the Caribbean

Ecotourism, its predecessors and allied cultural and historical activities have traditionally occupied a marginal position in the Caribbean tourism sector, as evident in Weaver's (1993) depiction of the three-tiered Caribbean tourism hierarchy (Figure 10.2). A major problem with this construct, however, is the confinement of ecotourism to the alternative tourism realm, which was criticised above as being an unduly restrictive relic of the 1980s. Moreover, it does not reflect the realities of contemporary Caribbean ecotourism, as discussed below. Another problem is the depiction of historical, environmental and cultural alternative tourism dimensions as separate tourism products.

Figure 10.3 updates this model by positioning ecotourism along the entire alternative tourism (AT)–mass tourism (MT) continuum, with diversionary ecotourism tending to locate at the MT end of this axis, and comprehensive and regional ecotourism tending towards the AT pole. The MT and AT dimensions blend into each other to recognise that there is no clear line between the two. Parallel to this ecotourism axis is an ACE spectrum that takes into consideration the hybrid character of many so-called 'ecotourism' products. While maintaining the fundamental three-tier structure of the older model, Figure 10.3 also modifies the status of activities in the top two tiers. The VFR (visiting friends and relatives) tourism of the second tier, for example, is shown as being more AT-oriented than business tourism, since such tourists typically stay in guesthouses or the homes of their social contacts, while business tourists are more hotel-oriented. In the highest or third tier, the beach-resort sector is shown as being somewhat larger than the cruise-ship sector, which in turn has an

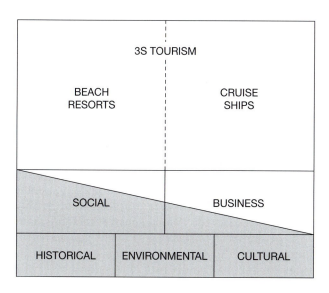

Figure 10.2 Reflexive status of contemporary Caribbean ecotourism (from Weaver 1993. Reproduced with kind permission of Kluwer Academic Publishers)

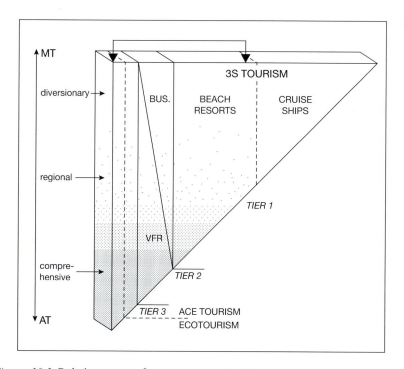

Figure 10.3 Relative status of contemporary Caribbean ecotourism

even stronger affiliation with mass tourism. A final important characteristic of Figure 10.3 is the arrow which joins tier one to tier three, thereby
acknowledging the expanding linkages between ecotourism and the
conventional 3S tourism sector. The following sections sequentially
examine the comprehensive, regional and diversionary manifestations of
Caribbean ecotourism.

Comprehensive ecotourism

Given the dominance of mass tourism in the region, it is not surprising
that only a few insular Caribbean states or dependencies have pursued or
are considering pursuit of a comprehensive ecotourism strategy. Candidates
include St Vincent, Montserrat and Dominica. In the case of St Vincent,
Duval (1998) attributes the case for comprehensive ecotourism to the
natural resource base of the main island (e.g. a small number of black-
sand beaches and an extensive landscape of forested mountains) as well
as capital and capacity limitations that dissuade the establishment of more
intensive types of tourism development. An influential leftist political tradition, moreover, has discouraged mass tourism development in St Vincent.
James Mitchell, a prime minister of the country in the early 1970s, gained
widespread attention from his 'To Hell with Paradise' *cri de coeur* against
3S tourism (Mitchell 1972). Such rhetoric has since diminished, but government and tourism industry support for smaller-scale, alternative types of
tourism, including ACE tourism and ecotourism, has remained strong. The
slogan 'The Natural Place to Be', used in the 1980s for promotional
purposes, is one example of this support (personal communication, Ministry
of Foreign Affairs, Tourism and Information). The situation of Montserrat
is both curious and tragic. This small British dependency had characteristics of and official support for alternative tourism and ecotourism during
the early 1990s, due to similar factors as found in St Vincent (Weaver
1995). An unexpected eruption of the dormant Chances Peak volcano in
1995, however, necessitated the abandonment of most of the island and
brought about the severe reduction in tourist activity. Ironically, a small
geologically based ecotourism sector has since emerged, focused on the
island's otherwise unwelcome volcanic activity.

Dominica offers the best opportunity in the region to examine the case
for comprehensive ecotourism. Prior to the early 1970s, the government
of Dominica attempted, without success, to establish a mass 3S tourism
industry on the island. Working against this objective were several factors,
including the lack of white-sand beaches, inadequate infrastructure, a long
wet season, vulnerability to hurricanes, and political instability. The release
of a wildly unrealistic consultant's report in 1971 appears to be the catalyst that prompted a fundamental rethinking of Dominica's tourism
strategy. Henceforth, the 3S resource liabilities of mountain, rainfall and
forest would be promoted as assets for environmentally and culturally

oriented tourism. Instead of attempting to emulate islands such as Barbados that were much better suited to 3S tourism, the Nature Island of the Caribbean would celebrate and emphasise its uniqueness as a relatively undeveloped nature-based destination. Promotional references to the island's purported 365 waterfalls during the 1980s offer an interesting counterpoint to Antigua's campaign that claimed the existence of 365 beaches. Aside from the marketing emphasis on the natural environment, the case for deliberate comprehensive ecotourism includes the establish-ment of a National Park network and its development on a limited scale as a tourist attraction, legislation ensuring a significant level of Dominican ownership in the accommodation sector, the presence of several small ecolodge-type facilities that appear to be in accord with basic ecotourism principles, and a relatively small stayover tourist intake induced by the decision not to expand either of the island's two small airports (Weaver 1991). Moreover, a significant complementary ACE tourism component is evident in the presence of high-profile attractions related to the remnant Carib indigenous population (Slinger 2000).

Other evidence, however, indicates that the ecotourism credentials of Dominica merit greater scrutiny. There is no evidence to date, for example, that ecotourism in Dominica is being formalised or professionalised to any significant extent through the introduction of educational initiatives, accred-itation schemes for operators, or other quality-control mechanisms that would help to ensure that the sector's core criteria are actually being fulfilled. Its superficial and ad hoc adherence to these criteria suggests that Dominica may more accurately be described as a 'proto-ecotourism' desti-nation that is not actually pursuing deliberate alternative tourism (Weaver forthcoming: b). More disturbing are developments within and beyond tourism that could undermine ecotourism. For example, while stayover visitation has increased modestly from 46,000 in 1991 to 70,000 in 2000, cruise-ship passengers have increased during the same period from 65,000 to 240,000 (Caribbean Tourism Organization (CTO) 2002). This reflects the government's investment in cruise-ship tourism as a way of stimu-lating medium-term economic development, given this sector's smaller capital requirements and lower marketing costs (TransAfrica Forum 2000). Questionable initiatives that have been pursued beyond tourism include the aborted formation of large-scale free port zones in the north of the island in the early 1970s, the more recent establishment of Dominica as an offshore banking, finance and internet gaming centre, and the creation of an Economic Citizen Programme that facilitates the acquisition of Dominican citizenship for a given amount of money. Equally disturbing, given ecotourism's emphasis on a viable natural environment, was the granting of a mining licence in 1996 to the international conglomerate BHP which would allow development of sensitive areas near the main National Park, and the alliance of Dominica with Japan in the International Whaling Commission to overturn the ban on commercial whaling. Weaver (forthcoming: b) interprets this pattern of pursued economic innovation as evidence of Dominica's desperation to attain economic development by whatever means possible in the face of severe resource limitations. Ecotourism, by this logic, is but one more expedient, opportunistically pursued as a default option following the inability to cultivate mass 3S tourism. If accurate, this does not bode well for Dominica's future as a viable comprehensive ecotourism destination. More charitably, this could simply indicate a comprehensive ecotourism strategy that is moving towards the more financially lucrative 'comprehensive-soft ecotourism' quadrant in Figure 10.1 (i.e. quadrant 4), given that most cruise-ship-passenger activity on land is still focused on a small number of natural and historical attractions rather than beaches or retail activity.

Regional ecotourism

Regionally based ecotourism strategies in the Caribbean assume that the destination-state or dependency will continue to derive most of its tourism-related income from conventional 3S tourism, but that certain regions of the state or dependency have a resource base that is more suited to ecotourism or ACE tourism (Weaver 1993). Peripheral islands are one type of venue where this option is commonly pursued, and Saba, a constituent island of the Netherlands Antilles, is one of the best Caribbean examples (though the degree of autonomy accorded to its tourism authority is such that a case for comprehensive ecotourism could also be argued). Ecotourism on Saba is focused mainly on the Saba Marine Park, which surrounds the 13 square kilometre island. A lack of beaches has prevented the estab-lishment of a 3S resort sector, and the terrestrial tourism industry instead emphasises small guesthouses, ecolodges and walking trails. Another strong Caribbean example of regional ecotourism on a peripheral island is St John, one of the US Virgin Islands. Here, a strong ecotourism reputa-tion is associated with Virgin Islands National Park, which accounts for 60 per cent of the island's land along with much of its coastal waters, and the luxury Maho Bay Camps. Among other states and dependencies, peripheral islands include ecotourism and ACE tourism as components of a slower-paced and smaller-scale tourism that contrasts with the prod-ucts offered on the main island. Examples of this watered-down ecotourism approach include Barbuda (relative to Antigua), Bonaire (Netherlands Antilles), Carriacou (Grenada), Cayman Brac (Grand Cayman Island), Marie Galante (Guadeloupe), Nevis (St Kitts), Tobago (Trinidad) and the Island of Youth (Cuba). At the archipelagic scale, tourism planning in the Bahamas designates the Family Islands (or Out Islands) in a similar manner, while encouraging intensive 3S tourism on New Providence and Grand Bahama Islands.

Mountainous and relatively remote interiors constitute a second common setting for regional ecotourism in the Caribbean. In this respect, large islands and smaller volcanic islands may be cited, in particular where high-order protected areas such as National Parks help to protect portions of the relevant landscapes. Cuba offers one of the best examples of large island mountain-based regional ecotourism, though it may also be atypical due to the country's idiosyncratic political and economic circumstances. Four interrelated characteristics that distinguish Cuba in this regard are (a) a strictly regulated central planning model, (b) the adoption of alternative, small-scale technologies as a way of coping with the US trade embargo, (c) the role of political and community-based tourism as corollaries to ecotourism, and (d) the inclusion of environmental restoration as part of the ecotourism product. All these characteristics are evident in the Las Terrazas area in the far west of Cuba, which is internationally known for the Moka Ecolodge (Honey 1999). Initiated by an influential member of Castro's inner circle, ecotourism in this part of Cuba is integrated into efforts to promote sustainable rural development by diversifying the local economy and restoring a large area of degraded hillside (CTO 2002). Concurrently, the project is intended to provide foreign visitors with a favourable impression of Cuba's socialist system. More typical of the region are the Dominican Republic, Jamaica, Puerto Rico and Trinidad, where forested mountains accommodate a small number of guesthouses and 'ecolodges' in the vicinity of protected areas. Coastal wetlands constitute a less common regional ecotourism venue in the Caribbean, with examples including the Zapata Peninsula in Cuba, the Caroni Swamp in Trinidad, and the Negril Great Morass in Jamaica. As with Dominica and the peripheral islands described above, the lack of formal regulations or education in these mountain- and wetland-based regional destinations suggests the presence of proto-ecotourism.

Diversionary ecotourism

Diversionary ecotourism occurs on all islands where mass 3S tourism is already well established. Several examples, as described below, suffice to illustrate the breadth of products that fall under this rubric, which essentially involves the participation of resort and cruise ship clients in short-term soft ecotourism opportunities within the vicinity of the resort or cruise ship. The Caribe Hilton, a major resort hotel in Puerto Rico, provides a private bird sanctuary and hiking trail on its 7-hectare peninsular property near San Juan. The Punta Cana Resort and Club, a large integrated resort in the Dominican Republic, has established a privately owned 400-hectare Nature Reserve on its property, a Biodiversity Lab that accommodates research students and faculty, and an offshore artificial reef that attracts divers. The Coral Canoa Beach Hotel and Spa, also in the Dominican Republic, operates a conservation programme to preserve the rhinoceros iguana. In Barbados, the Casuarina Beach Club offers opportunities for stargazing and for observing nesting sea turtles along the beach.

The nearby Coconut Court Beach Resort provides a Marine Education Programme, and is planting indigenous vegetation throughout its property to attract native fauna. Clientele in these and other Barbadian coastal resorts have the opportunity to take short excursions to public protected area remnants such as the Graeme Hall Swamp and the Turner Hall Woods. Finally, La Cabana beach resort on Aruba offers one-day cross-island 'safaris', as well as undersea observation on the *Atlantis* submarine. Most Caribbean resorts, in addition, provide opportunities for snorkelling and scuba diving, although the extent to which this type of activity should be included as a form of ecotourism remains a contentious issue (Cater and Cater 2001; Weaver 2001a). In all cases, diversionary ecotourism is culti-vated to diversify the 3S tourism product without distracting significantly from the core nature of that product or the associated provision of services and amenities.

Discussion and conclusion

In all its manifestations, ecotourism in the Caribbean remains a marginal component of the Caribbean tourism product when quantified as a propor-tion of all visitor activity. However, because of the tendency of mass 3S tourism to concentrate within a very small area of any particular state or dependency, ecotourism and ACE tourism are actually far more dominant than the former as a direct user of space, in particular because of its regional manifestations. It is unlikely, moreover, that mass 3S tourism will super-sede ecotourism and ACE tourism in the mountainous interiors and other non-beach areas of the Caribbean, given the unsuitable resource base of these spaces. Fears that such areas will progress along the unsustain-able trajectory of the conventional destination life cycle model through 'development' and 'stagnation' (as per Butler 1980) are therefore largely misplaced. But this should not breed complacency. Most Caribbean 'ecotourism' activity is actually better described as proto-ecotourism due to the absence of educational, accreditation and certification schemes that would help to ensure adherence to the three core principles described above. Hence, this proto-ecotourism does not appear to be fulfilling its potential to assist in environmental preservation or community development in the rural areas of the Caribbean, with the possible and anomalous exception of Cuba.

Further concerns about the region's commitment to ecotourism are raised by the situation in Dominica, which seems to be pursuing proto-ecotourism as an expedient or default option whose viability is therefore threatened by parallel expedient-type developments within and beyond tourism. Eco-tourism initiatives such as Australia's NEAP II and EcoGuide accreditation schemes (Issaverdis 2001) are clearly needed in the Caribbean, though real-istically, funding limitations as well as political and geographical fragmen-tation and parochialism will impede the development and implementation of such schemes in the foreseeable future. Ironically, it may be that ecotourism at the present time is best exemplified by the diversionary opportunities associated with the beach resorts. The providers of these opportunities (which are indicative of sustainable mass tourism) are able to work within small, manageable areas on or near their own properties. They are also more likely to possess the financial resources to develop high-quality ecotourism products and to have on staff environmentally qualified personnel to maintain this quality, and, hence, high customer satisfaction. Where they themselves own the ecotourism venue (as on a private nature reserve), the owners have an especially powerful incentive to invest in environmental stewardship and sophisticated site-hardening technologies that facilitate this stewardship. The synergies implied by this relationship should be expanded by encouraging the more enlightened beach resort operators of the Caribbean to play a larger role in the dissemination of ecotourism opportunities more broadly throughout all Caribbean destinations, in cooperation with alternative tourism providers, protected area managers, local communities, environmental NGOs and other stakeholders. It is perhaps in this partnership between sustainable mass tourism and deliberate alternative tourism that the establishment of a viable and genuine ecotourism sector can be effected in the Caribbean.

References

Blamey, R. (1997) 'Ecotourism: the search for an operational definition', *Journal of Sustainable Tourism* 5: 109–30.

—— (2001) 'Principles of ecotourism', in D. Weaver (ed.) *The Encyclopedia of Ecotourism,* Wallingford, UK: CABI Publishing.

Butler, R. (1980) 'The concept of a tourist area cycle of evolution: implications for management of resources', *Canadian Geographer* 24: 5–12.

Butler, R. and Hinch, T. (eds) (1996) *Tourism and Indigenous Peoples,* London: International Thomson Business Press.

Caribbean Tourism Organization (CTO) (2002) *Caribbean Tourism Statistical Report (2000–2001 Edition),* St Michael, Barbados: Caribbean Tourism Organization.

Cater, C. and Cater, E. (2001) 'Marine environments', in D. Weaver (ed.) *The Encyclopedia of Ecotourism,* Wallingford, UK: CABI Publishing.

Clarke, J. (2002) 'A synthesis of activity towards the implementation of sustainable tourism: ecotourism in a different context', *International Journal of Sustainable Development* 5: 232–50.

Duval, D.T. (1998) 'Alternative tourism on St Vincent', *Caribbean Geography* 9: 44–57.

Fennell, D. (1999) *Ecotourism: An Introduction,* New York: Routledge.

Hinch, T. (2001) 'Indigenous territories', in D. Weaver (ed.) *The Encyclopedia of Ecotourism,* Wallingford, UK: CABI Publishing.

Honey, M. (1999) *Ecotourism and Sustainable Development: Who Owns Paradise?,* Washington, DC: Island Press.

Hunter, C. (1997) 'Sustainable tourism as an adaptive paradigm', *Annals of Tourism Research* 24: 850–67.

Issaverdis, J. (2001) 'The pursuit of excellence: benchmarking, accreditation, best practice and auditing', in D. Weaver (ed.) *The Encyclopedia of Ecotourism,* Wallingford, UK: CABI Publishing.

Lawton, L. (2001) 'Public protected areas', in D. Weaver (ed.) *The Encyclopedia of Ecotourism,* Wallingford, UK: CABI Publishing.

Lawton, L. and Weaver, D. (2001) 'Modified spaces', in D. Weaver (ed.) *The Encyclopedia of Ecotourism,* Wallingford, UK: CABI Publishing.

McArthur, S. (1998) 'Introducing the undercapitalized world of interpretation', in K. Lindberg, M. Epler Wood and D. Engeldrum (eds) *Ecotourism: A Guide for Planners and Managers,* vol. 2, North Bennington: The Ecotourism Society.

Mitchell, J. (1972) 'To hell with paradise: a new concept in Caribbean tourism.' Address to CTA (Caribbean Tourism Association) Press Conference, Haiti, 21 September.

Page, S. and Dowling, R. (2002) *Ecotourism,* Harlow, UK: Prentice-Hall.

Romeril, M. (1985) 'Tourism and the environment – towards a symbiotic relation-ship', *International Journal of Environmental Studies* 25: 215–18.

Slinger, V. (2000) 'Ecotourism in the last indigenous Caribbean community', *Annals of Tourism Research* 27: 520–3.

TransAfrica Forum (2000) *The Impact of Tourism in the Caribbean,* Washington, DC: TransAfrica Forum.

Wearing, S. and Neil, J. (1999) *Ecotourism: Impacts, Potentials and Possibilities,* Oxford: Butterworth–Heinemann.

Weaver, D. (1991) 'Alternative to mass tourism in Dominica', *Annals of Tourism Research* 18: 414–32.

—— (1993) 'Ecotourism in the small island Caribbean', *GeoJournal* 31: 457–65.

—— (1995) 'Alternative tourism in Montserrat', *Tourism Management* 16: 593–604.

—— (2001a) *Ecotourism,* Brisbane: John Wiley and Sons Australia.

—— (ed.) (2001b) *The Encyclopedia of Ecotourism,* Wallingford, UK: CABI Publishing.

—— (2002) 'The evolving concept of ecotourism and its potential impacts', *International Journal of Sustainable Development* 5: 251–64.

—— (forthcoming: a) 'Tourism and the elusive paradigm of sustainable develop-ment', in A. Lew, A. Williams and C. Hall (eds) *A Companion to Tourism Geography,* Oxford: Blackwell.

—— (forthcoming: b) 'Managing tourism in the island microstate: the case of Dominica', in D. Diamantis and S. Geldenhuys (eds) *Ecotourism Planning and Assessment,* London: Continuum.

Weiler, B. and Ham, S. (2001) 'Tour guides and interpretation', in D. Weaver (ed.) *The Encyclopedia of Ecotourism,* Wallingford, UK: CABI Publishing.

Wheeller, B. (1994) 'Egotourism, sustainable tourism and the environment – a symbiotic, symbolic or shambolic relationship', in A. Seaton *et al.* (eds) *Tourism: State of the Art,* Chichester, UK: John Wiley & Sons.

Wight, P. (2001) 'Ecotourists: not a homogeneous market segment', in D. Weaver (ed.) *The Encyclopedia of Ecotourism,* Wallingford, UK: CABI Publishing.

11 Tourism, environmental conservation and management and local agriculture in the eastern Caribbean

Is there an appropriate, sustainable future for them?

Dennis Conway

Introduction

Throughout their colonial regimens, the island societies of the eastern Caribbean have struggled with the economic imbalance between natural resources and population; the limited 'carrying capacity' of their small land areas, the 'population pressure' their ex-slave societies experience, the embedded nature of structural unemployment among the island's youth, the lack of diversification in the local labour markets. Plantation agriculture ruled for over a century or two of profitable exploitation, in large part because sugar-cane growing was ideal for Caribbean climates and soils. When other export crops such as bananas, arrowroot, nutmeg and citrus were substituted for sugar in the smaller volcanic islands of the eastern Caribbean, their production levels were still evaluated in economic terms with reference to their contribution to the island's Gross Domestic Product, rarely their environmental impacts. More recently, with tourism growing to be many islands' major industry, agriculture has fallen even lower in the 'pecking order', and agricultural production appears to be less and less relevant to plans for sustaining the economic health and prosperity of these island societies.

This chapter examines the troubled history of conflictual relationships between tourism's growth and development, coastal zone environmental conservation and management and local, small-scale agriculture in eastern Caribbean islands in general and in St Lucia, in particular. Not only are these three strongly interrelated, but their sustainable futures may be predicated on the progressive paths all three follow, or are directed along. Institutional mechanisms forging linkages between all those involved in local tourism development, local agricultural development and environmental protection and management need to be carefully thought out, appropriately articulated (narrated), and analytically informed, if this difficult, complex industry is going to contribute to the sustainable futures of small Caribbean island societies.

First, Caribbean tourism's problematic and increasingly destructive relationships with small island environments and ecosystems, especially their coastal zones, are discussed. An understanding of how easily environmental relationships are disturbed is important because of their significance for the very notion of sustainability and for the health and long-term future of the industry. By developing and constructing the tourist 'built environment', and thus displacing farms and agricultural land, tourism's rapid monopolisation of the coastal zone drives up land prices and rents beyond their rural use-value (McElroy and de Albuquerque 1990). Not surprisingly, the proportion of agricultural land (and overall production) declines under mass tourism's challenge; whether with, or without, state intervention(s).

Second, utilising the UN Food and Agriculture Organization's (FAO) comparative statistics, the changing agricultural scene during the post-colonial period of political economic change (1960s to 1990s) is examined, to assess the trends in agricultural decline and diversification as Caribbean tourism takes hold, fluctuates and 'booms and busts', but eventually gets through its growing pains to become a mature and diversified industry. Inferential substantiation of the influences of the growing tourism sector and its stimulation of food-crop production in some islands, while apparently absent in others, is possible. On the other hand, the positive signs of changes in local agriculture's capacity to supply local and tourist markets, restaurants and tables during the 1990–2000 decade cannot be fully substantiated as causal outcomes.

Third, the scale of inquiry is narrowed to focus on local agriculture's linkages with tourism in St Lucia, and this synopsis of the temporal record of local agriculture's linkages with St Lucia's tourism industry references the field research of Bélisle (1983), Momsen (1972, 1986, 1998) and Timms (1999). Going beyond these authors' empirical observations and conclusions, tourism's relationship with local agricultural production is viewed as an appropriate path of soft growth, as opposed to hard growth. The distinctions between development paths of hard growth and soft growth build upon Herman Daly's (1990, 1991) progressive thoughts, which argue that environmental sustainability and economic growth need not be opposing goals. Hence a re-evaluation of small-scale agriculture's contributions to environmental management, social capital development and rural community consolidation is in order. Though tentative, because the outcome will depend upon the political will and managerial skill of Caribbean people and their governments, my answer to the question – 'Is there an appropriate set of sustainable growth paths that Caribbean tourism and local agriculture can follow?' – is, 'Yes.'

The fragility of tourism–environment relations

The Caribbean tourist industry has grown to become a complex service industry providing an ever-widening set of offerings, way beyond the allure and attractions of 'sun, sand and sea'. Caribbean islands are the environments with the all-important comparative advantage of the 'three-Ss', but this global industry is primarily managed, directed and produced by a wide range of private enterprises, mostly external and non-local. These include a multifunctional and geographically dispersed system of travel agents, airline companies, hotel chains, tour operators, advertising agencies, and communications and financial service sectors.

In the island destinations, this global ensemble interacts with the environment in a two-way system of direct and indirect feedbacks and relations, but the principal objective of this multifaceted and diverse international industry is to derive profits from their service endeavours. The bi-national, two-way system has Caribbean coastal environmental resources providing one of the basic 'ingredients' – the natural and/or man-made *locale* and island setting – for the tourist to visit, enjoy and 'consume'. For its international visitors the Caribbean is an 'exotic' resource to be consumed 'on vacation', but it is generally packaged, marketed and promoted in metropolitan marketplaces; they also interact with the destination environment *en masse*, thereby imposing aggregate burdens on its social, environmental and physical landscapes. Daily visits from cruise ships can bring excessive short-term overburdens, if not managed and regulated. Mass tourism can intrude on the relatively fragile ecosystems of the Caribbean, producing a variety of unwanted by-products, which have to be disposed of, thereby modifying the environment and creating negative environmental externalities.

There are, of course, qualitative and quantitative differences in the distribution of environmental resources over space in Caribbean islands, which definitively determine tourist development at the regional, national and international levels (McElroy and de Albuquerque 1991). Obviously, the smaller the island, the more limited the environmental options will be, but qualitative valuations of the coastal zone's potential will also vary with its natural endowments. Compare, for example, Dominica's black-sand beaches to Antigua's coral-sand stretches; both are numerous, but differ markedly in tourist potential. It might well be to the industry's advantage to propose that an important consideration for sustainable tourism development should be the preservation of both the quality and quantity of island coastal (and interior) environments at levels acceptable to the consumers, the tourists (Briassoulis and Van der Straaten 1992). An equally important requirement for small island sustainable development should be that the quality and quantity of Caribbean islands' natural resource bases are preserved at levels acceptable to the hosts.

Tourism-related activities compete for the environmental resources of an island's coastal zone, both among themselves and with other local and national economic activities (agriculture, fishing, forestry, industry, commerce, transportation, residential building and urban infrastructure, among others). Changes in the physical, spatial and socio-economic structure of tourist resort areas, of coastal zones, and of offshore ecosystems, as well as the existence of several burdensome environmental problems caused by mass tourist negative externalities, cause many local land-use and community conflicts. The need for some form of conflict management (resolution or reduction) leading to a more desirable allocation of environmental resources, or the proactive interdiction of environmental preservation practices by local authorities, is a regulatory agenda Caribbean governments, NGOs and regional environmental organisations should be pursuing, if they plan to keep a sustainable tourist industry. Participatory planning and co-management models for community organisation of local environmental protected areas are some of the 'grassroots' institutional mechanisms recommended for Caribbean islands (Conway and Lorah 1995; Pugh and Potter 2001). The jury is still out, however, on whether such progressive formulas can take root in the post-colonial societies of the eastern Caribbean, where hierarchical, top-down models of regulation and government overview are the enduring legacies (Pugh 2001).

A very important *direct* environmental impact of mass tourism concerns the industry's increasing demand for receptor services that will take care of unwanted by-products – solid waste, waste water, toxic effluents, for example. As early as the 1980s, Archer (1985) was warning us of the immediate offshore pollution that discharges from hotels along the over-built south coast of Barbados (see also Conway 1983). More recently, the cataloguing of environmental destruction by Island Resources Foundation and Caribbean Conservation Association, among others, effectively shows that in many islands tourism's economic profitability has taken precedence over environmental conservation and management (Island Resources Foundation 1996). Once an area becomes a tourist 'place', its resources undergo changes primarily because they are used (and often over-used), either directly for the production and consumption of the tourist product and/or indirectly by modernising activities linked to the tourist-related ones. The extent and intensity of the modifications depends upon two interrelated groups of factors: (a) the type of tourism activity (mass, selective, alternative), socio-economic, 'socio-cultural' and behavioural characteristics of tourists, the intensity and spatio-temporal distribution of environment use, and the strength of linkages between activities; (b) the social and biophysical characteristics of the islands' ecosystems – natural endowments, economic and social structure, forms of local political organisation and levels of tourism development or 'tourism style' (see McElroy and de Albuquerque 1991; UNEP 1982).

In addition, there are indirect impacts on the built environment and its infrastructure caused by industrial and service sector activities indirectly related to tourism; for example, the incubation and growth of a local handicraft industry, increased trade, local food and beverages production, increased entertainment venues, and greater commercial activity. McElroy and de Albuquerque's (1991, 1996) assessments of Caribbean small island tourism styles certainly give full recognition of the 'disconnect' between mass tourism's unregulated growth and environmental conservation. Their advocacy that small islands should plan to maintain low-density styles, and focus on alternative tourism offerings, rather than embrace mass tourism and its inevitable environmental consequences, is now widely accepted.

Not before time, environmental conservation issues were thrust into the limelight, in large part as a consequence of the Rio Earth Summit of 1992, Agenda 21, and the United Nations' Programme of Action (POA) for the Sustainable Development of Small Island Developing States (SIDS), which was formalised in Barbados in 1994. The latter advocated a plan for Sustainable Tourism Development in Small Island Developing States in which the principles of tourism and environmental management were spelled out. It identified action items at the national, regional and international levels which would ensure the viability of the tourism sector and its harmonious development with the cultural and natural endowments of small-island developing states (POA/SIDS 1994). The Caribbean Tourism Organization (CTO) had an equally broad perspective on how to ensure a sustainable tourist industry in the Caribbean. In 1998, the CTO Board gave high priority to environmental and socio-cultural considerations by endorsing a Sustainable Tourism Development Strategy and Plan of Action for the Caribbean. In particular, the development of community-based tourism development, in which local communities and stakeholders involve themselves in stimulating and managing their own national and local tourism development, was advocated as a means for ensuring the industry's sustainability (CTO 1998; Youngsman 1998). Most recently, the Association of Caribbean States held a Conference on Sustainable Tourism in 2001, in which the regional coordination of environmental planning and the promotion of Caribbean tourism would be mutually encouraged, planned and implemented. One important sector which needs to be included in any sustainable plan for a healthy and well-developed tourism industry in which environmental concerns are integrated with economic issues is, of course, tourism's linkages with local agriculture. It is to this that we now turn.

Agriculture's declining role in the region, with tourism helping agricultural resurgence, selectively

In the Caribbean region as a whole, there have been dramatic declines in agriculture's contribution to the national accounts of most small islands. Between 1960 and 1980, declines of 10 to 20 per cent in agricultural production affected large and small countries alike (Axline 1986). Declines were more common than turnarounds during the next twenty years, from 1980 to 2000, as many Caribbean countries' agricultural production continued its downward path. Notably, production levels were considerably lower in 2000 than 1960, for many small islands – Antigua and

Barbuda, the British Virgin Islands, Guadeloupe, Martinique, St Kitts and Nevis, St Lucia, St Vincent and the Grenadines, and the US Virgin Islands. A few, it should be noted, either registered increases in agricultural production by 2000 (the Bahamas, Dominica and Grenada), or maintained their productive output, registering little growth or decline since 1960 (Barbados and Jamaica). Some of the largest islands also registered considerable declines from their 1960 levels (the Dominican Republic, Haiti, Puerto Rico and Trinidad and Tobago), so size doesn't qualify as a discriminator of agricultural decline – the sector has shrunk, region-wide.

Across the region, and regardless of productivity declines or increases, there were also appreciable reductions in rural employment, during the 1960–2000 period. The most dramatic drops in this traditional employment sector occurred in the French overseas departments of Guadeloupe and Martinique and in Puerto Rico; with employment numbers plummeting at decadal rates of 20 per cent. During this same period Barbados experienced decadal declines of 19 per cent. Less precipitous declines were experienced in the Bahamas at 8 per cent per decade and Trinidad and Tobago at 4 per cent per decade; but some larger islands (e.g. the Dominican Republic, Jamaica) maintained their rural employment levels, irrespective of their productivity performances. Only Haiti and Belize increased their rural employment numbers by 2000.

McElroy and de Albuquerque (1990) in their examination of small farming sector performances in a set of small Caribbean islands differentiated in terms of tourism penetration – Antigua and Barbuda, Montserrat, St Kitts and Nevis and the US Virgin Islands – found that there has been a consequential loss of agricultural land to tourism-related development, as well as relative and absolute declines in the agricultural sector's productivity and output, which they attribute to small farming inefficiency. Where tourism grew rapidly, as in Antigua and Barbuda and St Lucia, the agricultural sector declined dramatically, with some labelling the substitution as 'direct displacement' (Hudson 1989; Weaver 1988). There was, at the same time, a suggestion that tourism and modernisation's impacts were most destructive and detrimental to agriculture's fortunes during the earlier decades of more rampant, unregulated growth of mass tourism (i.e. the 1950s and 1960s) than in more recent decades (Bryden 1973; McElroy and de Albuquerque 1990; Richards 1983).

Contrary to these consensual assessments of decline, alienation and retrenchment of agriculture, the temporal trends of agricultural land use differed somewhat. In the small islands of the Commonwealth Caribbean there has been a trend towards small-scale production, and the proliferation of smallholdings (most less than 10 acres) where family members produce domestic fruits and vegetables, and supplement their incomes with off-farm employment in the non-agricultural sectors such as services, informal sector, tourism (Weir's Agricultural Consulting Services 1980). To be sure there were declines in agricultural land-use areas by 2000 in Grenada,

Jamaica, St Kitts and Nevis and the US Virgin Islands, but elsewhere, modest gains in agricultural land use have been experienced in some smaller islands. And there have been appreciably larger gains in Cuba, the Dominican Republic, Haiti and Trinidad and Tobago (Table 11.1).

Structural realignments in most Caribbean islands accompanying modernisation, diversification and the growth of tourism, from the 1960s through to the year 2000, have certainly brought about dramatic declines in rural employment levels and declining fortunes in overall agricultural productivity for the region as a whole (thereby, concurring with McElroy and de Albuquerque 1990). Agriculture's rapid decline, however, has been most marked in the wholesale abandonment of acreages, growing traditional export staple commodities – sugar, bananas, coffee, nutmeg, among others. Urbanisation, residential development and tourist infrastructural development have contributed to this alienation of agricultural land, especially in island coastal zones (see Lorah 1995; Island Resources Foundation 1996). Realising the inevitability of plantation agriculture's declining prominence, governments, agricultural extension agencies and regional development agencies have promoted the redirection of agricultural production, either to non-traditional agricultural exports or to domestic and tourist market provision (Barker 1985; de Albuquerque and McElroy 1983; Wiley 1998).

If we examine the most recent patterns of production of crops grown more specifically for domestic, tourist and regional markets, then the hoped-for, tourism-generated turn-around of agriculture performance in the Caribbean can be inferentially surmised, though the positive picture doesn't apply everywhere (Table 11.2). Comparing 1990 and 2000 year totals, the Dominican Republic, Haiti and Cuba are the three Caribbean islands dominating the regional picture in terms of production of fresh fruit and vegetables; and, as it happens, accounting for the declines in overall production. Certain tourist islands, on the other hand, appear to have revitalised the production of many different crops used domestically and in tourist establishments; namely, Barbados, Jamaica and the French overseas departments of Guadeloupe and Martinique. Notably, non-traditional crops such as lettuce, cucumbers and tomatoes are being produced in greater quantities in these islands, as well as across the region. On the other hand, Puerto Rico's production levels of these same crops has dropped significantly; despite the island's large tourism sector. Fresh fruit and fresh vegetable production levels in the smaller islands mirror the patterns of salad vegetable production; but mangoes and papayas appear to be two successful Caribbean delicacies that have enjoyed notable increases of production from 1990 to 2000. Other tourist islands, like St Lucia, St Vincent and the Grenadines, Antigua and Barbuda and Grenada show more modest increments of domestic crop production from 1990 to 2000, though each shows a decline in the production of one or two specific crops. It is heartening to observe that the diversity of this production of

Table 11.1 Trends in agriculture in the Caribbean

	Agricultural production net per-capita PIN 89–91			Agricultural employment (thousands)			Agricultural land use (thousands of hectares)		
	1961	1980	2000	1960	1980	2000	1961	1981	2000
Antigua and Barbuda	120.1	102.4	95.1	n.a.	n.a.	n.a.	10	11	12
The Bahamas	94.4	39	115.8	9	5	6	10	11	13
Barbados	109.9	123	96.4	24	11	6	19	19	19
Belize	54.9	94.4	142.6	12	17	24	79	97	139
British Virgin Islands	310.6	133.8	75.1	n.a.	n.a.	n.a.	6	8	9
Cuba	99.2	92.3	61.1	897	879	785	3,550	5,938	6,665
Dominica	63.6	50	86.4	n.a.	n.a.	n.a.	17	19	17
Dominican Republic	123.6	108.8	89.5	635	682	606	3,082	3,517	3,696
Grenada	76.7	117	90.8	n.a.	n.a.	n.a.	22	16	12
Guadeloupe	192.6	123.5	104.2	42	24	7	58	59	49
Guyana	152	132.9	188.7	n.a.	n.a.	n.a.	1,359	1,715	1,726
Haiti	127.4	128.9	86.2	1,709	1,797	2,187	1,255	1,403	1,400
Jamaica	109.7	95.7	106.9	276	298	264	533	497	503
Martinique	131.5	69.6	116	40	17	8	34	38	33
Montserrat	48.3	69.9	303	n.a.	n.a.	n.a.	5	2	3
Netherland Antilles	233.3	290.9	143.6	n.a.	n.a.	n.a.	6	8	8
Puerto Rico	168.5	107.7	76.1	162	59	34	616	467	291
Suriname	47.8	104.6	72.4	n.a.	n.a.	n.a.	n.a.	n.a.	n.a.
St Kitts/Nevis	135	141.8	103	n.a.	n.a.	n.a.	20	15	10
St Lucia	87.6	70.2	63.4	n.a.	n.a.	n.a.	17	20	19
St Vincent/Grenadines	83.6	72.2	72.4	n.a.	n.a.	n.a.	10	12	13
Trinidad and Tobago	161.9	145.1	107.1	61	46	50	102	127	133
US Virgin Islands	533.6	145.1	89.2	n.a.	n.a.	n.a.	12	16	10

Source: Adapted from FAOSTAT (2002).

crops for domestic and tourism consumption is being upheld in many islands, increasing in a few, and not dropping precipitously in any; though Cuba's and Puerto Rico's records are the most uneven in this respect (Table 11.2).

Agriculture's future in so many small Caribbean islands is both beguiling and a paradox of uncertainties to governments and planners, who now view tourism development as a given – whether an 'evil' or a 'blessing' – in today's globalising world. Tourism might displace plantation agriculture (and the production of export staples) as the economic driver, but it was also expected to stimulate agricultural production, and hopefully revive it. At first, it seemed that the hoped-for benefits were illusory. In large part, this came about because the proponents, who advocated that Caribbean farmers would be stimulated to expand and diversify their production to meet the tourist sector's needs, failed to acknowledge the strength and resilience of the structural limitations and debilitating effects of the plantation system and its social and economic legacies. The structural limitations of this sector of the island's economic landscape stems from both internal and external influences (Momsen 1998), but they were endemic. Not surprisingly, it has taken decades for Caribbean island societies to emerge from plantation agriculture's stranglehold on agricultural modes of production and livelihood (Beckford 1972; Best 1971), and the accompanying lacklustre performance of the small-farming sector (Brierley and Rubenstein 1988; Gomes 1985).

Traditional practices, imprinted since slave-plantation times, had always favoured imported food stuffs, so self-sufficiency was never a realistic objective, and the domestic provision of food was rarely a major objective of Caribbean agriculture. Furthermore, such colonial and post-colonial legacies were not the only constraining factors. The externally oriented and directed international tourist industry, especially the larger hotel chains and corporations, have vested interests in maintaining profitable throughputs of overseas imported foods and beverages (Britton 1991; Bryden 1973). In this profitable marketing strategy, they have been aided and abetted by the Caribbean *comprador* elite of merchant proprietors who have always profited from oligopoly control of their island's import licensing procedures. This alliance of merchant and international capital has been a dominant feature of Caribbean small island economies (Beckles 1990; Conway 1990; Watson 1994).

Today, however, this core–periphery relationship is undergoing redefinition, or at least is being refined, so that intra-regional commerce is competing with external imports, and food import substitution is occurring in tourist islands such as the US Virgin Islands with neighbouring island producers of fruit and vegetables replacing US suppliers (see de Albuquerque and McElroy 1983). Elsewhere in the region, the more favoured (and fertile) Windward island small-farming sectors – Dominica, for example – are providing fresh produce for the tourist enclaves in the

Table 11.2 Trends in agricultural production of crops for local and tourism cuisine in the Caribbean (1990–2000) (metric tonnes)

	Avocados		Cantaloupe		Gr. Chillies		Cucumbers		Eggplant		Fresh Fruit		Grapefruit	
	1990	2000	1990	2000	1990	2000	1990	2000	1990	2000	1990	2000	1990	2000
CARIBBEAN	240,913	144,268	74,434	62,335	69,862	40,151	68,304	69,900	8,868	11,840	128,677	121,350	406,218	293,029
Antigua and Barbuda			650	750	40	60	200	200	230	250	6,500	6,500		
The Bahamas			400								4,700	400		11,715
Barbados	470	470				900	350	1,300			1,200	1,300		
Cayman Islands	3	13												
Cuba	9,000	7,500	23,000	34,000	42,630	12,748	45,000	40,000	1,400	1,700	55,000	50,000	331,157	188,810
Dominica	695	500					1,961	1,650					15,600	20,600
Dominican Republic	162,620	81,736	39,572	16,000	19,640	14,560	3,000	2,500	4,748	6,700			2,800	3,200
Grenada	1,677	1,750	4,072	3,270			54	110	300	180	1,646	1,000	2,006	2,000
Guadeloupe	86	280	2,500	2,900	70	90	2,500	3,918	1,000	800	1,031	1,050	166	604
Haiti	57,000	45,000											9,500	12,750
Jamaica	3,200	4,000			2,137	4,438	8,115	13,500	79	360	55,000	57,000	40,000	42,000
Martinique	1,066	363	2,240	2,390			4,200	4,000	30	50			150	110
Puerto Rico	4,286	1,950	2,000	3,025	4,763	6,500							261	300
St Kitts and Nevis						60					1,200	1,300		
St Lucia	500	396			350	380		80			2,400	2,800	310	2,580
St Vincent/Grenadines													415	460
Trinidad and Tobago					219	400	2,859	2,575	1,081	1,800			3,853	7,900

Table 11.2 continued

	Limes/Lemons		Lettuces		Mangoes		Papayas		Pumpkins		Roots/Tubers		Tomatoes	
	1990	2000	1990	2000	1990	2000	1990	2000	1990	2000	1990	2000	1990	2000
CARIBBEAN	130,756	92,361	13,555	17,481	608,138	534,038	59,564	77,793	123,354	135,763	2,002,005	2,336,893	330,489	487,992
Antigua and Barbuda	270	220			1,500	1,300			230	220	275	255	190	170
The Bahamas	1,400	8,147									1,150	942	7,500	3,352
Barbados				450					174	1,000	6,691	7,445	335	650
Cuba	61,130	18,698			72,476	45,536	39,909	43,790	56,000	50,000	617,680	953,826	164,960	153,883
Dominica	2,585	1,000	150	170	4,030	1,900			552	810	32,517	26,550	170	200
Dominican Republic	8,500	8,500	4,600	1,598	190,000	180,000	14,000	22,500	15,965	22,000	232,142	250,822	117,491	285,630
Grenada	489	470	53	70	1,760	1,850	162		226	260	3,545	4,050	46	60
Guadeloupe	800	1,608	2,500	3,146	1,050	1,023	45	55	210	220	18,294	19,800	3,300	3,071
Haiti	24,500	25,000	600	600	300,000	250,000					771,000	769,340	2,800	6,000
Jamaica	24,000	24,000	2,355	3,257	4,000	5,000	3,861	8,248	26,249	42,000	232,276	234,635	14,261	20,941
Martinique	862	270			500	198					20,850	22,220	2,600	6,100
Puerto Rico	3,735	1,811	3,100	7,800	6,336	17,375	1,587	3,200	20,412	11,000	26,282	10,390	15,649	5,000
St Kitts and Nevis											1,410	1,050	80	80
St Lucia	270	382			24,000	28,000					11,080	11,216		
St Vincent/Grenadines	850	880			1,800	1,250			420	480	15,940	12,500		
Trinidad and Tobago	1,300	1,300	197	390	430	430			2,907	7,700	10,705	11,635	993	2,728

Source: Adapted from FAOSTAT (2002).

drier and less productive Leeward Islands (Wiley 1998). Within the Caribbean Community (CARICOM), processed foodstuffs – preserves, jams, juice, spices, dried fruits – which are manufactured by local corporations in Trinidad and Jamaica, are finding market niches both within the region and beyond. Agriculture may never reach its former levels of dominance as a primary sector, but there appear to be welcome signs that it is not quite 'dead and buried', except perhaps in Puerto Rico (see Tables 11.1 and 11.2).

Tourism–local agriculture linkages in St Lucia

Past research has addressed linkages between the tourist industry and the domestic agricultural sector in terms of hotel characteristics; namely size, ownership and maturity of hotel. In 1971, Momsen examined the major constraints on both production and marketing that may have negatively affected an increase in the domestic supply of foodstuffs in St Lucia (Momsen 1972). Momsen found demand from hotels to be generally strong. However, 70 per cent by value of food had to be imported, and 70 per cent of hotels complained that the uncertainty of local supply was a major hindrance to increased use of local produce. In addition, Momsen found a discrepancy between the four largest foreign-owned hotels, which relied extensively on imported foodstuffs, and small locally owned hotels, which imported only 33 per cent of their food by value. Momsen's updated survey of St Lucia in 1985 examines changes in the use of locally produced foodstuffs by the hotel sector over time (Momsen 1986), and found improvements in a falling percentage of imported food used by the largest hotels (58 per cent). This was attributed to changing attitudes about serving local cuisine, to increased government assistance to producers for improving supply, and to the protective influence of import restrictions.

Following up on Momsen's leads, Timms's (1999) research uncovered seven main forms of hotel–local agriculture linkages utilised in St Lucia; while acknowledging that more may exist. Listed in descending order of potential stimulation and transformative effects on the local agricultural sector, they were the following: (1) formal contract, (2) cooperative marketing, (3) informal linkage with individual producer, (4) wholesalers, (5) agents, (6) informal linkage with vendors and (7) self-supply. The seven forms were then examined and compared; with analysis focusing on how the linkage communicates supply and demand, stimulates increased production, and in some cases facilitates the transformation of the local agricultural production system through access to agricultural inputs. Local small hotels and guest houses still used more local suppliers, but an encouraging sign was the increased use of local vendors and suppliers by large, well-established hotels. Large foreign import bills were favoured by several middle-sized hotels which had only recently been built, but with maturity and better local knowledge, even this category of hotels was increasing its local inputs. The multiplier effects of tourism appeared to be working in St Lucia, although it was by no means a story of unqualified success for local agriculture. There were plenty of bottlenecks in supply linkages, hoteliers and their staffs maintained flexibility in their contractual arrangements, and local agricultural cooperatives appeared to be avoided rather than welcomed (Timms 1999). This said, the hoped-for linkages between tourism and local agriculture in places like St Lucia, the US Virgin Islands and Barbados now appear to be sustained, and sustainable.

'Soft growth' rather than 'hard growth'

Economic growth, generally measured in terms of GNP, can be divided into two components. The first is hard growth (quantitative, physical) which involves an increase in size and scale, the second is soft growth (qualitative, non-physical) based on improvement, efficiency and reaching a fuller, better state. The first is severely constrained by natural limits; the second is potentially sustainable (Daly 1990, 1991). The distinction between the two is essential to any conceptualisation of sustainable development. Hard growth is based on increasing the amount of natural resources exploited (i.e. each year cutting more pulp wood than the last). This intensifies pressure on the natural resource base by increasing the flow of matter and energy through the economy.

Soft growth is not directly tied to the natural resource base (Daly 1990). Instead, it relies on increasing efficiency (i.e. producing more paper by reducing waste in paper manufacturing, recycling, improving the quality of paper produced and charging more for it, etc.) or producing services. The difference between these two components in the sphere of production is analogous to the difference between 'growth' and 'development'. According to Daly (1991: 402), 'growth is quantitative increase in physical scale, while development is qualitative improvement or unfolding of potentialities. An economy can grow without developing, or develop without growing, or do both or neither.'

There is reason to believe that current levels of economic production already approach the natural limits of some resource-based economic sectors at the global scale; strong evidence exists that these limits to hard growth have been surpassed in many island regions (Vitousek *et al.* 1995).[1] Following Robert Goodland (1992) we can adapt Daly's idea to the Caribbean island context and posit that the insular economy is an open, growing subsystem contained within a finite ecosystem. In such a model, the scale of the economic subsystem is determined by a combination of population size and per capita resource use. Under capitalist development, the economic subsystem relies on the global ecosystem for two purposes: as a source for all the material inputs fuelling the economy, and as a sink for all waste the economy produces (Goodland 1992). An island ecosystem's ability to absorb waste and provide natural resources for hard growth is therefore limited. Paradoxically, as the demand for natural resources and for environmental sinks grows, the environment's ability to absorb waste and provide raw materials declines. When this occurs over an extended period of time, harvesting rates of renewable resources surpass regeneration rates, and non-renewable resources are rapidly depleted (Daly 1990, 1991). Sustainability will only be achieved if the size of the economic subsystem remains within the ability of the ecosystem to assimilate it (Goodland 1992).

Given the finite nature of island ecosystems, it is very disturbing that the concept of an optimal scale of the economic subsystem for such ecosystems is not a concern of current macroeconomic theory. The economy is simply assumed to grow forever (Daly 1991). This untenable axiom is still current, despite the work of Repetto (1987a, 1987b), which forcefully demonstrates that many forms of economic (hard) growth actually leave us poorer in the long run. They destroy potentially renewable resources and/or create negative externalities that actually outweigh the short-term, or medium-term, economic gains produced by the unsustainable productive enterprise. Island ecological systems, whether mass-tourist dominated or low-density styled, deserve conservation and management institutional development, in which hard growth maxims are replaced by soft growth policies and practices.

Conclusions

During the last two decades, significant changes in Caribbean tourism–environment–agriculture relationships have occurred. Concerning hotelier–local agriculture linkages there appears to be a positive shift in the formation of supply-networks between local food and beverage producers and some segments of the hotel industry. In part, this appears to be accompanying a maturing of island tourist industries in places such as Barbados, Jamaica, the Virgin Islands and St Lucia, and in part it is helped by the growth of regional–transnational hotel corporations, like Sandals, and the growing global cosmopolitan character of alternative tourism offerings (Telfer and Wall 1996).

In addition, today's increasingly stark (and gloomy) picture of small island vulnerability, not to mention the pressures of today's globalising world and the threats of neoliberal 'free-marketeering', have forced Caribbean governments and their agricultural producers to move away from their entrenched reliance on export staple commodity production and to produce alternative foodstuffs for regional markets and develop local specialisations. The recently concluded 'banana wars', between the US and the EU, all-too-vividly demonstrated the eastern Caribbean farmers' vulnerability and helplessness in the new WTO deregulated regime of commerce and global trade of basic commodities and staples. Small island exporters were not to be given any 'third chances' in the negotiations for the maintenance of long-standing preferential trade agreements, such as the Lome III. Not surprisingly, intra-regional (CARICOM) trade in agricultural products has developed a new vigour, and fresh-food production has been taken up by large- and small-scale farmers alike. Some are using new intensive cultivation technologies, such as the hydroponics cultivation of fresh lettuce in Barbados, for example; others are relying upon tried and tested small-farming intensification methods (Brierley 1991; Hills 1988), and local cooperative associations, as well as governmental organisations, have been successful in organising the marketing and trade of these foodstuffs both domestically and intra-regionally (de Albuquerque and McElroy 1983; Momsen 1998; Timms 1999).

As well as a growing demand for local cuisine in the maturing hotel and restaurant business sectors of Caribbean islands, there is a growing market for local Caribbean foods among overseas enclave communities (Conway 1999/2000). The globalisation of food consumption habits (Goodman and Watts 1994), and the introduction of more exotic foods and cuisine into the metropolitan cultures of the tourist's home, has led to the hoped-for stimulation of agricultural diversification and production of Caribbean foods. Today, domestic and non-traditional export commodities – mangoes and mango chutney, limes and lime pickle, habañero peppers and pepper sauce, for example – now find profitable outlets both in the local tourist industry, and overseas in the tourists' metropolitan supermarkets.

In several, though not all, Caribbean islands' rural and small-town sectors, non-agricultural economic expansion and local agricultural growth have accompanied tourism and modernisation, though more as an indirect, complementary effect than as a result of the sector's direct economic influence. At the community and household levels, handicrafts, food-production, cafés, bars and eating establishments, boarding houses, bed and breakfast accommodation, provide much-needed income for the entrepreneurial-minded. Being petty-commodity producers, such small informal sector operations have cost structures, economic relations and motivations based on familial and/or household relations (Wheelock 1992). They therefore supplement household incomes, add flexibility to the survival strategies open to members, and help maintain rural farming endeavours, including their 'Antillean gardens', food-forest plots and smallholdings (Berleant-Schiller and Pulsipher 1986; Brierley 1991; Hills 1988; Hills and Iton 1983).

Soft growth is being generated, and the scope and scale of production is balanced and maintained to fit within the economic realm of the household or small enterprise, and its labour and capital inputs. In addition, such micro-economic 'flexible' responses to marginality in island rural communities are likely to have influential (and quite possibly progressive) social repercussions. In these complementary and ever more valuable activities, the balance of social power and economic decision-making shifts from the formal to the informal (complementary) economy, or at least the addition of such activities broadens the bases of negotiation within the household.

They broaden rural household and family economic roles from activities dominated by men to incorporate activities managed (and dominated) by women as significant and important to household survival and reproduction (Kinnaird and Hall 1994; Momsen 1994).

Note

1   Conservatively, human activities appropriate 25 per cent of the *global* net primary product and 40 per cent of the potential *terrestrial* net primary product. Humans also affect much of the other 60 per cent of terrestrial net primary product, sometimes heavily (Vitousek *et al.* 1995).

References

Archer, E. (1985) 'Emerging environmental problems in a tourist zone: the case of Barbados', *Caribbean Geography* 2(1): 45–55.

Axline, W.A. (1986) *Agricultural Policy and Collective Self-sufficiency in the Caribbean*, Boulder, Colo.: Westview Press.

Barker, D. (1985) 'New directions in Jamaican agriculture', *Caribbean Geography* 2(1): 56–64.

Beckford, G. (1972) *Persistent Poverty*, New York: Oxford University Press.

Beckles, H.M. (1990) *A History of Barbados: From Amerindian Settlement to Nation State*, Cambridge: Cambridge University Press.

Bélisle, F. (1983) 'Tourism and food production in the Caribbean', *Annals of Tourism Research* 10(4): 497–513.

Berleant-Schiller, R. and Pulsipher, L.D. (1986) 'Subsistence cultivation in the Caribbean', *New West Indian Guide* 3(1): 1–40.

Best, L. (1971) 'Size and survival', in N. Girvan and O. Jefferson (eds) *Readings in the Political Economy of the Caribbean*, Kingston: New World Group.

Briassoulis, H. and Van der Straaten, J. (1992) 'Tourism and the environment: an overview', *Tourism and the Environment: Regional, Economic and Policy Issues*, Dordrecht: Kluwer Academic Publishers.

Brierley, J.S. (1991) 'Kitchen gardens in the Caribbean, past and present: their role in small farming development', *Caribbean Geography* 3(1): 15–28.

Brierley, J.S. and Rubenstein, R. (eds) (1988) *Small Farming and Peasant Resources in the Caribbean*, Winnipeg: Department of Geography, University of Manitoba.

Britton, S. (1991) 'Tourism, capital and place: towards a critical geography of tourism', *Environment and Planning D: Society and Space* 9: 451–78.

Bryden, J. (1973) *Tourism and Development: A Case-Study of the Commonwealth Caribbean*, London: Cambridge University Press.

Caribbean Tourism Organization (CTO) (1998) *Caribbean Tourism Statistical Report, 1998*, St Michael, Barbados: Caribbean Tourism Organization.

Conway, D. (1983) *Tourism and Caribbean Development*, Hanover: University Field Staff International Report, No. 27.

—— (1990) *Small May Be Beautiful, But is Caribbean Development Possible?*, Hanover: University Field Staff Reports. Latin America, 1990–91/No. 13.

—— (1999/2000) 'The importance of migration for Caribbean development', *Global Development Studies*, Winter 1999–Spring 2000, 2(1–2): 73–105.

Conway, D. and Lorah, P. (1995) 'Environmental protection policies in Caribbean small islands: some St Lucia examples', *Caribbean Geography* 6(1): 16–27.

Daly, H.E. (1990) 'Sustainable growth: an impossibility theorem', *Development: Journal of Society for International Development* 3(4): 45–7.

—— (1991) 'Sustainable growth: a bad oxymoron', *Environment and Carcinogenic Reviews* 8(2): 401–7.

de Albuquerque, K. and McElroy, J. (1983) 'Agricultural resurgence in the United States Virgin Islands', *Caribbean Geography* 1(2): 121–32.

FAOSTAT (2002) Agriculture. Online. Available: <http://apps.fao.org/page/collections?subset=agriculture> (accessed February 2003).

Gomes, P.I. (1985) *Rural Development in the Caribbean*, London: C. Hurst; New York: St Martin's Press.

Goodland, R. (1992) 'The case that the world has reached its limits: more precisely that current throughput growth in the global economy cannot be sustained', *Population and Environment* 13(3): 167–82.

Goodman, D. and Watts, M. (1994) 'Reconfiguring the rural or fording the divide? Capitalist restructuring and the global agro-food system', *Journal of Peasant Studies* 22(1): 1–49.

Hills, T. (1988) 'The Caribbean food-forest: ecological artistry or random chaos?', in J.S. Brierley and H. Rubenstein (eds) *Small Farming and Peasant Resources in the Caribbean*, Winnipeg: Department of Geography, University of Manitoba.

Hills, T. and Iton, S. (1983) 'A reassessment of the "traditional" in Caribbean small agriculture', *Caribbean Geography* 1(1): 24–35.

Hudson, B. (1989) 'The Commonwealth Eastern Caribbean', in R.B. Potter (ed.) *Urbanization, Planning, and Development in the Caribbean*, London: Mansell.

Island Resources Foundation (IRF) (1996) *Tourism and Coastal Resources Degradation in the Wider Caribbean*, St Thomas, USVI: Island Resources Foundation.

Kinnaird, V. and Hall, D. (eds) (1994) *Tourism: A Gender Analysis*, Chichester: John Wiley.

Lorah, P. (1995) 'An unsustainable path: tourism's vulnerability to environmental decline in Antigua', *Caribbean Geography* 6(1): 28–39.

McElroy, J. and de Albuquerque, K. (1990) 'Sustainable small-scale agriculture in small Caribbean islands', *Society and Natural Resources* 3: 109–29.

—— (1991) 'Tourism styles and policy responses in the open economy–closed environment context', in N.P. Girvan and D. Simmons (eds) *Caribbean Ecology and Economics*, Barbados: Caribbean Conservation Association.

—— (1996) 'Sustainable alternatives to insular mass tourism: recent theory and practice', in L. Briguglio, B. Archer, J. Jafari and G. Wall (eds) *Sustainable Tourism in Islands and Small States: Issues and Policies*, London and New York: Pinter.

Momsen, J.H. (1972) 'Report on vegetable production and the tourist industry in St Lucia', Calgary: Department of Geography, University of Calgary.

—— (1986) *Linkages Between Tourism and Agriculture: Problems for the Smaller Caribbean Economies*, Seminar Paper, No. 45, Department of Geography, University of Newcastle upon Tyne.

—— (1994) 'Tourism, gender and development in the Caribbean', in V. Kinnaird and D. Hall (eds) *Tourism: A Gender Analysis*, New York: John Wiley.

—— (1998) 'Caribbean tourism and agriculture: new linkages in the global era?', in T. Klak (ed.) *Globalization and Neoliberalism: The Caribbean Context*, Lanham, Md.: Rowman & Littlefield.

POA/SIDS (1994) *Report of the Global Conference on the Sustainable Development of Small Island Developing States*, Bridgetown, Barbados, 25 April–6 May, United Nations, Sales No. 94.I.18, Chapter 1, Resolution 1, Annex II.

Pugh, J. (2001) 'Deconstructing participatory environmental planning: dispositions of power in Barbados and St Lucia', unpublished dissertation, Department of Geography, Royal Holloway College, University of London.

Pugh, J. and Potter, R.B. (2001) 'The changing face of coastal zone management in Soufriere, St Lucia', *Geography* 86(3): 247–60.

Repetto, R. (1987a) 'Creating incentives for sustainable forest development', *Ambio* 16(2–3): 94–9.

—— (1987b) 'Population, resources, environment: an uncertain future', *Population Bulletin* 42(2): 1–43.

Richards, V.A. (1983) 'Decolonization in Antigua: its impact on agriculture and tourism', in P. Henry and C. Stone (eds) *The Newer Caribbean*, Philadelphia: Institute for the Study of Human Issues.

Telfer, D.J. and Wall, G. (1996) 'Linkages between tourism and food production', *Annals of Tourism Research* 23(3): 635–53.

Timms, B. (1999) 'Linkages between domestic agriculture and the hotel sector in St Lucia', unpublished thesis, Department of Geography, University of Indiana.

UNEP (1982) 'Tourism', in UNEP: *The World Environment, 1972–1982*, Dublin: Tycooly International Publishers.

Vitousek, P.M., Loope, L. and Andersen, H. (1995) *Islands: Biological Diversity and Ecosystem Function*, Berlin and New York: Springer-Verlag.

Watson, H.A. (1994) *The Caribbean in the Global Political Economy*, Boulder, Colo., and London: Lynne Rienner.

Weaver, D.B. (1988) 'The evolution of a "plantation" tourism landscape on the Caribbean island of Antigua', *Tijdschrift voor Economische en Sociale Geographie* 79(5): 313–19.

Weir's Agricultural Consulting Services (1980) *Small Farming in the Less Developed Countries of the Commonwealth Caribbean*, Barbados: Caribbean Development Bank,

Wheelock, J. (1992) 'The household in the total economy', in P. Ekins and M. Max-Neef (eds) *Real-life Economics: Understanding Wealth Creation*, London and New York: Routledge.

Wiley, J. (1998) 'Dominica's economic diversification: microstates in a neoliberal era?', in T. Klak (ed.) *Globalization and Neoliberalism: The Caribbean Context*, Lanham, Md.: Rowman and Littlefield.

Youngsman, M. (1998) 'Tourism and the environment: sustaining the sector', *Caribbean Handbook, 1998/1999*, Tortola, British Virgin Islands: FT Caribbean (BVI).

12 Community participation in Caribbean tourism

Problems and prospects

Simon Milne and Gordon Ewing

Introduction

This chapter focuses on local community participation in the planning, development and ownership of Caribbean tourism. In recent years all of the tourism-oriented nations in the region have joined, to varying degrees, the global move towards planning and developing more sustainable forms of tourism (Wilkinson 1997; ACS 1998). An integral component of any sustainable development strategy must be the protection and improvement of quality of life in communities influenced by the industry. It is now generally accepted by commentators that, unless residents are empowered to participate in decision-making and ownership, tourism development will not reflect community values and will be less likely to generate sustainable outcomes (Milne 1998; Timothy 2002).

The chapter begins with a brief review of the inclusion of community in recent Caribbean-wide tourism plans and policies. While local communities have often been viewed as passive actors in past tourism development, we outline (1) the recent regional growth in interest in public participation initiatives, and (2) the development of tourism products that actively integrate community into the visitor experience. Case-studies drawn from Cuba, Bonaire and Jamaica are then introduced to exemplify the range of problems and prospects facing those attempting to increase public participation in the tourism development process.

In the case of Cuba, we highlight the degree to which national political agendas and structures hinder participation as the nation attempts to develop alternatives to mass tourism. A short review of tourism planning and development in Bonaire during the 1990s then reveals some of the difficulties that can arise in achieving stakeholder collaboration and under-standing – even in a setting that is broadly supportive of participatory approaches. The final case set in Jamaica highlights the role that information and communication technologies (ICT) can begin to play in facilitating community participation, stakeholder understanding and small business development in tourism. In conclusion we review some of the key lessons that can be learnt for the rest of the region.

The rise of community participation in Caribbean tourism

Historically, tourism development in the Caribbean has taken place in the context of a relatively weak state and marginalised local populations (Erisman 1983; Pattullo 1996). Policy has tended to flow 'top-down' and large tourism companies have been the dominant players in determining industry outcomes. As a result of these factors wealth has been poorly distributed (Bryden 1973), land-use and other types of conflict have been frequent (Hills and Lundgren 1977), and little attention has been paid to enhancing the sustainability of tourism at the local community level (Freitag 1996). All of this has occurred in the context of the constant nego-tiation, between the state and overseas corporations, of the policy and practice under which tourism resources and infrastructure will be devel-oped. Communities have been conspicuous by their absence from these negotiations (Hudson 1996; Pattullo 1996).

The role played by communities in the planning and development process has increased in recent decades as the broader Caribbean region and the nations within it have adopted economic development plans that incorpo-rate, to varying degrees, the desire to create a more sustainable tourism industry (Poon 1990; Weaver 1991; de Albuquerque and McElroy 1992; Wilkinson 1997; Duval 1998). The Organization of American States (OAS) has, for example, placed sustainability at the centre of tourism policy and related technical assistance, and in 1997 the hemisphere's tourism ministers adopted the *San Jose Declaration and Plan of Action on the Sustainable Development of Tourism* (OAS 1998).

The San Jose Plan proposes initiatives to be executed by different part-ners interested in the sustainable growth and development of tourism. Central to these initiatives is the need to enhance community understanding of, and input into, the tourism development process. Indeed the incorpo-ration of community input into the planning process has become almost a mantra – repeated on a regular basis at regional workshops and symposia on tourism development throughout the region.

The Association of Caribbean States' (ACS) Special Committee on Tourism (1998), in its own Plan of Action for the creation of a more sustainable tourism industry in the region, stresses that community partic-ipation is central to the future of the industry:

> The creation of a zone of sustainable tourism in the Caribbean will only be realised to the extent to which the communities that receive or will receive tourists find that they benefit from tourism. It is obvious that the positive effects derived from tourism must go well beyond the mere creation of jobs for members of the community. It is therefore essential that communities be in the forefront of the planning of tourism activities and become important actors in its development.
>
> (ACS 1998: 9)

Such participation must underlie any meaningful attempts to create a 'new model' for tourism development, a model based on

> a series of indicators whereby one could determine the sustainability of the entire system. The continuous estimation and analysis of these indicators would enable the decision-makers to have timely information in order to devise strategies, to prioritize programs and establish policies that will tend to maintain the tourism activities and the limits established for sustainability.
>
> (ACS 1998: 37)

Such declarations stem from the realisation that if the tourism industry is to perform effectively for communities and nations in the region it must not only generate long-term economic benefits, it must also mesh with the needs and desires of local people, be characterised by some degree of local ownership and control, and not destroy resident quality of life (Timothy 2002; O'Riordan 2002).

It is important to note here that this increasing interest in community participation has not simply been born out of the desire to move to a new paradigm of more sustainable approaches to planning and development. The fact that 'community' is a growing part of 'alternative' tourism products being developed throughout the world means that local involvement at all stages of the tourism development process is particularly vital (Milne and Ateljevic 2001). The need to enhance community buy-in to tourism and other economic development projects is also driven by broader directives from increasingly pressured aid donors from North America and Europe (Boyce 2002).

Put simply 'you can't have your cake and eat it too'; if governments want to incorporate community culture, heritage and local environs into a more diversified tourism product – then community must be an active rather than passive participant in the development process. Likewise foreign investment-led tourism development can no longer ignore local issues and priorities. Companies can no longer assume that a deal with the national or provincial government means that local and regional interests will disappear from the agenda, communities are increasingly willing and able to confront and undermine public sector and corporate objectives in the face of threats to their quality of life.

At the same time those researching tourism and the development process must come to terms with the complexity and changing nature of 'community' and its relationship with other stakeholders. Definitions of a homogeneous community wholly committed to a unified political and economic objective are increasingly seen as misleading (Telfer 2002). Simplistic views of small-scale, under-resourced and disempowered communities pitted against multinational companies possessing an overwhelming corporate ethos and rationalistic mentality are proving to be increasingly unrealistic, as is the argument that these two groups engage only in a unidirectional relationship (Milne 1998).

It is also important to stress that participation is a multifaceted process. Thus Gray (1989) and Sanoff (2000) stress the fact that there are several phases in the collaborative/participatory process, the dynamics of which evolve as participants move from one stage to the next. Gray (1989) identifies three phases of collaboration and participation: (1) problem-setting in which the nature of the challenge is diagnosed; (2) direction-setting wherein some type of consensus is achieved; (3) structuring which is concerned primarily with implementation and programming. While local involvement may occur in early phases it needs to be spread throughout the whole process, from problem setting through to implementation, if it is to be truly participatory (Sanoff 2000).

We now present three cases of local community involvement and participation in the Caribbean which reflect the different stages outlined by Gray (1989). National and local cases from Cuba reveal how broad-based political and structural constraints can still prevent even the most basic meaningful participation from occurring. A summary of Parker's (2000) review of collaboration in tourism planning in Bonaire highlights the difficulties associated with moving through to the implementation phase. The Jamaica case then presents some tools and approaches that may actually enhance broad-based participation in problem and direction setting and, just as importantly, increase the ability to move through to Gray's implementation phase.

Cuba

In principle, community participation in shaping decisions about Cuban tourism development should be widespread. After all, a central tenet of Marxism-Leninism is that power is in the hands of the proletariat. The Cuban Minister of Foreign Affairs (2001) puts it as follows: 'We are striving to achieve for our children levels of equality, social justice and community participation that have not been reached in any other society.' In reality initiatives of an economic nature are almost invariably promulgated by the Politburo to be implemented locally. While central government sees a role for community participation in some spheres, such as matters involving the family or health, participation is more a matter of engaging the public in centrally initiated programmes, than of the actively seeking input into design and implementation (Rosendahl 1997).

While meetings of locally elected Communist party politicians do take place at the local *asamblea municipal*, one commentator has likened such gatherings to a 'Greek drama in which everyone has his or her given place and role' (Rosendahl 1997: 143). Meetings are not aimed at enabling local people to shape the form of economic and social development in their district. Rather they provide a forum for disagreement and criticism, a sounding board that allows party leaders to keep in touch with grass-roots concerns. When it comes to questions of what sort of tourist development should take place, and what products should be developed, local participation is rarely, if ever, sought (Rosendahl 1997: 143).

Such limitations to public participation can have their advantages in times of economic crisis. As prices for its major foreign currency earner, sugar, fell following the collapse of the Soviet Bloc, Cuba had to build up tourism as the new mainstay of a faltering economy (Salinas 1998). To increase rapidly the number of visitors, large infusions of capital were needed to build more and higher-quality hotels, airports and local infrastructure. The government was forced to take the unprecedented step of allowing foreign capital to invest in hotel joint ventures with government-owned hotel chains. Planners identified pristine beaches on the northern keys of Cayo Coco and Cayo Guillermo in Ciego de Avila province, and a range of other locations, and permitted rapid development of the areas. As a result the number of overseas visitors rose rapidly – reaching 750,000 in 1995, a million the following year, and 1.8 million by 2000 (Espino 2001).

The lack of participatory processes not only allowed such developments to be 'fast-tracked', but also enhanced the ability of central government to ensure that all visitor spending occurs in state-owned agencies. By reducing the ability of the local population to own elements of the industry, the government can stall the development of two classes of citizen – one with direct access to foreign currency and the other without.

People who are self-employed in the tourism industry are called *cuentapropista*. The main domains of such dollar-based activities are home-based, family-run restaurants (*paladares*), room rentals in private homes (*casas particulares*), and private taxi services (*transportistas*). *Casas particulares* represent important competition for hotels (Rodriguez 1997). Prior to the introduction of the Decree-Law No. 171 on private home rentals in May 1997, it was estimated by housing and immigration authorities that about one-fifth of vacationers stayed in private homes, and two years later another source estimated the figure had climbed to one-third (Zúñiga 1999 as reported in Henken 2001). A monthly tax on *casas particulares* now applies regardless of how often a room is let, many owners of such rooms seeing this as a way for the government to run them out of business (Henken 2001).

Pressures to increase levels of public participation and ownership are, however, growing as the government attempts to diversify the tourist product away from traditional enclave mass tourism towards niche products including: health, cultural and nature tourism. These are products that often bring visitors in closer proximity to residents and the communities they live in (Salinas and Estevez 1996; Salinas 1998; Honey 1999). One area identified to have great potential for nature-based and cultural tourism is the Viñales Valley in the western province of Pinar del Rio. This area of outstanding karst landscape, in which traditional methods of agriculture have survived unchanged for several centuries, was recently designated as a national park, and declared a World Heritage Site by UNESCO in 1999. There are only three Cuban-owned hotels in the valley, with a total of 250 rooms. They are built to modest standards and were refurbished long before the joint venture era. The local village, Viñales, puts on a weekly *Noche del Patio Decimista* where a popular local form of improvised poetry, *decimas*, is put on for locals and tourists. The state-run *Casa de Cultura* supports such activities and there is also a museum, art gallery, library and bookstore open to tourists (Thivierge 2001).

There are, however, other forms of tourist services offered in the area: in addition to over twenty *casas particulares*, there are also guided tours provided by 'volunteers' (Thivierge 2001). In one case an elderly woman escorts small parties of tourists around a botanical garden she maintains at her home. The state has offered to buy it and turn it into a formal tourist attraction, but she has refused. Another attraction is the small rural community of *Los Aquaticos*, which boasts local water with curative powers with a volunteer escorting small groups and providing translation if needed. A group of Cuban speleologists also offers to guide tourists through the Santo Tomás cave system for a nominal charge.

The state remains ambivalent towards community involvement in tourism in the district. Unlike the *cuentapropistas*, volunteers have no legal status. The money they receive, usually in the form of gratuities, does no more than defray some of their living costs. The fact that these activities are tolerated by the authorities, despite not being licensed, reflects the small scale of the operations, the age of the providers and the fact that they are not a threat, either ideologically or financially. The *casas particulares* on the other hand continue to face real difficulties in coping with punitive government licensing and taxation policies.

One example of nascent local participation comes from La Moka Ecolodge, which was built in 1994 adjacent to the community of Las Terrazas in the Sierra del Rosaria. While this community-managed hotel was the brainchild of the Minister of Tourism, the local community was consulted about the development. After a series of community meetings, a team of psychologists and sociologists carried out a detailed house-to-house survey, soliciting opinions on all aspects of tourism (Thivierge 2001). The community is scheduled to repay the Ministry's $6 million investment over 15–20 years and 40 per cent of the hotel's profits go into a community development fund overseen by the neighbourhood Committee for the Defense of the Revolution, with a further 10 per cent going directly into the community's health clinic.

Bonaire

Parker's (2000) analysis of collaboration in tourism policy in Bonaire presents another face of stakeholder participation in Caribbean tourism. The total population of the island is only 15,000, and yet by 1998 it was receiving over 62,000 visitors (compared to 25,000 in 1986). Many of the visitors are drawn to the country by the Bonaire Marine Park which was established in 1979. Growing visitor numbers, and increasing concerns about the negative environmental impacts of the industry, led to the need for a multi-stakeholder tourism plan to be developed. There were two clear themes to emerge – the need to have sustainable tourism growth, and the need to protect the environment and local quality of life. Two leadership clusters – commercial and environmental – drove this process, with hoteliers and dive operators focused on filling rooms and dive trips, and environmentalists focused on building a water-treatment plant to protect the reef.

A collaborative committee comprised of these key groups produced a report in 1993 which presented a vision of Bonaire based on slow, planned and controlled growth of the dominant sector of the economy – dive tourism. According to Parker this represented an important turning point – people could see each other's perspectives and the links between the natural environment and sustained economic growth. While participatory processes were engaged in and indeed encouraged it is interesting to note that approximately two dozen networked individuals, most of them key players in the industry and related private and public sectors, dominated the process. As Parker (2000: 85) notes:

> Instead of establishing a system that might allow all stakeholders to meet together on an ongoing basis in some type of umbrella group or peak association, collaboration occurred primarily through the mechanisms of overlapping membership, joint participation and periodic general meetings.

While communication occurred in ad hoc situations, and a wide range of stakeholders were included, the process became somewhat exclusive and made it difficult to create a more institutionalised and sustainable approach to future tourism development. Thus despite what appeared to be some emerging common ground between different stakeholder groups on the need to emphasise the island's environmental product, and to steer a course away from the mass tourism emphasised in nearby Aruba and Curaçao, implementation was not achieved. A slow-down in industry growth in the late 1990s exacerbated the situation with the local industry once again focusing on marketing and increasing visitor numbers rather than emphasising sustainability.

Jamaica

Jamaica has certainly had its fair share of past criticism over its lack of participatory planning and community involvement in tourism (see e.g. Hudson 1996; Pattullo 1996). Nevertheless there are important ICT-based initiatives emerging on the island and elsewhere that offer the potential for improvement (see also Kenny 2002).

The Association of Caribbean States (1998: 16) stresses the role that technology can play in achieving more sustainable tourism development. ICT and the related area of community informatics (CI) offer the opportunity to establish new channels to connect residents, tourism operators, planners and other stakeholders, facilitating a shared understanding of how different stakeholders view planning and development issues (Milne and Mason 2001). In the case of Jamaica a number of ICT-based initiatives have been developed to enhance participation in the tourism development process and to increase the ability of locally owned businesses to make their presence felt on the global tourism stage. The initiatives build on the key elements of community informatics – a discipline that emphasises the design and delivery of technological applications to enhance community development, and improve the lives of residents (Gurstein 2000). Commentators identify three strategies for using CI as an enabler of community development: as a marketing tool for local business; as a mechanism to bring together a range of 'linked' resources of value to improving quality of life; and as a distributed network that can assist the creation of new relationships and economic linkages (O'Neil 2002: 82).

Established in 1997 the Jamaican Sustainable Development Network Programme (JSDNP) provides a virtual meeting place that aims to 'service the sustainable development information needs of all sectors of society' (see www.jsdnp.org.jm). It is part of a broader global SDNP initiative by the United Nations Development Programme and has, as its core objectives, the need to:

1   Introduce and connect stakeholders to sources of information on sustainable development utilising the Internet and other ICT.
2   Develop appropriate information services to support the implementation of local and national development plans.
3   Provide information on Jamaica's environment and social and economic development needs to national and international communities (quote from www.jsdnp.org.jm/aboutus.htm).

A significant element of the JSDNP is the development of sustainable tourism projects in Jamaica. These take a number of forms and are funded by a range of agencies. The focus is the need to stimulate tourism that creates direct benefits for communities and which diversifies the national tourism product. The emphasis is on nature and heritage and the involvement of community in the planning and development of tourism.

One example is the Coastal Water Quality Improvement Project (CWIP) (www.jsdnp.org.jm/susTourism-proActive.htm). The project, started in 1998 with the assistance of the Canada Jamaica Green Fund, aims to support community organisations, NGOs and private sector organisations to develop and manage economic activities that can assist in improving coastal water quality and generate employment and/or business opportunities. Examples of such an approach may be education initiatives focused on locals and tourists. This initiative is supported by the development of community cybercentres.

To date, six cybercentres have been established – with the Montego Bay Marine Park in the west of the island, and the Caribbean Coastal Area Management programme in Lionel Town, Clarendon, being the two centres with the greatest relevance to the CWIP. These cybercentres have three major functions: (1) to train community members to use computer technology and the Internet as a means of communication and gaining information; (2) to provide broad-based access to the Internet; (3) to coordinate the gathering and dissemination of community-based information (see www.jsdnp.org.jm/cybercentres.htm).

These initiatives are, in turn, closely linked to the Community Tourism Secretariat of the JSDNP which focuses directly on local tourism development initiatives. The core aims include (see www.jsdnp.org.jm/cs3.htm):

1   To provide administrative, marketing and other resources for communities.
2   To assist in creating and presenting an alternative tourism product – focused on cultural and environmental exchange between 'hosts' and 'guests'.
3   To support programmes that protect and preserve environment, heritage and culture.
4   To foster and encourage economic linkage to communities.
5   To provide educational programmes for communities in sustainable tourism.
6   To assist government agencies in conveying policies as widely as possible and in embracing community input and concerns.

Mandeville (with a population of 100,000), in the parish of Manchester in the centre of Jamaica, has become the 'capital' of community tourism and is the centrepiece of attempts to prevent the largely unplanned development that has marred both Negril and the North Coast by increasing levels of local participation in planning and ownership of the industry (see www.jsdnp.org.jm/cs2.htm).

ICT is also being used in Jamaica and the rest of the Caribbean to enhance the performance of locally owned small businesses. The Small Tourism Enterprises Project (STEP 2003) of the OAS is using the Internet as a tool to facilitate marketing, finance and environment management aspects of small enterprises in the region, with website-based 'toolkits' providing information necessary for small enterprises to improve their operations (see www. caribbeaninnkeeper.com/aboutstep.asp). Assistance on the technology front revolves around five components (see www.caribbeaninnkeeper.com):

1   Basic computer training.
2   Walk-in Resource Centres. In each participating country, including Jamaica, the centres provide a computer workstation, a digital camera and scanner for website development, and other resource material for use by small-enterprise owners.
3   Caribbeaninnkeeper.com. This website caters to small hoteliers and serves as a virtual walk-in centre. It includes a resource library, demonstration software and links to other small business sites.
4   Technology support for small hotels. Cable and Wireless provides eligible small hotels with discounted ISP access in countries that it services.
5   Caribbeanexperiences.com. This Internet-based marketing portal enables small hotels that meet certain standards to be able to take advantage of advertising and booking via the Internet.

While it is still early days there is no doubt that these initiatives being adopted in Jamaica and elsewhere in the region offer an important tool to assist in attempts to make sure that participation is carried through initial design phases through to implementation of tourism products and developments.

Conclusions

The importance of community participation in the tourism planning and development process is increasingly recognised by Caribbean governments, NGOs, industry and local residents. The cases presented here show that there are several problems to be overcome before participation can really develop as a sustainable force in tourism development in the region.

In the case of Cuba it is clear that a paternalistic style of government for over forty years has meant that the majority of Cubans have come to adopt a passive attitude to involvement in local issues, believing on the whole that they can have little influence on local decision-making. However, the need to diversify the traditional tourism product and reach into higher-yield segments, like nature and health tourism, mean that the government cannot continue to avoid engaging with community in a meaningful manner.

The case of Bonaire reveals that even in environments that are politically supportive of participatory and collaborative planning, changes in the external business environment can threaten fragile alliances and frameworks. At the same time the nature of internal processes can leave a legacy of failure in terms of maintaining participation through to the final implementation phase. Decentralised and informal systems of interaction may not stand the test of time and can lead to the exclusion of key groups (Parker 2000). A durable institutional framework is really needed to keep participation on track and it is clear that careful attention must be paid to the power differentials that exist between different stakeholder groups.

This brings us to the critical point raised by a number of commentators – how do we ensure not only that local involvement and participation occurs, but that it can be sustained in such a way that it leads to effective development outcomes? The Jamaican examples provide insights into the ways in which ICT may be harnessed to achieve this elusive outcome. Nevertheless this does not change the fact that on the Internet, just as in traditional communities, citizens must be motivated if they are to participate. Whilst the Internet changes the capacity and quantity of information that is available, Bimber (1998: 138) asserts that 'it is not yet clear that it will also change motivation and interest, let alone cognitive capacity'.

For the Caribbean, and elsewhere, it remains clear that, regardless of political systems, evolving stakeholder relationships and access to information technology, participation will not occur in a sustainable fashion unless people themselves have the will and interest to take control over their own destinies.

Acknowledgements

We wish to acknowledge the input of our students who have worked in the region in recent years: Wolfgang Haider, Paul Ridoutt, Beverley Mullings, Dominique Brief, Pascale Thivierge and George Vincent

References

Association of Caribbean States (ACS) (1998) *Plan of Action*, 3rd Meeting of the Special Committee on Tourism, Antigua and Barbuda, 9 and 10 November.

Bimber, B. (1998) 'The Internet and political transformation: populism, community and accelerated pluralism', *Polity* 31: 133–60.

Boyce, J.K. (2002) 'Unpacking aid', *Development and Change* 33: 239–46.

Bryden, J. (1973) *Tourism and Development: A Case-Study of the Commonwealth Caribbean*, Cambridge: Cambridge University Press.

Cuban Ministry of Foreign Affairs (2001) Speech delivered by the Minister of Foreign Affairs, Felipe Roque, at the 57th Session of the U.N. Commission on Human Rights, Geneva, Mar. 27. Online: <http://www.iacenter.org/cuba_032701.htm>.

de Albuquerque, K. and McElroy, J.L. (1992) 'Caribbean small-island tourism styles and sustainable strategies', *Environmental Management* 16: 619–32.

Duval, D.T. (1998) 'Alternative tourism on St Vincent', *Caribbean Geography* 9: 44–57.

Erisman, H.M. (1983) 'Tourism and cultural dependency in the West Indies', *Annals of Tourism Research* 10: 337–62.

Espino, M. (2001) 'Cuban tourism: a critique of the CEPAL 2000 report', in *Papers and Proceedings of the 11th Annual Meetings of the Association for the Study of the Cuban Economy*.

Freitag, T.G. (1996) 'Tourism and the transformation of a Dominican coastal community', *Urban Anthropology* 25: 225–58.

Gray, B. (1989) *Collaborating: Finding Common Ground for Multiparty Problems*, San Francisco: Jossey-Bass.

Gurstein, M. (2000) 'Community informatics: enabling community uses of information and communications technology', in M. Gurstein (ed.) *Community Informatics: Enabling Community Uses of Information and Communications Technologies*, London: Idea Group.

Henken, T. (2001) 'An analysis of Decree-Law No. 171 on private home rentals', in *Papers and Proceedings of the 11th Annual Meetings of the Association for the Study of the Cuban Economy*.

Hills, T. and Lundgren, J. (1977) 'The impact of tourism in the Caribbean: a methodological study', *Annals of Tourism Research* 4: 248–67.

Honey, M. (1999) *Ecotourism and Sustainable Development: Who Owns Paradise?*, Washington, DC and Covelo, CA: Island Press.

Hudson, B. (1996) 'Paradise lost: a planner's view of Jamaican tourist development', *Caribbean Quarterly* 42: 22–31.

Kenny, C. (2002) 'Information and communication technologies for direct poverty alleviation', *Development Policy Review* 20: 141–57.

Milne, S. (1998) 'Tourism and sustainable development: the global–local nexus', in C.M. Hall and A. Lew (eds) *Sustainable Tourism: A Geographical Perspective*, Harlow: Longman.

Milne, S. and Ateljevic, I. (2001) 'Tourism, economic development and the global–local nexus: theory embracing complexity', *Tourism Geographies* 3: 369–93.

Milne, S. and Mason, D. (2001) 'Tourism, web-raising and community development', in P.J. Sheldon, K.W. Wober and D. Fesenmaier (eds) *Information and Communication Technologies in Tourism (ENTER 2001)*, New York: Springer Wein.

O'Neil, D. (2002) 'Assessing community informatics: a review of methodological approaches for evaluating community networks and community technology centres', *Internet Research: Electronic Networking Applications and Policy* 12: 76–102.

Organization of American States (OAS) (1998) *Inter-American Program for Sustainable Tourism Development*, Washington, DC: OAS.

O'Riordan, T. (2002) 'Protecting beyond the protected', in T. O'Riordan and S. Stoll-Kleemann (eds) *Biodiversity, Sustainability and Human Communities: Protecting Beyond the Protected*, Cambridge: Cambridge University Press.

Parker, S. (2000) 'Collaboration on tourism policy making: environmental and commercial sustainability on Bonaire, NA', in B. Bramwell and B. Lane (eds) *Tourism Collaboration and Partnerships: Politics, Practice and Sustainability*, Clevedon: Channel View Publications.

Pattullo, P. (1996) *Last Resorts: The Cost of Tourism in the Caribbean*, London: Cassell.

Poon, A. (1990) *Prospects and Policies for Caribbean Tourism to the Year 2000*, London: Commonwealth Secretariat.

Rodriguez, E. (1997) 'Orden en casa', *Granma International*, 8 June: 2.

Rosendahl, M. (1997) *Inside the Revolution: Everyday Life in Socialist Cuba*, Ithaca, NY: Cornell University Press.

Salinas, E. (1998) 'Turismo en Cuba: desarrollo, retos y perspectivas', *Estudios y Perspectivas en Turismo* 7: 137–64.

Salinas, E. and Estevez, R. (1996) 'Aspectos territoriales de la actividad turistics en Cuba', *Estudios Geográficos* 57(223): 327–50.

Sanoff, H. (2000) *Community Participation Methods in Design and Planning*, New York: Wiley.

Small Tourism Enterprises Project (STEP) (2003) *STEP Newsletter* January 1(2).

Telfer, D.J. (2002) 'The evolution of tourism and development theory', in R. Sharpley and D.J. Telfer (eds) *Tourism and Development: Concepts and Issues*, Clevedon: Channel View Publications.

Thivierge, P. (2001) 'Tourism, stakeholder networks and sustainability: the case of Viñales Valley, Cuba', unpublished MA thesis, Department of Geography, McGill University.

Timothy, D.J. (2002) 'Tourism and community development issues', in R Sharpley and D.J. Telfer (eds) *Tourism and Development: Concepts and Issues*, Clevedon: Channel View Publications.

Weaver, D.B. (1991) 'Alternatives to mass tourism in Dominica', *Annals of Tourism Research* 18: 414–32.

Wilkinson, P.F. (1997) *Tourism Policy and Planning: Case Studies from the Commonwealth Caribbean*, Elmsford, NY: Cognizant Communications.

Zúñiga, J. (1999) 'Preocupación por alojamiento de turistas en casas particulares', *CubaNet News*, 10 September. Online: <www.cubanet.org/Cnews/y99/sep99/10a2.htm>.

13 Tourism businesses in the Caribbean

Operating realities

*K. Michael Haywood and
Chandana Jayawardena*

Introduction

The viability of tourism, as the Caribbean's primary industry, is ultimately determined by the performance and profitability of a myriad of private and public sector enterprises, such as restaurants, retailers, hotels, resorts, attractions or destination management organisations. To the uninitiated and inexperienced, the intensity of effort that goes into managing these businesses, whether they are simple or complex, is quite profound. It is little wonder that the famous free market economist, Milton Friedman, once argued, 'the business of business is business' – there is little time to do anything else, so you are better to focus all efforts and resources on achieving your mandate.

Unfortunately, as other authors in this book so poignantly emphasise, and as many businesspeople are grudgingly realising, focusing simply on the demands of the business is insufficient. The critical skill is navigating highly public, ethical and social concerns, and overcoming restraints that are far more subtle than those encountered in standard business practice. Increasingly tourism businesses find themselves caught between seemingly contradictory goals: satisfying the investors' expectations for progressive earnings growth, the consumers' desire for fabulous (Caribbean) experiences, and the community's demands for derived benefits and social and environmental responsibility. Realising these goals is complicated by the fact that the industry is intensely competitive, and sensitive to the slightest turbulence in climatic change, incidents of crime, technological improvement, social unrest or seasonal market demand. Anticipated cash flows can flounder and fall; as a consequence, profitability from one year to the next is far from being a certainty.

This chapter, therefore, provides perspective on the challenges associated with operating tourism enterprises, and in creating sustainable value both from tourism over time, and for the broad constituency of stakeholders, each of whom has, or should have, a vested interest in the systematic competitiveness of the industry. With emphasis based on creating appreciation for the complexities facing tourism enterprises achieving profitability, the chapter begins by briefly reviewing how Caribbean nations can ill afford to allow tourism to become a commodity-like industry, which ultimately becomes a race to the bottom. Undifferentiated and poorly planned tourism development hastens obsolescence; diminishes market value; intensifies competitiveness; reduces incomes; and destroys any hope of sustainability.

The second section examines the major activities in which tourism enterprises must excel if the industry is to survive. It provides a starting point for the industry to determine whether there is alignment between the internal activities of tourism enterprises and the value-creating strategies designed to achieve desired results and outcomes. A simplified value chain framework is employed (Porter 1985; Heskett *et al.* 1997) with commentary on business activities based on correspondence with managers, the authors' knowledge, and working experiences in various Caribbean tourism businesses and institutions.

Tourism enterprise viability and the goal of industry sustainability

Visual evidence of the wealth captured within the industry is a common but misleading indicator of success. It is true that some tourism businesses require massive investment dollars. The grandeur (and often opulence) of these physical structures and assets is awesome, particularly in stark contrast to some surrounding communities. Surface appearances, however, may belie the facts surrounding fiscal performance, particularly as visitor counts and expenditures have been on the decline in many island nations. Based on the cost structures of larger businesses, financial break-even points are quite high and declines in visitor counts cause anguish. Indeed, to meet guest-service requirements it is becoming increasingly difficult to improve productivity and find cost savings in operating and maintenance budgets. Fixed costs, moreover, are rarely negotiable – failure in the obligation to pay, and businesses are placed in jeopardy, with receivership and possibly bankruptcy imminent. And, if dollars are still being repatriated to the offshore investment community, they often represent a form of debt or interest repayment.

While to those who argue that tourism favours the select few, it may be welcome news that any loss of opportunity to earn a constant stream of profits is small consolation given the exploitation of a country's natural and cultural resources, and the extent of investors' private wealth. Appreciating the financial woes facing many tourism enterprises in the Caribbean may not shatter perceptions about tourism as a 'profit zone', but at least an appreciation of the realities and challenges facing tourism businesses might reshape and broaden the current debate on sustainable tourism. For example, Caribbean nations have already been devastated by agriculture as a no-profit zone. The airline industry is another major example of an industry that does not produce profit, and, in the process, compromises the viability of the Caribbean industry through diminished airlift capacity. If and when tourism is added to the roster of looming no-profit-zone industries in the Caribbean, deteriorating economics will certainly follow. Hence the question: if tourism becomes a no-profit industry will it render obsolete the notion of sustainability?

Answers to this question invariably challenge the assumption that businesses are the basis of the economy and the primary source of modern society's wealth; however, it seems safe to assume that success in business will remain of significant consequence to national and regional well-being for some time – economically, socially, politically and environmentally. Perhaps the reason why this will be so coincides with the emergence of the natural-systems model of post-modernism, on which sustainability is built (Hawken *et al.* 1999). It has become abundantly clear that the core characteristic of a healthy system is that the varied parts interact with each other through relationships of mutual independence. The significance of this observation should be obvious. The future of tourism businesses as a whole is threatened by environmental neglect, particularly if relationships are neglected. If, on the other hand, they are fostered it promises to be one of the major determinants of tourism development, competitive advantage and corporate survival in the decades to come. Since most Caribbean nation-states depend on tourism, the loss of the potential restorative power and flow of resources that could (and should) flow from tourism to support societies and their natural resources would be devastating. This benign reference to the organic cycle of life and death of businesses is self-evident, yet it would be perilous if the value-enhancing and wealth-generating capacity of an entire industry were allowed to fade.

The competitiveness challenge for, and perhaps the survival of, Caribbean destinations demands strategic tourism management responses. These responses are in play throughout the Caribbean (CTO 2002), but it is not unusual to find that the strategic focus requires resolution of a quandary: is the primary determinant of a country's ability to create wealth – and its concomitant spread in social and economic benefits – access to resources (Wernerfelt 1984; Barney 1991), pursuit of activities (Porter 1985), or position within the industry structure? A government's or NGO's attempt to choose one of these explanations over the other generally proves fruitless. There are continuous reciprocal relationships between industries, among enterprises within industries, and between enterprises and their environment, with competition shaping capabilities and capabilities in turn shaping competitive positions (Henderson and Mitchell 1997). Even this statement is benign as social-political environments and systems of governance must be taken into account. As such, it is widely acknowledged that tourism depends on the cooperation of extended enterprises – a global network of interrelated businesses, government agencies and other stakeholders that create, sustain and enhance a country's or an organisation's value-creating capacities. In other words it is inter-organisational relationships and institutional mechanisms as well as resources, transactions and activities that are the ultimate sources of wealth.

For decades Caribbean tourism has relied on the myriad of tourism's natural and man-made, tangible resources to drive the industry and create value. In today's global economy, however, competitive advantages, even for the tourism industry, are increasingly being built on relationships, knowledge, people, brands and systems – intangible assets that take centre stage. Tourism resources in the Caribbean may be unparalleled, but these resources *per se* do not create value (Porter 1996). Rather, value creation results from the activities in which the resources are applied. Primary and supportive activities, which are the basic units of competitive advantage, are embodied in internal business processes that comprise its value chain (Porter 1985); that is, collectively these activities comprise systems where value is created by transforming a set of inputs into more refined outputs – the creation of visitor experiences, for example. Since tourism is a classic service industry, the systematic composition of these processes varies from typical goods-producing firms, and creates unique challenges in the management of service chains (Heskett *et al.* 1997). Furthermore, if hotels are unable to fill their rooms over any given time period the potential revenue lost cannot be recouped. Hotel rooms, like airline seats, are perishable. Capacity management, therefore, becomes a high-priority task, especially as room revenue is the most profitable.

With markets located some distance from Caribbean destinations, visitors need to be reassured that the value of the Caribbean experience will exceed the acquisition costs (monetary and psychological) plus the price paid. Visitors expect and buy results; even tourism's other critical stakeholders expect results. When tourism is managed for results (problem solution, economic wealth, profits, preservation), strong direction is provided both in the implementation of effective strategies and the achievement of outstanding performance.

The most critical dimensions which lead to the achievement of these results in the Caribbean tourism industry are the value of the products, services and experiences delivered to visitors. Visitor satisfaction, repeat visitation and positive word-of-mouth (in essence, loyalty) are the important results that must be managed. Since employees in the tourism industry, as well as the host community, deliver these services, expectations for hospitable behaviour and competency must be evident. The capability for delivering results must be demonstrated. In tourism, therefore, employee satisfaction, loyalty and productivity must be included in the result-equation. Similarly, since tourism utilises communities at large, impacts must be managed. Again tourism must be managed for results – results that hopefully are in alignment with a country's articulated vision for tourism, which gives priority to its citizens and endowments.

Since results are achieved through activities, the artful development of successful and sustainable strategies also requires alignment between the internal activities of tourism enterprises and results. In simple terms these internal processes can be separated into four business processes: visitor management processes, operational processes, enterprise and tourism development processes, and community and environmental processes. All of these are important and must be performed well by every organisation within the industry. But as outlined below, the delivery of results to visitors is easily compromised when organisations are confounded by restraints that make it difficult to excel in any process that will have an effect on visitors and their overall experience.

As industry clusters, tourism destinations create value in interactive ways. The business strategy literature has identified these as networks or constellations (Normann and Ramirez 1993, 1998), value systems (Porter 1996), and value configurations (Stabell and Fjeldstad 1998). In other words, the focus of strategic analysis need not be a company, or even an industry, but the value-creating system, within which different economic and non-economic actors – suppliers, businesses, customers, environmentalists, legislators, planners, and so on – work together to produce (or destroy) value. In this case, the key strategic task is the (re)configuration of roles and relationships among this constellation of participants in order to mobilise the creation of value in new forms and by new players (Normann and Ramirez 1993: 66). As a consequence a destination's offering is a result of a complicated set of activities between, and relationships among, participants within the value-creating system.

These observations lead to recognition that Caribbean nation-states, which are extremely vulnerable to the vicissitudes of tourism as a dominant industry, must pay greater attention to the logic of value and its configuration. It has significant implications for the crafting and implementation of strategy. For example, as value is created within a country's or a region's tourism constellation, competition is not simply between tourism enterprises, but between the tourism offerings or experiences, which are, in turn, the result of the cooperation among tourism enterprises (Brandenburger and Nalebuff 1996; Morgan 1998). Furthermore, as the tourism offerings become more complex and varied, so do the relationships necessary to produce them. Stand-alone strategies are no longer viable, and most attractive offerings imply strategies of networking by different participants. Therefore, the only true source of competitive advantage is to conceive and implement an entire value-creating system. This implies, among other things, that tourism enterprises must continually reassess and redesign their competencies and relationships based upon a thorough understanding of how visitors may create value for themselves within the destination. Evidently tourism leaders will have to become very astute in managing the value network if destinations are to reap the rewards (Haywood, forthcoming).

Assessment of operational and value-creating activities and resources

In order to legitimise and appreciate the growing need for value-based strategies in conjunction with both the operational, value-chain issues facing tourism enterprises, and the state of the tangible and intangible resource base within the Caribbean, what follows is a general overview of concerns. A modified value chain is used as a heuristic device to describe and assess a range of activities that are required enhance value, particularly to visitors to Caribbean destinations. While no attempt is made to map the value chain, resource considerations and relationships are considered. The intended outcome from such an assessment is the discovery of new ways for the industry to begin functional upgrading and to leverage greater value from the Caribbean tourism experience – value which is important to a wide variety of stakeholders:

- visitors, as there is an absolute need for Caribbean destinations to remain market driven;
- employees, as there is an absolute need to get them engaged and appropriately compensated;
- the community, as there is an absolute need for local commitment and resource sustainability; and
- shareholders, as there is an absolute need to reward their investment and willingness to assume risk.

Initially developed by Porter (1985), value chains normally identify an industry's primary and support activities. For the sake of simplicity these have been encapsulated into four processes, as previously identified: visitor management processes, operational processes, enterprise and tourism development processes, and community and environmental processes. Because each of these processes can be aligned with a major theme or objective, this is stated along with associated activities, many of which are discussed in general terms. Included is reference to both associated tangible and intangible assets and resources.

Visitor management processes ('enhance visitor value')

Successful destinations seek to attract, and establish lasting relationships with, an appropriate number and type of visitors. As a service industry this objective is accomplished through a broad range of marketing and branding activities that spill over into operating and development processes. Building more intimate relationships with visitors in the Caribbean has suddenly become a more challenging endeavour. Since 11 September 2001, visitation to the Caribbean has fallen dramatically, with occupancies during the following year reported to be down 18 to 35 per cent. ADR (average daily rates), and RevPAR (revenue per available room) were also reported to be significantly lower (Deloitte and Touche 2002). With new and emerging destinations, economic slowdown occurring in many tourism generating markets, cutbacks in travel due to security concerns, the desire to stay closer to home intensifying, and increases in tourism advertising occurring in most destinations, recovery is expected to be painfully slow. Exacerbating the problem are real exchange rates and an apparent decline in expenditure per visitor. While a recent study of price elasticity of tourism demand in the Caribbean has revealed substantially lower elasticities than previously reported (Rosenweig 1988) it appears as if many Caribbean countries have been quite successful at differentiating their island experience. But as the authors conclude: 'The value (of long-run income elasticity) suggests that, even if tourism overall is a luxury good, Caribbean tourism is perhaps less so' (Maloney and Rojas 2001: 8).

The importance of keeping rate structures and price fluctuations in check has never been more critical. In lower demand situations many businesses operate close to break-even points. Until visitor traffic increases occur, higher operating costs are likely to be reflected in lower profits. The unfortunate aspect of being locked into a price-sensitive/tight-margin business is the lack of ability or interest in re-investment. Further differentiation has to be found if the constraints around price responsiveness are to be lifted. But the capital available to accomplish the associated tasks may be difficult to obtain.

In order to break out of the commodity trap, market research needs to determine what services customers want, where, when, and at what prices. By learning more about why they want these services, and how they want them delivered, new business models and enhanced service capabilities can be introduced. Similarly, better understanding of service evaluation processes and visitor loyalty needs to be acquired. Unfortunately insufficient time and money is being dedicated to visitor knowledge and its management, either by businesses or governments (Haywood, forthcoming). The Caribbean Tourism Organization, based in Barbados, does an admirable job, but its mandate and resources are limiting.

Industry standards are also floundering. The following remarks made by John Bell, Executive Director of the Caribbean Hotel Association, are worth heeding:

> We have to fight for everything we have. Our biggest market, Florida, is also our biggest competitor. We have to pay an awful lot more attention to our standards. Too often we are found wanting and we need to overcome the difference between service and servitude. Failing to do so, we can look at the business dwindling even more. . . . It's time for professionalism, de-politicising tourism, and managing the industry as a business.
>
> (*Black Britain* 2002)

Responses to the inherent problems have to be multifaceted, and will be addressed in relation to discussion on each to the internal business processes. Obviously marketing activities (for example, advertising, branding, market research, etc.) are pivotal. The prohibitive costs of reaching and appealing to offshore and highly segmented markets – for example, leisure, adventure, convention, incentive, honeymoon, gaming and eco – cannot be undertaken solely by individual businesses. Government assistance is required. In the latter part of 2002, the Caribbean Hotel Association Charitable Trust, a public/private sector affiliation of major hotel chains, airlines and credit card companies, joined forces with both CARICOM (Caribbean Community) and non-CARICOM countries, and revived their US$16 million advertising campaign, 'Life Needs the Caribbean', to market and promote the Caribbean as a single destination.

What effect this campaign will have is difficult to tell. In the past there has been a failure to live up to advertising's promises. Results from a fairly recent survey (Karma Centre for Knowledge and Research in Marketing 1998) reveal six areas of concern (scoring high on importance and low on satisfaction): crime rate, cleanliness of food and water, local price levels, affordability of accommodations, dining and restaurants, affordability of flights. Visitor satisfaction exceeded expectations in four 'high worth' areas: general sporting activities, snorkelling/scuba/diving, beach/sunbathing/swimming, and dance clubs/bars.

If the survey results are accurate, and visitor satisfaction wavers, visitor loyalty is unachievable. To be truly successful brands must be 'relevant' and 'resonate' with customers, if they are to achieve 'brand strength' (Bedbury 2002). Even John Bell admits something is amiss (*Black Britain* 2002). He points to all the linkages – operational processes, enterprise and tourism development, and community and the environment. Clearly the problem is complex, but, as he points out, a first-order priority is to resolve the structural deficiencies in the relationships among the stakeholders.

Visitor management must also pay more attention to dependency on intermediaries such as tour operators, travel agents, airlines, cruise and other transportation modes, Internet and online travel sites. Each of these intermediaries, and still more, get involved and influence visitors' Caribbean experience. They either strengthen or weaken the Caribbean or destination's brand, and the value creation process. Of major concern are those organisations who, by fiat, can limit the number of visitors or accessibility to the islands. Airlines and cruise lines have caused, and will continue to cause, grief at times. Even tour operators negotiate from positions of strength when they purchase blocks of rooms at resorts. Managing these relationships is stressful to say the least. Indeed, when Accor Hotels closed down properties in Guadeloupe and Martinique for a while, many sensitive issues were aired openly in the French press. Negative publicity over the years has also been caused by civil unrest or destructive hurricanes. All these crises influence brand touchpoints that indirectly make impressions about a destination, country or the entire Caribbean region.

Brand touchpoints – 'all the different ways that a brand interacts with, and makes an impression on, customers, employees, and other stake-holders' (Davis and Dunn 2002: 58) – also include 'purchase touchpoints' in which visitors are exposed to the destination, revel in the delights of experiences, bask in the warmth of Caribbean hospitality, or recoil at service *faux pas*. All of these, including adjustment to a different place and pace of life, have to be managed if the customer-value proposition is to be maximised. Because management of many of these touchpoints really falls under operational processes, and tourism and product development, one might assume that the interdependency and interplay among all these processes and activities is far from seamless. Marketing, operations and development occupy organisational silos. Breaking through the bureau-cracy in some businesses in the Caribbean can be quite a challenge.

This glimpse into one process, visitor management, reveals that tourism in the Caribbean is built on essential, though nuanced, blocks of resources and assets (private and public, tangible and intangible, naturally or humanly inspired), as well as a set of activities used to create visitor experiences and measurement systems that record success. It is now possible to see how assets relate to processes, how processes connect to outcomes, how measurements relate to market results and how relationships are critical determinants of the creation of value. This is, essentially, the DNA of tourism, but every destination adds its own flavour and tempo that are difficult to capture on paper and hard to fathom if you are an outsider. For example, expatriate managers, new to the islands, may arrive equipped to understand and appreciate the requirements of guests, but may lack appre-ciation of typical Caribbean customs and may have difficulty in fostering local relationships that might serve to untangle supplier relationships or motivate employees. Earning trust and confidence takes considerable time and lots of effort.

Operational processes ('achieve operational excellence')

There are few tourism enterprises that do not set out to achieve some form of operational excellence. Taking its cue from visitor management, the accompanying strategy for this theme or objective usually places emphasis on cost control, standardisation, quality improvement, efficiency and improved cycle time of operating processes, and excellent supplier rela-tionships.

Operating a tourism business in the Caribbean can be decidedly expensive. Few Caribbean countries are self-sufficient, and most items required to build, equip and operate resorts and other enterprises have to be imported. Transportation costs, and sometimes tariffs, are added to shipments of supplies and equipment. Items get short-shipped, lost or stolen in transit, and may take considerable time to obtain between ordering and receiving. As such, some items may have to be ordered in bulk or larger quantities; inventory costs may be high and possible spoilage may be incurred, particularly for perishable food items. Sourcing locally, if possible, is often complicated by the fact that the capacity to produce in sufficient quantity on a day-to-day basis may be impossible, without some form of long-term contractual obligation or investment.

A particularly sensitive issue in the Caribbean is the low economic multiplier that is derived from tourism; inter-industry linkages are encouraged, particularly with agriculture and fisheries, but government assistance is definitely required in this realm. Because the requirements of, and between, the sectors are often not compatible, cooperation can be onerous. The hotel and restaurant sector, however, is often encouraged to create menus and recipes using local produce and product; the success stories are plentiful. With length of stays shortening, however, most visitors are not always amenable to altering their eating habits. Food choice and meal selection are hard to dictate when the guest is paying the bill and wary of becoming ill in a foreign country. As the previously referenced visitor survey indicated, the issues of cleanliness, and food and water quality, are uppermost among visitors' concerns. Sensitivity to price, however, need not be an issue if meal experiences provide outstanding value.

The tourism industry in the Caribbean is labour intensive. From an employment perspective tourism suffers from an image of offering low-skilled positions that are often labelled 'McJobs'. This characterisation is unfair as tourism provides Caribbean people with tremendous employment opportunities that otherwise would not exist. In other words there is a high degree of dependency on the industry, and when it falters job losses mount and family incomes fall. Certainly some of the jobs may not require the application of high-level skills, but most tourism businesses take their employment mandate very seriously. The link between visitor satisfaction and service value is profound (Heskett *et al.* 1997). Visitors' strong value-orientation is served through the provision of services and service process quality. In other words it is employees who determine whether visitors will receive the results they expect. The interactive nature of most tourism jobs depends on well-trained, competent and highly motivated staff, though some managers in the region acknowledge difficulty in finding such employees. Consequently, more and more businesses are undertaking their own training programmes. The payback in terms of visitor loyalty can be immense, if calculated; but this type of measurement rarely happens.

Tourism jobs are demanding, as the industry operates around the clock on a day-in day-out basis. Obviously, constant handling of the demands of visitors away from their familiar surroundings can be tiresome and tedious; pay rates may not be high by European or North American standards, but those who benefit by receiving tip income can do well financially.

Many Caribbean nations do not tax income, so take-home pay is likely adequate. Unions are active in some countries, and relationships are known to sour. Bermuda, for example, has been hit hard by strikes.

Attracting the Caribbean's brightest and best to the industry is difficult as there is a prestigious aura, as well as higher remuneration, associated with positions in government, the professions and financial services. In the past, senior level and management positions within the industry tended to be filled by expatriates with considerable international experience. Intervention by government to stop this practice, and availability of home-grown talent, has now opened up the senior industry ranks to well-educated and experienced Caribbean nationals. There is tremendous local talent, and the industry has fostered its own group of well-known entrepreneurs and industry leaders. There is a need, though, to foster more entrepreneurial talent within the industry, but support from the financial community and micro-credit organisations is not as evident as it should be.

Productivity issues, many associated with age-old, inherited, bureaucratic practices from colonial days, are still rampant; however, modern managerial practices and service process systems are being implemented in the larger operations. This is due, in part, to sophisticated management systems of chain and franchise hotels and professional hotel management companies that operate throughout the Caribbean. Nevertheless, productivity- and efficiency-related problems are common; they can be linked to poorly designed service-delivery systems that conform to outmoded standard procedures, limit employee latitude and create unnecessary service errors. Obviously visitor satisfaction can be compromised through extremely annoying and frustrating requests, time delays and inconveniences. Given the financial squeeze and tight margins that continue to plague tourism businesses, there is tremendous opportunity to improve service-delivery systems.

As mentioned, operating costs are high, particularly for small enterprises. Room taxes, import taxes, customs and stamp duties, and airport departure taxes for visitors, are the bane of their existence. A recent study (1999) by the Organization of American States (OAS) reported that utility costs are higher than in the US, and that insurance premiums may be unaffordable resulting in lack of insufficient liability coverage. Small businesses are also marginalised due to insufficient knowledge of accounting and human resource management, including training. They are less likely to be part of their national tourism organisations and, hence, destination marketing programmes may not work for them.

Enterprise and tourism development processes ('become world class')

In attempts to grow and become world class, most destinations focus attention on the theme of development. Building the destination characteristically has been based around the construction of glamorous physical structures, attractions and amenities. Expensive infrastructure improvements normally have to accompany this growth. Many Caribbean destinations, however, suffer from inadequate tourism-related infrastructure.

Since sustainable competitive advantage is the implicit or explicit goal that virtually every Caribbean tourism destination is attempting to reach, then three things must occur. First, visitors must perceive a consistent difference between a destination's offerings and those of competing destinations, and that difference must occur in one or more key purchasing or visiting criteria, that is, in one or more of the attributes that shape the purchase decision. Second, the difference must come from a resource and/or a capability gap between the favoured destination and its competitors. Third, the differentiation and the capability gap must endure over time. Because of the increasing diversity or segmentation of a destination's markets – for example, adventure, nature-based, convention, gambling – competitive advantages are becoming more difficult to identify. To have strategic value the differentiation must be of significant interest to large numbers of visitors who will continue to visit on a regular basis. The constant requirement to fill rooms (a capacity issue) and the three indicators of success – visitation rates, length of stay and expenditure levels – have a tendency to divert attention away from identifying and measuring differentiation strategies.

Typically, Caribbean countries rely on their natural resource base, culture and tourism superstructure for this differentiation. But is this form of differentiation sufficient given the variety of sun, sea and sand destinations that exist around the world? In a nutshell the answer is 'no'; the key criteria are only partially related to the destination's cultural and natural characteristics. Selection or visitation criteria are usually attributes that affect the way potential visitors perceive the destination and the destination experience (the desired results and value anticipated), their access to it and the costs associated with the trip and sojourn. As John Bell (*Black Britain* 2002) indicated, sustainable competitive tourism will be denied unless Caribbean nations seriously address all the attributes that determine these perceptions.

Paradoxically the easiest and yet most costly attribute that is constantly addressed has to do with the physical product. Caribbean destinations have become very astute in appealing to and attracting investors to spur development in the traditional sense. This focal pursuit creates a situation in which tourism has become 'supply-led'. Hosts of incentives, tax breaks and other concessions are designed to attract investment dollars and the leading brand-name hospitality and tourism conglomerates. Of course, these efforts can have immense payoffs: new and affluent visitors, tremendous brand synergy, attention from the international press, new jobs and tax revenues. The payoff is the anticipated creation of new wealth. Development fostered by well-known brands is an attention-grabber. It can spur complementary or even competitive ventures. The snowball effect, if it occurs, can be both rewarding and risky. Proposed developments may not be in keeping with the differentiation required; economic, social and environmental impacts are hard to anticipate, though always accompanied with a positive spin; and everything rides on the expectation of a surge in demand and the overall success of tourism. As the industry becomes more dominant, however, exposure to downside risks can be heightened; there are few counter-balances when unpredictable swings in seasonal as well as business cycles occur (Haywood 1998).

Understandably financial institutions are becoming increasingly wary about supply-led hotel and tourism developments. They are more fastidious in conducting due diligence before lending capital, and are more inclined to underwrite for well-known brands that have a proven track record, than they are for independent investors and operators. Hotel and attraction development is relatively capital intensive, and requires long-term capital suitably structured, to enable it to deal with the vagaries of natural and man-made disasters and business cycles. Unfortunately the capital markets in the Caribbean are still very much in a state of infancy, but, as a major investor in Caribbean tourism, the Caribbean Development Bank is anxious to do more. However, it needs to find ways to diversify its risk, and to attract a broader spectrum of investors.

Access to capital is particularly a major problem for small businesses in the Caribbean. Consequently, hotel properties may be poorly maintained, and not up to standard (OAS 1999). Weather-related damage can be a severe problem, so quality problems can be exacerbated. In order to address the financing problem, the Organization of American States is in the process of raising funds for support of the Small Tourism Enterprise Project (STEP) and the Caribbean Experience brand to assist owners of small unbranded hotel properties in the Caribbean region. There is also talk of establishing 'microcredit' organisations to encourage entrepreneurial development; and the Caribbean Development Bank, in association with the Caribbean Association of Industry and Commerce, held a major symposium in 2002 to discuss private and public sector collaboration to make the Caribbean private sector more internationally competitive.

As a service industry, the capability and competency of employees is of critical importance. Governments throughout the Caribbean realise that the Caribbean experience is showcased through its people. The opportunities brought about by tourism development, however, cannot be realised until the capability of people is developed. It must become a priority. The service profit chain (Heskett *et al.* 1997) provides the justification: Profit and growth are linked to visitor loyalty; visitor loyalty is linked to visitor satisfaction; visitor satisfaction is linked to the value of services provided; service value is linked to employee productivity; employee loyalty is linked to employee satisfaction; and employee satisfaction is linked to quality of life, both in the workplace and in the community. Employee capability is being achieved through the implementation of occupational skill standards, certification and credentialing programmes through the cooperation of the Organization of American States, the Caribbean Tourism Organization, the Caribbean Financial Services Corporation, the Caribbean Development Bank, the Caribbean Hotel Association, the Caribbean Hospitality Training Institute, and many other public institutions and private enterprises. Most Caribbean countries have hospitality and tourism programmes in place at all academic levels. Training and education for senior managerial levels does exist, but many businesses still suffer from a deficiency of individuals who possess the appropriate management skills.

The momentum behind improving capabilities in all areas (brand value, visitor relations, management capabilities, alliances, technology, employee relations and environment and community issues) must be maintained if achievement of competitive advantage is to be sustained. As suggested, the tourism industry in the Caribbean is learning the hard way that the value inherent in the Caribbean experience is beginning to migrate (Slyworthy 1996). Detection of this phenomenon has been complicated by the fact that the early stages of this migration are often silent and subtle. The identity and priorities of visitors are shifting in ways that are not immediately apparent to the incumbents in the industry. Nevertheless detection is a prerequisite to sharing in the next cycle of value growth. Turning once again to market research activities, hypersensitive competitive radar screens and strategic understanding of the various visitor markets are required to pick up on the shifting patterns of change, or on any disillusionment that may be occurring

Community and environmental processes ('be a good corporate citizen')

As an industry that generates volumes of visitors that descend on, and utilise, large portions of the land mass and surrounding waters of small, fragile tropical isles, the industry struggles with the objective of being or becoming good corporate citizens and neighbours. Industry associations are leading the charge. Some of their visions for tourism include: 'To be the safest, healthiest and happiest of comparable destinations in the world' (Caribbean Epidemiology Centre); 'The Caribbean cares' (Caribbean Hotel Association); 'Caribbean development through quality tourism' (Caribbean Tourism Organization). The Caribbean, in particular, depends on the pristine conditions of its coastal zones – fabulous beaches, clear water and healthy marine life. Islands are fragile, closed ecosystems with limited resources. Water is a precious commodity in short supply, and the industry tends to consume an inordinate proportion of it. For tourism to flourish, it needs to be available into the foreseeable future.

Making the case for the sustainability of tourism enterprises dependent on a healthy natural and physical environment has required tremendous effort. There are comprehensive programmes of action for attaining sustainable development in the works. The Caribbean Healthy Hotels Project, for example, 'takes a holistic approach to improve the health and hygiene conditions for guests and staff in hotels, leading to decreased liability, improved profitability and sustainability of tourism and economic development'. The Caribbean Alliance for Sustainable Development (CAST) is another endeavour that offers education and training on sustainable tourism. It represents a partnership between the Caribbean Hotel Association, Green Globe 21 and the International Hotel Environment Initiative.

Sustainable development was identified as the most critical issue in the most recent Caribbean Strategic Plan (CTO 2002). The plan makes the case for a more holistic view of tourism, including a new appreciation of the fact that tourism is more than the sum of all arrivals. Instead, issues of sustainability are paramount and the plan highlights the need for consideration of the natural environment, security and health, community involvement and for a sustainable source of funding for tourism development programmes, among others (CTO 2002).

Conclusion

The sustainability of the Caribbean tourism industry depends on finding and developing constructive business relationships with Caribbean society. There seems to have been some reluctance to become so attuned. To a large extent this is due to the tremendous effort required in coping with the inordinate demands of the business/tourism world. Gradually tourism executives and leaders operating in the Caribbean are beginning to move from 'value-neglect' postures to 'value-attuned' responsiveness. Tourism cannot be allowed to become another no-profit industry. If the 'Life Needs the Caribbean' advertising campaign is to be truly effective, it must be credible. In other words, it must be relevant to, and resonate with, visitors before, during and after their visits. Credibility and trust will only occur when individual organisations and the industry as a whole work collectively to improve all the activities and processes associated with visitors, operations, development, community and environment. In the process the industry will need to become better integrated, value-attuned and oriented to collaborative relationships.

In a connected industry such as tourism, it has become clear that technology, alliances, brand value, employee relations, customer relations and innovation are increasingly driving value propositions. Physical assets do not encapsulate as much value as they once did. Most brand-name hotel chains, for example, have shed or spun off their real estate and now operate as professional management companies. In other words, greater proportions of investments are going into intangible resources. As such determination of the market value of these firms now requires the additional quantification of the value of intangible assets (Kaplan and Norton

1996). While an increasing number of corporations are engaging in these measurements, it seems appropriate that destinations follow suit. All tourism activities or processes should be measured against results expected by all stakeholders. The development of broadly based industry and tourism enterprise scorecards is a task that definitely requires research and attention. Tourism organisations need to measure relationships between individual links in the service profit chain through meaningful scorecards, and to utilise leading as well as lagging indicators to measure the results associated with each major activity performed in the value chain (Heskett *et al.* 1997). With the scorecards designed to incorporate measures of progress in terms of the visions and strategies designed to 'enhance visitor value', 'achieve operational excellence', and 'be a good corporate citizen', the basis for achieving sustainability of the tourism industry in the Caribbean is likely to be more successful. The performance and profitability of tourism enterprises, therefore, is a multifaceted undertaking that provides the basis for achieving the goals of sustainability and competitiveness of the Caribbean tourism industry.

References

Barney, J. (1991) 'Firm resources and sustained competitive advantage', *Journal of Management* 17: 99–120.

Bedbury, S. (2002) *A New Brand World*, New York: Viking Penguin.

Black Britain (2002) Caribbean Tourism Troubles. Online: <http://www.blackbritain.co.uk> (accessed 21 October 2002).

Brandenburger, A.M. and Nalebuff, B.J. (1996) *Co-opetition*, Cambridge: Harvard Business School Press.

Caribbean Development Bank (2002) 'Competitive private sector development: an imperative for the future – report on proceedings', Barbados: Caribbean Development Bank.

Caribbean Tourism Organization (CTO) (2002) *Caribbean Tourism Strategic Plan*, Barbados: Caribbean Tourism Organization.

Davis, S.M. and Dunn, M. (2002) *Building the Brand-Driven Business*, San Francisco: Jossey-Bass.

Deloitte and Touche (2002) 'Hotel benchmark survey – Caribbean and Latin America', London: Deloitte and Touche.

Hawken, P., Lovins, A. and Lovins, L.H. (1999) *Natural Capitalism: Creating the Next Industrial Revolution*, Boston: Little, Brown & Co.

Haywood, M. (1998) 'Economic business cycles and the tourism cycle concept', in D. Ionnides and K. Debbage (eds) *The Economic Geography of the Tourism Industry: A Supply Side Perspective*, London: Routledge.

—— (forthcoming) 'Boosting the competitiveness of the Caribbean tourism industry: application of intellectual capital and knowledge management' , in C. Jayawardena (ed.) *Caribbean Tourism: Visions, Missions and Challenges*, Kingston, Jamaica: Ian Randle Publishers.

Henderson, R. and Mitchell, W. (1997) 'Introduction', *Strategic Management Journal* 18: 1–4.

Heskett, J.L., Sasser, W.E. and Schlesinger, L.A. (1997) *The Service Profit Chain*, New York: The Free Press.

Kaplan, R.S. and Norton, D.P. (1996) *The Balanced Scorecard: Translating Strategy into Action*, Boston: Harvard Business School Press.

Karma Centre for Knowledge and Research in Marketing (1998) *Caribbean Tourism Survey*, Montreal: McGill University Press.

Maloney, W.F. and Rojas, G.V.M. (2001) *Demand for Tourism*, New York: The World Bank.

Morgan, B.W. (1998) *Strategy and Enterprise Value in the Relationship Economy*, New York: Van Nostrand Reinhold.

Normann, R. and Ramirez, R. (1993) 'From value chain to value constellation: designing interactive strategy', *Harvard Business Review*, Jul./Aug.: 65–77.

—— (1998) *Designing Interactive Strategy*, New York: John Wiley.

Organization of American States (OAS) (1999) *Caribbean Small Hotel Sector*. Online: <http://207.61.131.81/hoi_sect.htm> (accessed 12 January 2003).

Porter, M.E. (1985) *Competitive Advantage*, New York: The Free Press.

—— (1996) 'What is strategy?', *Harvard Business Review*, Nov.–Dec.: 61–78.

Rosenweig, J. (1988) 'Elasticities of substitution in Caribbean tourism', *Journal of Development Economics* 29: 89–100.

Slyworthy, A.J. (1996) *Value Migration*, Boston: Harvard Business School Press.

Stabell, C.B. and Fjeldstad, O.D. (1998) 'Configuring value for competitive advantage: on chains, shops and networks', *Strategic Management Journal* 19(5): 413–37.

Wernerfelt, B. (1984) 'A resource-based view of the firm', *Strategic Management Journal* 5: 171–80.

14 What makes a resort complex?

Reflections on the production of tourism space in a Caribbean resort complex

Tim Coles

Introduction

In the past two years geopolitical conditions have impacted seriously on the travel and tourism trade. As several commentators have noted, the events of 11 September 2001 have precipitated both short-term and more persistent long-term spatial restructuring and reorganisation of demand (Goodrich 2002; Mills 2002; TBP 2002). Within Europe, popular destinations were hit hard in the autumn and winter as North Americans chose not to make long-haul trips. Conversely, some British holiday resorts experienced a late season increase in demand as domestic visitors chose to travel in the United Kingdom rather than risk even short-haul flights to European short-break and early winter-sun destinations (Vasagar 2002; Coles 2003). The Bali bombing and, more recently, the events leading up to, and attending, the second Gulf War have forced business and independent travellers to reappraise their plans in the far and near East (Lawrence 2003). In the case of south-east Asia, the recent outbreak of SARS (Severe Acute Respiratory Syndrome) has compounded already negative public opinion in the western world about travel to the region (*Observer* 2003).

Where elsewhere in the world there appears to be uncertainty and above-average risk is perceived, in the British travel press at least the Caribbean has been identified as a safe haven, a relatively low-risk destination region for visitors from the 'global north'. As other chapters in this collection demonstrate, the Caribbean is not an internally homogeneous region from a tourism development perspective (Pattullo 1996). While the islands of Jamaica, Antigua and Barbados have long-standing and well-deserved reputations as tourist playgrounds, in recent years destinations in the northern Caribbean such as Cuba and the Dominican Republic have become increasingly popular with European visitors. Cuba's appeal is grounded in perceptions that it has a good climate; it is relatively affordable and cheap once there; and that it offers one of the last chances in the world to gaze on the kitsch spectacle of a fading communist regime. More importantly perhaps, it has relatively unspoiled environments when compared to established European resorts, and there has been significant recent investment in new, high-quality resort hotels that compete with eastern Caribbean competitors on price and non-price terms. Many similar attributes help explain the appeal of the Dominican Republic. With rising prices in the Mediterranean and greater competition for services and amenities, the Dominican Republic is perceived to offer much greater value for money for sun-seekers than 'traditional' southern European honey pots, notwithstanding the lingering doubts over, and persistent cruel jibes about, the standard of food hygiene (Mawer 2002).

Both destinations are brought to market in Europe primarily through package deals. More often than not these take the form of standardised one- or two-week trips to resort hotels, many of which are embedded in larger resort complexes (Howlett 2002; Mawer 2002). These enclaves offer the visitor the two simultaneous advantages of insularity and exposure (cf. Edensor 2001; King 1997; Shaw and Williams, forthcoming). Within the resort 'bubble' there is the reassurance of security allied with an enhanced level of facilities and the tantalising potential of what lies beyond. There is a feeling of being 'elsewhere', but with the sureness born of a comfortingly predictable level of service delivery in what Edensor (2001) calls 'enclavic space'. External exposure to environment and society is available, but it is always possible to exercise control and limit contact by retreating back into the resort 'bubble'.

This chapter presents a micro-level study of a resort complex on the Atlantic, northern coast of the Dominican Republic and teases out its wider implications. Resort hotels dominate discourse on the critical issues facing tourism in the Dominican Republic (Kermath and Thomas 1992; Freitag 1994; Cater 1996; Pattullo 1996). While previous studies have explored the development of resorts and resort hotels with their concomitant outcomes (Kermath and Thomas 1992; Freitag 1994), this chapter adopts an alternative tack. It presents a supply-side analysis of the mediation of the tourism experiences on offer in resort hotels and complexes; in short, it explores the recreational opportunities and products that are presented to the denizens of 'enclavic spaces', and thereby it seeks to unravel some of the processes behind the mediation of tourists' vacation experiences. Opportunities and experiences both within and beyond the boundaries of resort space are documented as means of charting the production of holidays on the island. Exploratory and suggestive in scope, this chapter focuses less on the specifics of visit episodes and decision-making *per se* (cf. Mercado and Lassoie 2002). Rather, its aim it to expose wider messages about the development and usage of tourism spaces as well as the modes of tourism production and consumption both on the island and in the region.

Tourism production and consumption: choices, structures and agency

One of the most provocative ideas to appear over the past quarter of a century is Ritzer's so-called 'McDonaldization' thesis (see Ritzer 1998, 2000). At its most basic it contends that contemporary consumers have been faced with (and continue to encounter) a façade of choice when, in fact, consumption has become increasingly globalised and standardised. The rise and spread of the fast-food chain is indicative of a rationalising modernity (Meethan 2001: 75); that is, the application of Fordist principles to fast food is reflective of more omnipotent global forces that serve to discipline production and consumption (Barber 1995; Ritzer 1998; Lee 2003). Services have become mass produced and highly standardised for routine mass consumption. Similar modes of rationalisation and organisation are evident in such diverse sectors as fashion (Lee 2003), higher education and tourism (Ritzer 1998). Ritzer and Laska (1997) denote the trends in the tourism industry as 'McDisneyization'. In spite of the appearance of greater choice at the start of the twenty-first century, what people really want is vacations that are highly predictable, highly efficient, highly calculable and highly controlled (Ritzer and Laska 1997: 99–100). In their search for the familiar, the reassuring and the predictable, tourists consume in their masses standardised forms of vacation, above all, in the form of the package deal. Security and consumption have become intertwined to such an extent that standardisation is almost inevitable. Holidays, like other forms of 'service commodities', are replicated and reduced to relatively banal and mundane configurations. Consumer drivers as well as industry triggers provide the strong impulses in this movement.

Notions of mass production of tourism products and experiences have been challenged. The principal criticisms revolve around their inherent reductionism when, in fact, the world is increasingly characterised by diversity and plurality. For Meethan (2001: 75), the thesis shows a poor understanding of the nature of tourism markets and segments. Instead of a single tourist to which the standardised product can appeal, there are several *tourisms*. Tourism has become increasingly characterised by a multitude of consumers, each of whom has quite individual tastes, interests and contexts (Urry 1990). In a similar vein, Chang's (1999) work exposes the limitations behind reducing the role of local agencies, and hence the tourist, to that of a mere passive receiver of products and product-related information. Tourists are not automatons, or Fordist dupes willing to consume uncritically the opportunities dutifully served to them by tour operators and travel agents. Rather, they make sophisticated decisions about their holidays, born of informed choices. Consumers have become increasingly empowered by greater wealth, leisure time (to search out opportunities), and information in the form of the worldwide web and the plethora of guidebooks, newspaper travel supplements, travel programmes on television and radio and the like. Major vacations represent significant allocations of discretionary expenditure and time. As investments, they can represent a considerable portion of the (annual) household budget, and as such consumers have a vested interest in seeking out the experiences most appropriate to their individual contexts. Finally, in this respect, some criticism surrounds how Ritzer's thesis conceptualises, or, more accurately, largely overlooks, tourism spaces and the production of place. Aspects of the thesis appear incongruous when contrasted with the emerging accepted orthodoxy that the production of place in tourism is the outcome of the collision between global forces and local conditions, which are specific in time and space (see Milne and Ateljevic 2001; Green 2001; Chang *et al.* 1996; Chang 1999).

McDisneyization may not be a perfect set of postulates, but it does have several redeeming aspects. Increasingly, the largest tour operators have been widening their scope both spatially and organisationally. In advertisements major chains remind readers of just how many destinations worldwide they may be found in, and how invariably high is their quality of service irrespective of where in the world one visits. Strategies of horizontal, vertical and even diagonal integration among major transnational tour operators allied to the rationalisation of operation have been pursued as a means of developing market share, enhancing price competitiveness, and responding to global market imperatives. Standardisation allows costs to be driven down, an approach that has helped businesses respond to the intense competitive pressures in the challenging global tourism market in recent years. Homogenisation to a high standard also engenders predictability of service provision, and hence can enhance consumer confidence in the quality of experience s/he will have. In this respect, Butler's (1990: 40) observation is particularly apposite: namely, that

> many people seem to enjoy being a mass tourist. They actually like not having to make their own travel arrangements, not having to find accommodation when they arrive at a destination, being able to obtain goods and services without learning a foreign language, being able to stay in reasonable, in some cases considerable comfort, being able to eat reasonably familiar food, and not having to spend vast amounts of money or time to achieve these goals.

In spite of the panoply of alternative forms of tourism and experiences, the outcome of positive agency can be reductionism when, in fact, it is frequently implied that agency results in the search for diversity and individuality.

These two sets of arguments represent two sets of polar opposites. For some commentators, such a banal dialectic is incapable of adequately dealing with the complexities of the real world (Torres 2002: 112). Instead, a synthesis or a spectral approach is more appropriate. As Shaw and

Williams (2002: 242) have noted, even Ritzer (1998: 148) conceded that, in reality, positions of relative homogenisation and heterogeneity exist rather than absolutes. For him, 'both the McDisneyized tourist and the post-tourist exist, but neither gets at the truth of tourism. What exists are concepts that allow us to understand things about tourism that we may not have understood before.' Torres's (2002) reading of tourism development in Cancun is testament to this, suggesting that elements of Fordism and post-Fordism are evident in the production and consumption of tourism space. These are, however, complemented by the obvious presence of an intermediate form. Neo-Fordist production, or mass customisation, is present whereby there is some (albeit limited) flexibility in the mass production of vacations (Torres 2002: 113). Torres's conclusion is that Cancun, which is also a Caribbean destination, is a 'predominantly Fordist mass tourism resort ... [and] within this predominant mass tourism paradigm, there is a growing diversity and flexibility'. Post-Fordist and neo-Fordist elements are present, and such flexibility positions Cancun well to respond to global shifts in tourism and to diversify its product mix. In effect, there is greater product differentiation and more specialisation in an otherwise Fordist mode of production, and new experiences and segments are thus fostered by high volumes of production and exploitation on a massive scale (see also Ritzer and Laska 1997; Shaw and Williams 2002).

In what remains, the production of tourism in 'enclavic space' within the Dominican Republic is explored within this framework. Two questions resonate strongly: first, are the same trends to be observed in other parts of the Caribbean, and particularly in (smaller) island destinations; and second, does the production of tourism have particular policy implications for the Dominican Republic?

Perfecting paradise: tourism development in the Dominican Republic

The Dominican Republic has a population of nearly 8 million people (Fuller 1999), and it comprises the eastern two-thirds of the island formerly known as Hispaniola, which is the second largest in the Caribbean archipelago. It is separated by a north–south border from Haiti in the west.

Tourism has become a major feature in the economy alongside agriculture (sugarcane, coffee, cocoa and tobacco) and trade in manufactured goods in the fifty-two industrial free-trade zones. Cameron (2000: 245) reports that tourism generates receipts in excess of US$2 billion, contributes 20 per cent of GDP and supports 5 per cent of the labour force through 50,000 direct and 110,000 indirect jobs. In 1999 there were 2.1 million arrivals excluding returning Dominicans from the diaspora (Cameron 2000: 246). Europeans and North Americans dominate arrivals. According to the World Tourism Organization (WTO 2001), they comprised 1.064 million and 805,000 arrivals respectively of 2.703 million in total in 1998. In 1997 most visitors arrived from the United States (402,039) followed by Germany (328,860) and the United Kingdom (216,970) (Fuller 1999). The average length of stay was ten nights in 1998, and this translated into average occupancy rates for serviced accommodation of 69.7 per cent (WTO 2001).

The growth of tourism has been a relatively modern phenomenon, with the main impetus behind development in the mid-to-late 1970s (Freitag 1994; Fuller 1999). Rapid increases have been recorded in the number of accommodation units and bed-spaces. Between 1987 and 1999 total bed-spaces alone multiplied over fourfold from 11,400 to 49,623 (Cameron 2000: 245). Accommodation is not evenly distributed across the country (Figure 14.1). Hotel projects have been mainly polarised in the Puerto Plata area (in the north); the Samana peninsula; on the east coast (Punta Cana-Bavaro-El Cortecito); in Santo Domingo, the capital; and the adjacent towns of Boca Chica and Juan Dolio on the south coast (Mawer 2002; Cameron 2000; Freitag 1994).

As Latzel and Reiter (2002) enthuse, it is the nature and quality of the natural environment that it is the principal selling-point of the island: the coastline, the beaches, coves and bays, and the crystal blue waters of the Caribbean (and Atlantic); tropical (rain)forest, scrub and cactus; mountains and white water (see also Mercado and Lassoie 2002). Cater (1996) notes that the diversity and preservation of the environment may offer the Dominican Republic opportunities to develop more ecologically responsible and sustainable modes of consumption (cf. Cameron 2000; Mercado and Lassoie 2002). Notwithstanding this great allure, paradise has been paved to cater for, and to stimulate, the influx of (foreign) visitors. While there are plans to increase the number of bed-spaces by a further 50 per cent, nearly half of current hotel rooms are sold on an all-inclusive basis (Cameron 2000: 245–6). These are mainly in resort hotels, a formula which is widely replicated across the Caribbean (Freitag 1994; Pattullo 1996). Many of the resorts are operated by transnational corporations with strong pan-Caribbean presences. For instance, Occidental Hotels and Resorts group has over eighty hotels in fifty-eight destinations in thirteen countries (Antigua, Aruba, Costa Rica, Dominican Republic, France, Mexico, Morocco, Spain and Tunisia (Occidental Hotels and Resorts 2003)), while Hotetur is part of the My Travel group, for whom this facia provides a brand presence in mainland Spain, the Canary Islands, the Ballearics and the Dominican Republic (Hotetur 2003). Both chains have multiple presences in the Dominican Republic: Occidental-Allegro has ten establishments on the island with six in the Puerto Plata area alone, while Hotetur has a presence in both Santo Domingo and Puerto Plata. Other operators have formed chains limited to, or highly concentrated in, the Dominican Republic. Nevertheless, in light of the evidence, Fuller (1999) continues to maintain that, unlike other states in the Caribbean, the level of domestic ownership of resort hotels is relatively strong. She reports that in 1987

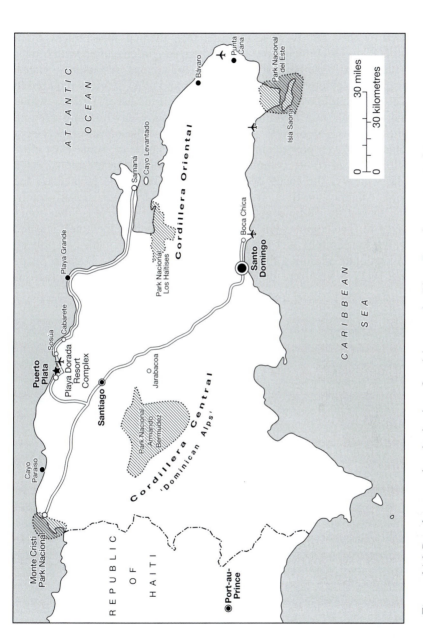

Figure 14.1 Popular excursion destinations for visitors to the Playa Dorada resort complex

only 21 per cent of hotel rooms were foreign-owned, compared to a region-wide average of 63 per cent (Freitag 1994: 540; Pattullo 1996: 20).

There are often difficulties in attempting to delineate and demarcate the term 'resort'. In the case of the Dominican Republic resort hotels are characterised by their relatively large scale; their themed appointment, often in styles (Plate 14.1); their carefully landscaped gardens, often with beach-front access; their plethora of services such as shops, restaurants, stages, pools, sports, nightclubs, casinos, golf clubs and other services required to service the complete vacation; and their 'all-inclusive tariffs', which include all meals and beverages as well as many daytime and evening activities such as water sports and discos. Many of these resort hotels exist as isolated enclaves punctuating the coastline; however, some exist clustered together as resort complexes (Freitag 1994). For example, Playa Dorada, which is the focus of the discussion below, is located on the northern, Atlantic coastline. It contains thirteen archetypal resort hotels, set out around an 18-hole golf course designed by Robert Trent Jones (Table 14.1). The complex has a cellular nature; each of the hotels is gated and policed separately, and the entire complex is gated with two policed entry/exit points adjoining the main highway from Puerto Plata to Sosua (Figure 14.2). Several are owned by transnational organisations such as

Plate 14.1 A resort hotel in the Playa Dorada complex themed in a colonial plantation-type style

Table 14.1 The resort hotels in the Playa Dorada resort complex, Dominican Republic

Name	Operator	Size/Remarks
Ahmsa Paradise Beach Club and Casino	Ahmsa	436 rooms
Caribbean Village Club on the Green	Allegro Resorts	336 rooms in 45 2-storey villas
Gran Ventana Beach Resort	Victoria Hotels	510 rooms
Hotetur Dorado Club	Hotetur	165 rooms
Hotetur Villas Dorada	Hotetur	244 rooms
InterClubs Fun Royale and Tropicale	InterClubs	352 rooms in 2- or 3-storey blocks
Jack Tar Village Puerto Plata Beach Resort and Casino	Allegro Resorts	291 rooms, casino
Occidental Flamenco Beach	Occidental Hoteles	582 rooms and suites
Occidental Playa Dorada	Occidental Hoteles	500 rooms, casino
Playa Naco Golf and Tennis	Naco	418 in 3-storey blocks
Puerto Plata Village	PP Village	380 rooms in cottages
Rumba Heavens	Coral Hotels	192 rooms and suites
Victoria Resort	Victoria Hotels	190 rooms

Source: Adapted from Cameron (2000: 139–41).

Occidental Allegro and Hotetur, while major international tour operators market, and deliver tourists to, several of the hotels within the complex (see Tables 14.2 and 14.3). The hotels vary in their standard of appointment from 3- to 5-star but, as discussed below, they basically offer the same set of experiences (both in-resort and beyond-resort) to visitors because of the savings to be gained by corporate replications of delivery by the accommodation providers and tour operators.

Although these clusters are not perhaps of the magnitude of the mega resorts found more widely in the global 'pleasure periphery' such as parts of south-east Asia, East Africa, Latin America and the Pacific (King 1997; Torres 2002; Shaw and Williams, forthcoming), resort hotels and resort complexes represent a principal component in the state's tourism development strategy. Not surprisingly, they also generate many of the critical issues facing tourism. For instance, the state invested US$76 million in the Puerto Plata region in the period 1974 to 1982 to facilitate tourism (cited in Freitag 1994). At the start of this decade, at a conservative estimate, Playa Dorada alone contained approximately a tenth of all bed-spaces in the island (cf. Cameron 2000). A fall in the popularity of this format of vacation experience or in the particular destination will have a major impact on the state and local economy, as the recent 'food poisoning' episode testifies (Mawer 2002). Similarly, the majority of international visitors inhabit these institutional resort hotels and complexes for the limited time they are in the country. While resorts engineer jobs, wealth and taxation, common complaints have surrounded the level of earnings for local

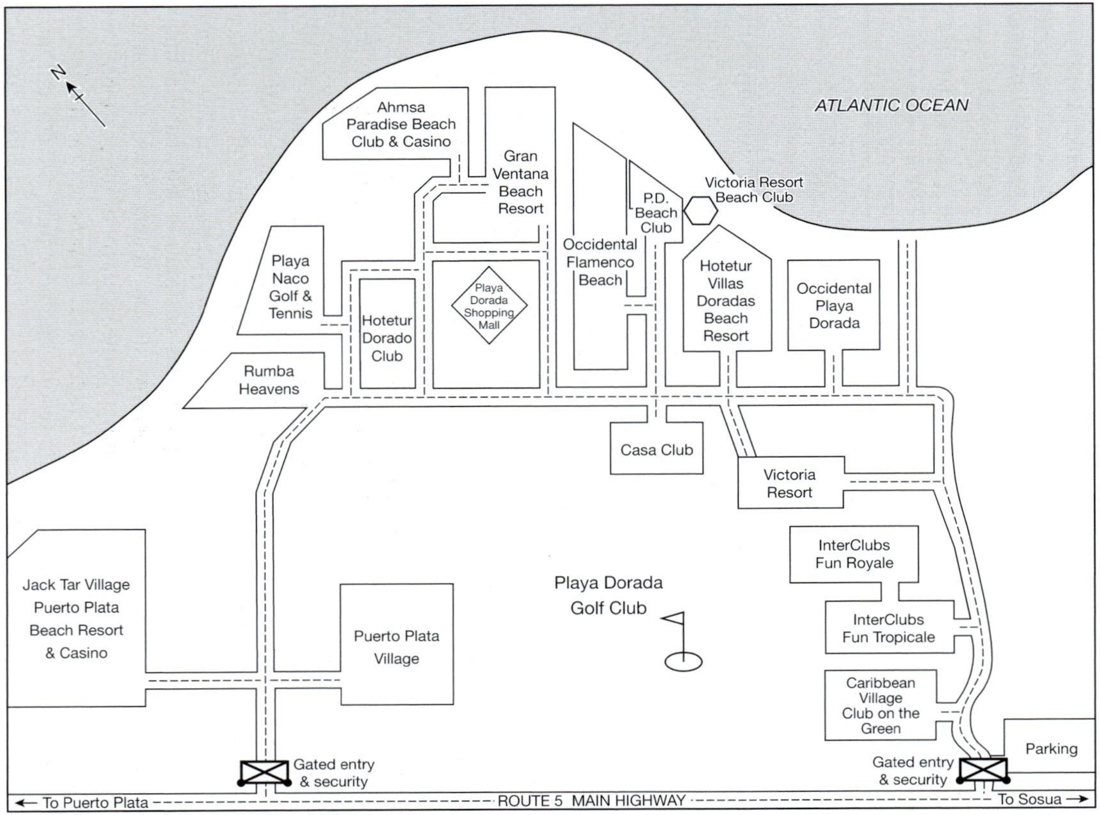

Figure 14.2
The Playa Dorada resort complex at Puerto Plata in the Dominican Republic

people; limited spending and restricted supply from the local economy (Mawer 2002); and the flow of capital offshore. In a study of Luperon to the west of Puerto Plata, Freitag (1994) notes that, rather than encouraging regional economic diversification, enclave tourism has notable limitations and restrictions as a development tool. The creation of resort enclaves restricts social interaction between tourists and local people (Fuller 1999), with other concomitant social costs such as restricted access to the beaches (Fuller 1999), prostitution (Sanchez Taylor 2001), and the spread of HIV/ AIDS (Forsythe *et al.* 1998). Finally, there are environmental concerns, which centre on such issues as waste disposal, water delivery and power supply (not least for local people for whom there are often power cuts). More widely, Anderson (2003) also points to the inherent ephemerality of tourism and tourists, and the long-term financial risk from environmental degradation and the destruction of unique ecosystems.

What emerges is that, in general, tourism consumption in the Dominican Republic is highly disciplined, regimented as it is by the structural framework supplied by the resort hotel format, and mediated by the transnational accommodation providers and tour operators. In the next section, the specific opportunities, experiences and products on offer inside the 'enclavic space' of Playa Dorada will be audited and analysed. This is informed by an inventory compiled from the posters, files and brochures deposited in the resort hotels by local representatives of the transnational tour operators, and from the publicity materials of local tour companies with concessions in the complex. Unfortunately, customer numbers, which would have allowed a more complete picture of consumption to have been painted, were not forthcoming from the service providers.

Beyond the bubble? Decisions, decisions!

In spite of the exhortations of guidebooks, independent travel around the island is difficult for all but the most determined and footloose travellers. Instead, most visitors attend all-inclusive resort hotels. Many take advantage of the opportunities provided by their tour operator/package provider 'on their doorstep' as well as the panoply of in-resort activities and facilities. Shortly after arriving in their resort hotels, tourists are invited to orientation meetings by the local, 'on-site' representatives of their tour operators where they are offered the chance to book activities, excursions and tours beyond the basic price in addition to those ordered and paid for at home. They are regaled with stories, popular myths and marketing hyperbole as a means of driving purchases that will enhance the quality of their holiday experience. Anecdotal evidence from tour operators' representatives suggests that parties tend to book one to two tours each, irrespective of whether they are staying for one week or two. More adventurous (!) visitors seek out the shopping mall in the central plaza where local independent tour operators are keen to impress on them the quality of their tours, their price competitiveness and their qualitative advantages over the 'official' products offered via the transnational tour operators' representatives.

In-resort activities

In their resort hotels, the array of choices offered visitors has been harmonised. Standard prescriptions embraced by the all-inclusive tag include: swimming, tennis, windsurfing, boating, snorkelling, mini-golf, (beach) volleyball, shooting, basketball, water polo, aerobics, sea-kayaking, gym and fitness, table tennis, horse riding and archery. This portfolio is intended to imply a choice of leisure activities to punctuate visitors' time spent around the pool or on the beach. Shooting and archery are restricted to the over-eighteens. Beach volleyball is played by the 18–30s, while basketball is the preserve of teenage guests. Equally, additional fees are payable for some activities such as motorised water sports (jet skis), para-sailing, riding waterborne dirigibles ('banana boats'), scuba diving, golf, billiards and snooker and floodlit tennis. As the travel journalist Howlett (2002) notes, this is not an untypical selection when compared to resorts in other destinations on the island.

Excursions, tours and 'experiences'

Further choices exist in the types of tours visitors can take to develop a wider appreciation of the Dominican Republic. Should they wish to, tourists can, in fact, exercise a high degree of independence. They may extend their search for opportunities beyond those preordained by their tour operator. Car-hire facilities do exist, as do opportunities to hire taxis and drivers by the day. Instead, given the perceived difficulty and inconvenience associated with these options, and the advice regarding the complexity of travel offered in some guide books, following Butler's approach many visitors instead are willing to submit themselves to the pre-produced tours and experiences.

In a similar manner to the operation of tour concessions in Thailand (Shepherd 2003), the external experiences offered in orientation meetings are not operated directly by the transnational tour operators themselves. Rather, representatives of tour operators appear to operate as booking agents for approved local tour companies. Transnational operators enter into commercial collaborations where they believe the experiences on offer by the local providers are of potential interest to their guests, and where the local businesses are able to provide assurances about their health and safety measures and records. Further private, independent (i.e. 'non-aligned') tour companies also function from the central shopping mall in Playa Dorada. Hotel guests are not restricted by their transnational tour operators from buying tours from the independents.

Table 14.2 describes the choice of tours offered to visitors to the Playa Dorada complex in the summer of 2002. Beyond this façade, choice was rather more limited than would appear, and it was rationalised and choreographed by the tour operators and their agents. Wherever an international tour operator had guests throughout the complex, it offered them the same portfolio. As far as this author could ascertain, there was no differentiation in the opportunities made available to guests depending on the standard of their accommodation and, hence, the cost of their holiday. Furthermore, although the tours varied in scope and spatial reach, the same basic formulae were repeated. As Table 14.3 shows, products may be classified into five broad thematic groups depending on the experiences they provided, albeit with fluid boundaries. For instance, in the most extreme hybrid form of excursion, a day's soft adventure in the Dominican Alps, while more obviously including a short jeep safari, optional horse-riding and sightseeing at Jarabacoa Falls (which appeared in *Jurassic Park*), it also includes sightseeing in the country's second city, Santiago de los Cabaillos, a visit to a cigar factory and a shopping stop at a ceramics factory in La Feria. The majority of holidays last a week or a fortnight. With ample time allocated to sitting around the pool and/or sunbathing on the beach, most visitors do not reserve a large number of excursions (see Howlett 2002). Once more, anecdotal evidence from tour operators' representatives suggests that tourists tended to book one of the sun, sea and sand day trips first and foremost (Plate 14.2), which, if required, was then typically supplemented by a soft adventure tour, which allows (limited) exposure to the hinterland.

As Table 14.2 would suggest, there were variations between the transnational tour operators in the types of tours and experiences they were willing to market to their guests. Nevertheless, a common core of products was made available by all the majority of operators with some more diverse ones offered by just a few of the operators. Independents delivered the same tours as the transnational tour operators with additional opportunities for tourists to tailor-make their experiences. Mimicry and price drove the rationalisation of tours and excursions. In the case of the transnational tour operators, their product portfolios were on public show not just to their guests, but also to the guests of their global counterparts. As such, there was pressure on each transnational operator to provide the same types of excursion and levels of appointment in order to underscore the quality of their vacations. Equally, this logic also drove price convergence among the transnational tour operators (Table 14.4). For independents, mimicry offers the opportunity to undercut the prices of the tour operators. In interviews these operators claimed they provided a schedule to mirror the international tour operators' products. Some even argued that their intimate knowledge of the island added value to their tours over those of the tour operators. They claimed they were able to charge a lower price (for 'identical' experiences) because their mark-up was lower, and there

Table 14.2 Excursions and visits offered by major tour operators

Company to offer (principal market)

	My Travel (UK)	JMC (UK)	FTI Touristik (Germany)	I*2 Fly (Germany)	Jahn Reisen Germany (Germany)	Weltweit Reisen (Germany)	Neckermann (Germany)	Sunquest/Alba Tours (Canada)	Vacances ATH (France)	Evenements Reizen (Neths)	Holland Internat. (Belgium, Neths)	Sunjet (Belgium)	Helvetic Tours (Switzerland)	Berge and Meer (Germany)	Mary Vic Tours (Local operator)	C.B. Tours (Local operator)
Saona Island	X	X	X	X	X	X	X	X		X	X		X		X	X
Samana/Cayo Levantao/Los Haïtses	X	X	*	X	X	*	X	X	X	*	X		X	X	X	X
Cayo Paraiso	X	X	X	X	X	X	X	X	X	X	X			X	X	X
Playa Grande																
Catamaran	X	X	X	X	X	X	X		X	X	X	X	X	X	X	X
Dominican Alps	X	X	X	X	X	X	X	X	X	X	X	X	X	X	X	X
River rafting	X	X	X	X	X	X	X	X	X	X	X	X	X	X	X	X
Horse riding	X	X	X	X	X	X	X	X	X	X	X	X	X	X	X	X
Mountain biking	X							X							X	
Quad biking	X	X	X	X	X	X	X	X	X	X	X	X	X	X	X	X
Jeep safari	X	X	X	X	X	X	X	X	X	X	X	X		X	X	X
Truck safari	X	X			X	X		X						X		
Santo Domingo	X	X	*	X	X	*	X	X		X	X	X		X	X	X
Puerto Plata	X	X	X	X				X	X			X	X			
Puerto Plata/Sosua	X	X	X	X			X	X	X			X	X		X	X
Haiti	X		X	*	X	*			X	X	X	X		X	X	X

Source: Author's inventory of tour operators' in-resort marketing materials.

Note
All are day trips unless marked '*' where 2-day (1-overnight stay) options were also offered.

Table 14.3 Broad typology of experience/commodity in tours offered to visitors at Playa Dorada

Sun, sea and sand	Saona Island, Samana, Cayo Paraiso, Los Haïtses, Playa Grande, Catamaran trips
Culture and heritage	Santo Domingo, Puerto Plata, Sosua, Haiti
Active recreation	River rafting, Mountain biking, Horseback riding, Trekking
Soft adventure	Jeep safari, Truck safari, Quad biking
Mixed (soft adventure/ cultural heritage)	Dominican Alps

Plate 14.2 Here today, gone tomorrow? Tourists from the resort hotels of the North Coast stake their claim to a space on the 30m-long sand bar Cayo Paraiso in the North Atlantic

was no additional agency commission to be extracted by the international tour operator from their local collaborator. Conversely, the international tour operators implied the lower prices outside reflected poorer quality, and they questioned the safety of independents, not least as to whether insurance is included in their products and prices.

Indeed, risk appears to be an important determinant in the product mix presented to the tourist. British guests were not offered the opportunity to

Table 14.4 The cost of excursions

Company to offer (principal market)	My Travel (UK)	JMC (UK)	FTI Touristik (Germany)	I*2 Fly (Germany)	Weltweit Reisen (Germany)	Neckermann (Germany)	Sunquest/Alba Tours (Canada)	Evenements Reizen (Neths)	Berge and Meer (Germany)	Mary Vic Tours (Local operator)	C.B. Tours (Local operator)
Saona Island	215	216		219	235	216	215	199		180	210
Samana/Cayo Levantao/Los Haïtses	59		55/133	84	46/138	80	59	–	59	39	38
Cayo Paraiso	79	75	79	79	69	75	79	–	75	60	59
Playa Grande					32			–		31	30
Catamaran	69	65	70	72	73		69	–	69	60	59
Dominican Alps	59		53	59	49	55	59	–	49	39	38
River rafting	99	90	100	92	73	90	99	–	90	77	72
Horse riding	65	65	55	60	35	–	65	–	55	40	45
Mountain biking	45						45				
Quad biking	69	69	–	79	39	–	69	–	75	30	17
Jeep safari	75	78	60	69	44	69	75	60	60	65	35
Truck safari					75				75		45
Santo Domingo	49	58	53/–	59	49/153	58	49	–	55	45	43
Puerto Plata	22	30	25	37		30	22				
Puerto Plata/Sosua	35	38					35				30
Haiti			78	78/230	–/259			69	69	65	69

Source: Author's inventory of tour operators' in-resort marketing materials.

Notes

All prices are in US$; all for day trips unless marked with '/' and with 1- and 2-day prices given.

visit Haiti by the international tour operators unlike their German and Dutch counterparts. One reason for this may pertain to Haiti's status as a country to which the Foreign and Commonwealth Office (FCO 2003) advised (and continues to advocate) against travel on safety grounds. Should visitors have exercised their prerogative to travel, they may well have invalidated their travel insurance as policies often exclude coverage where travel is to countries not considered 'safe' by the FCO. The concept of safety as a structural constraint in the provision and consumption of opportunities was at the heart of a debate recently reported by Mawer (2002). One American owner of a bar and grill, who stood to benefit should tourists be prised out of the complex, complained that 'the reps in Playa Dorada imply it's not safe to come into the town, even though the island is just about the safest in the Caribbean'. Conversely, a local manager for Thomson retorted, 'Tourists stay within their hotels to eat and drink partly because they've already paid for the cocktails and meals, partly because there's not much to do outside. As a result, there is little incentive for locals to set up appealing bars and restaurants.'

Finally, choice was disciplined by the carrying capacity of the excursion sites (which may disappear at high tide! (see Plate 14.2)), and by the infrastructure required to service excursions. Travel times are not rapid along the main highways. For some of the more far-flung destinations such as Saona and Samana, alternative modes of transport such as air and speedboats have to be used. As Table 14.5 reveals, excursion operators of all types were involved in an intricate time–space budgeting exercise. In addition to limiting the number of times they repeated an excursion per week for profitability reasons, there is evidence that they have tried to engineer locational avoidance in the weekly rhythm of consumption. For instance, Air Tours visitors were offered Saona on Tuesdays and Fridays, while a local independent, Mary Vic Tours, took guests there on Wednesdays and Sundays. Similarly, Air Tours visitors could visit Cayo Paraiso on Mondays, Thursdays and Saturdays; Mary Vic's on Tuesdays and Thursdays; FTI Touristik's on Saturdays; and Berge and Meer's also on Saturdays. Samana and the Dominican Alps were similarly rationed as excursion sites.

Conclusion

In 2002, the *Observer* in the United Kingdom carried advertisements from the state tourism office of the Dominican Republic. Potential visitors were invited to experience 'A Land of Sensation'. A principal image of a deserted, pristine white sandy beach with palm trees was deployed with secondary views of a group white-water rafting (in the Dominican Alps), golfers against a backdrop of the ocean, and Christopher Columbus's house in the capital, Santo Domingo.

The inference from the advert is that, besides taking it easy on the beach under the hot Caribbean sun, the variety of activities offers holidaymakers

Table 14.5 A time–space mapping of excursion opportunities for tourists with selected tour operators

	Coco Tours	Mary Vic Tours	Berge and Meer	FTI Touristik

Monday	Samana
	Cayo Paraiso
	Puerto Plata
	Truck safari
		Samana
		Haiti
			Dominican
			  Alps
				Dominican
				  Alps
				Jeep safari
				Rafting

Tuesday	Saona
	Dominican
	  Alps
	Puerto
	Plata/Sosua
	River rafting
	Jeep safari
	Horse riding
		Cayo Paraiso
		Santiago/Jarabacoa
			Santo
			  Domingo
			Haiti
				Haiti
				Samona
				  Island
				Jeep safari
				Riding
				River rafting

Wednesday	Puerto Plata Sosua Jeep safari Truck safari	Samana Playa Grande Saona	Catamaran Jeep safari Riding	Puerto Rico Santo Domingo Jeep safari Riding
Thursday	Samana Cayo Paraiso Santo Domingo Jeep safari Puerto Plata/Sosua Horse riding	Cayo Paraiso Santo Domingo	Jeep safari Truck safari	Jeep safari River rafting
Friday	Saona Dominican Alps Sosua Jeep safari Truck safari	Santiago/Jarabacoa	Haiti Jeep safari	Haiti Jeep safari Riding
Saturday	Cayo Paraiso River rafting Truck safari	Samana	Cayo Paraiso Catamaran Jeep safari Riding	Samana/Los Haïtses Cayo Paraiso Jeep safari
Sunday	Jeep safari	Playa Grande Saona	Samana	–
Daily	Catamaran	Catamaran Jeep safari River rafting Horse riding Sea fishing Scuba diving	River rafting Quad bikes	Catamaran Quad bikes

Source: Author's inventory of tour operators' in-resort marketing materials.

a full and rich vacation experience. Beyond the three Ss, the Dominican Republic has much to offer, not least in terms of cultural (heritage) tourism and sports tourism. This is underscored by a supplementary commentary (appearing as a glossy advertisement in a Sunday UK broadsheet), which, one is left to assume, is a tourist's:

> Here, I feel free. Here, there's much more than the Caribbean Sun. Much more than a thousand golden beaches, much more than the rhythm of the Merengue. Here, you will get the feel of life and peace, feel the movement of history, the joy of our people, and the sounds of nature. Here, an entire country awaits you. A land of sensations.

Strictly speaking, there is much more to the Dominican Republic than the Caribbean sun and the golden beaches. However, there is perhaps not the freedom of activity that this marketing copy would suggest. Choices do exist but they are regimented within institutionalised networks of supply and spatial consumption patterns. More accurate is Cameron's (2000: 19–20) pronouncement that 'all-inclusives [are] designed to offer everything you could need for a beach holiday, with maybe a day trip or two to break the monotony'.

In terms of the wider implications of this study, there has been a willingness among commentators to marginalise Ritzer's ideas about the mass production of holiday experiences. While it is clear that holidays are not mass produced and mass marketed in the strictest, most inflexible Fordist sense, it is equally evident that the free and unconstrained choice of holiday and activities advocated by critics is similarly mythical in this part of the Caribbean. Instead, a third reading is required based on the precepts of structure and agency. The contemporary automobile industry perhaps offers a model for conceptualisation. As the pioneer of the production line, Henry Ford is supposed to have commented that 'any customer can have a car painted any colour that he [sic] wants so long as it is black'. Automobile production has progressed from such rigidity to a position of mass customisation; the purchaser is given the opportunity to choose engine size and type, body colour, accessories and other options. From the standard shell and fittings, the customer tailors the product to meet his or her needs; the producer then delivers the commodity. The process is not uncommon and clearly represents a more appropriate metaphor for the construction of holiday experiences for the denizens of Caribbean resort complexes. At home, the potential visitor is offered the basic model – a week or two weeks in the sun. From this starting point, options to embellish the product are first offered. This process starts with the type of board (half-board, full-board or all-inclusive), potential flight and transfer upgrades, and the opportunity to pre-book excursions. The final product is not completed until the visitor is in the resort. The optional extras included in the price and situated within the resort may be selected, while further opportunities exist to book excursions, tours and experiences through representatives and even independently through local suppliers.

Tourism in the resorts of the Dominican Republic is marked by an interesting duality. Production is dominated by an essentially neo-Fordist mode on the spectrum of production, and consumers are offered the choices one would recognise from mass customisation. Beyond the resort enclave, although the excursions are consumed in great numbers, this high volume is a function of low per capita uptake rates by tourists. Nevertheless, practically every tourist participates in at least one tour so that they can return home and proclaim, in effect, that 'they've seen the island' (Howlett 2002; Mawer 2002). Among tourists, there is a tendency to act in a relatively more Fordist manner, to restrict themselves in spite of the (limited) opportunities offered them. Unlike Cancun, the Dominican Republic has yet to develop more fully or more systematically post-Fordist alternatives. This duality has interesting policy implications for the state. On one level, best value is not being achieved for the state or local residents from a sector of the economy that has received considerable public funding. In order to secure greater benefits the state should explore ways of working with transnational operators to unlock the potential associated with encouraging foreign visitors to take even just one or two extra excursions. On another level, it has yet to develop more fully post-Fordist options, and hence the ability of the local tourism sector to respond to changing global imperatives must be questioned. As the negative publicity over food poisoning demonstrates, and as the recent crises elsewhere in the world have underscored, tourists are fickle and ephemeral, and they can switch their spatial allegiances quickly. Resorts date and have to be reinvented; transnational accommodation providers exit destinations if commercial realities require it.

The Dominican Republic is currently in vogue, but no destination can afford to be so complacent as to consider its success to be assured over the long term. Where in the past the state has recognised the possibilities afforded by 'mass' tourism and duly invested in it heavily, now is the time for it to consider supporting alternatives, in particular post-Fordist initiatives to diversify the destination product mix, and thereby to provide a degree of indemnification against the vagaries of the global marketplace. For Caribbean states like the Dominican Republic, if the economic advantages of tourism are to be secured in the long term, they must protect themselves against the vulnerability of reliance on a single mode of consumption. Auditing based on the Fordist spectrum may form an important first step in the appraisal process.

References

Anderson, J. (2003) 'Can tourism and conservation co-exist in the Caribbean?' Press Release from Conservation International, 24 April 2003. Online: <http://www.conservation.org/xp/news/press-releases/042303.xml> (accessed 1 May 2003).

Barber, S. (1995) *Jihad vs. McWorld. Terrorism's Challenge to Democracy*, New York: Ballantine Books.

Butler, R. (1990) 'Alternative tourism: pious hope or Trojan horse?', *Journal of Travel Research* 28(3): 40–5.

Cameron, S. (2000) *Footprint Handbook: Dominican Republic, First Edition*, Bath: Footprint Books.

Cater, E. (1996) 'Ecotourism in the Caribbean: a sustainable option for Belize and the Dominican Republic', in L. Briguglio, R. Butler, D. Harrison, and W.L. Filho (eds) *Sustainable Tourism in Islands and Small States: Case Studies*, New York: Pinter Press.

Chang, T.C. (1999) 'Local uniqueness in the global village: heritage tourism in Singapore', *Professional Geographer* 51(1): 91–103.

Chang, T.C., Milne, S. and Pohlmann, C. (1996) 'Urban heritage tourism: the global–local nexus', *Annals of Tourism Research* 23: 284–305.

Coles, T.E. (2003) 'A local reading of a global disaster: some lessons on tourism management from an annus horribilis in South West England', *Journal of Travel and Tourism Marketing* (forthcoming).

Edensor, T. (2001) 'Performing tourism, staging tourism: (re-)producing tourist space and practice', *Tourist Studies* 1(1): 59–82.

Foreign and Commonwealth Office (FCO) (2003) Travel – Country Advice: Haiti (updated 26 March 2003). Online: <http://www.fco.gov.uk> (accessed 1 May 2003).

Forsythe, S., Hasburn, J. and Butler de Lister, M. (1998) 'Protecting paradise: tourism and AIDS in the Dominican Republic', *Health Policy and Planning* 13(3): 277–86.

Freitag, T. (1994) 'Enclave tourism development: for whom the benefits roll?', *Annals of Tourism Research* 21(3): 538–54.

Fuller, A. (1999) 'Tourism development in the Dominican Republic: growth, costs, benefits and choices.' Unpublished Manuscript. Online: <http://www.kiskeya-alternative.org/publica/afuller/rd-tourism.html> (accessed 1 May 2003).

Goodrich, J.N. (2002) 'September 11, 2001 attack on America: a record of the immediate impacts and reactions in the USA travel and tourism industry', *Tourism Management* 23(6): 573–80.

Green, M. (2001) 'Urban heritage tourism: globalization and localization', in G. Richards (ed.) *Cultural Attractions and European Tourism*, Wallingford: CAB International.

Hotetur (2003) 'Hotetur: our history.' Online: <http://www.hotetur.com/> (accessed 1 May 2003).

Howlett, P. (2002) 'Joining the club?', *Guardian*, 6 April 2002. Online: <http://www.guardianunlimited.co.uk> (accessed 1 May 2003).

Kermath, B.K. and Thomas, R.N. (1992) 'Spatial dynamics of resorts. Sosua, Dominican Republic', *Annals of Tourism Research* 19(2): 173–90.

King, B. (1997) *Creating Island Resorts*, London: Routledge.

Latzel, M. and Reiter, J. (2002) *Insight Compact Guide: Dominican Republic*, 2nd edn, Singapore: APA Publications GmbH.

Lawrence, F. (2003) 'Travel companies slash prices in response to collapse in bookings', *Guardian*, 29 March 2003. Online: <http://www.guardian.co.uk/news/story/0,7445,925054,00.html> (accessed 1 May 2003).

Lee, M. (2003) 'One size fits all in McFashion', *Observer*, 4 May (Review): 1–3.

Mawer, F. (2002) 'Much easier to stomach', *The Daily Telegraph*, 14 January 2002. Online: <http://www.dailytelegraph.co.uk> (accessed 1 May 2003).

Meethan, K. (2001) *Tourism in Global Society: Place, Culture, Consumption*, Basingstoke: Palgrave.

Mercado, L. and Lassoie, J.P. (2002) 'Assessing tourists' preferences for recreational and environmental management programs central to the sustainable development of a tourism area in the Dominican Republic', *Environment, Development and Sustainability* 4(3): 253–78.

Mills, S. (2002) 'Tourism leaders meet in Paris to develop future for industry', Press Release circulated, World Travel and Tourism Council, 10 May.

Milne, S. and Ateljevic, I. (2001) 'Tourism, economic development and the global–local nexus: theory embracing complexity', *Tourism Geographies* 3(4): 369–94.

Observer, (2003) 'Flight prices hit rock bottom', *Observer*, 20 April (Escape): 4.

Occidental Hotels and Resorts (OH&R) (2003) 'Occidental Hotels & Resorts: who are we?' Online: <http://www.occidental-hoteles.com/home.html> (accessed 1 May 2003).

Pattullo, P. (1996) *Last Resorts: The Cost of Tourism in the Caribbean*, London: Cassell.

Ritzer, G. (1998) *The McDonalidization Thesis*, London: Sage.

—— (2000) *The McDonaldization of Society: New Century Edition*, Boston: Pine Forge Press.

Ritzer, G. and Laska, A. (1997) '"McDisneyization" and "Post-tourism": complementary perspectives on contemporary tourism', in C. Rojek and J. Urry (eds) *Touring Cultures: Transformations of Travel and Theory*, London: Routledge.

Sanchez Taylor, J. (2001) 'Dollars are a girl's best friend? Female tourists' sexual behaviour in the Caribbean', *Sociology* 35(3): 749–64.

Shaw, G. and Williams, A.M. (2002) *Critical Issues in Tourism*, 2nd edn, Oxford: Blackwell.

—— (forthcoming) *Tourism, Tourists and Tourism Spaces*, London: Sage.

Shepherd, N. (2003) 'How ecotourism can go wrong: the cases of SeaCanoe and Siam Safari, Thailand', in M. Lück and T. Kirstges (eds) *Global Ecotourism Policies and Case Studies: Perspectives and Constraints*, Clevedon: Channel View Publications.

Torres, R. (2002) 'Cancun's tourism development from a Fordist specturm of analysis', *Tourist Studies* 2(1): 87–116.

Travel Business Partnership (TBP) (2002) 'Comment: measuring the real impact of September 11', *Travel Market Monitor* 1 (April): 3–4.

Urry, J. (1990) *The Tourist Gaze: Leisure and Travel in Contemporary Societies*, London: Sage.

Vasagar, J. (2002) 'September 11 keeps Britons away from US', *Guardian*, 22 October. Online: <http://travel.guardian.co.uk/news/story/0,7445,816806,00.html> (accessed 1 May 2003).

World Tourism Organization (WTO) (2001) *Compendium of Tourism Statistics, 2001 Edition*, 21st edn, Madrid: WTO.

15 Hucksters and homemakers

Gender responses to opportunities in the tourism market in Carriacou, Grenada

Beth Mills

Introduction

In recent years governments of the small island states of the Caribbean have sought ways to substitute local agricultural products for imports and diversify domestic agricultural production, thus improving their overall balance of trade and strengthening local economies. Primarily as a result of global neoliberal economic policy, many island states have, since the early 1990s, used tourism and the niche marketing of agricultural products in order to compete in a somewhat unregulated international marketplace.

Tourism is obviously critical to the economies of the region (see McElroy, Chapter 3). Between 1996 and 2000 estimates of visitor spending in the Caribbean Community (CARICOM) nations alone rose from US$4,302.4 million to US$5,201.1 million (Caribbean Tourism Organization 2002). Concomitantly, government officials and analysts in the region have recognised the importance of linkages between tourism and agriculture (Brown 1974; Belisle 1980; Gomes 1993; Momsen 1986, 1998; see also Conway, Chapter 11). Local production can, in some instances, make an important contribution to the balance of payments in tourist-dependent economies.

With the establishment of the World Trade Organization as a regulatory institution for global trade policies, the special trade arrangements that the former colonies in the eastern Caribbean shared with individual European nations are to be eliminated or are in serious jeopardy. The general weakening of the state that many observers see accompanying the dominance of global capital, especially in the form of powerful transnational corporations and 'footloose' industries, implies that small governments have fewer resources to invest in grassroots economic development projects.

The growth of international tourism as an industry is part of the process of globalisation. Improved technologies, communication and transportation contribute to the international movement of people and ideas, including individuals travelling for entertainment and relaxation. It is logical to look for possible links between agriculture and tourism in the Caribbean in the hope of finding new paths for economic development given the current global economic climate. Upon closer investigation, however, the specifics of the process of creating linkages between the two sectors at a community level can be surprising.

The purpose of this chapter is to highlight how the complex social structures of a place creates impediments for linking agriculture with tourism, especially in the context of differences in gender responses to market opportunities. To illustrate this, the small island of Carriacou in the eastern Caribbean is discussed. The results reinforce the importance of looking deeply into the social structure and the nature of individuals' interactions before anticipating any given outcome from otherwise rational economic initiatives. It is clear from this example that institutions and agents in a specific locale will incorporate, reject and transform practices in response to ongoing globalisation processes.

The data from this chapter stem from a series of interviews conducted by the author in July and August of 1998 on Carriacou. The managers of all the guesthouses and hotels on the island were interviewed and a formal questionnaire was administered. More informal interviews were conducted with visitors who arrived by yacht as well as produce vendors who had stalls in the primary market area of the main town (Hillsborough). The fieldwork coincided with both the annual 'Regatta' and the peak tourist season, when the tiny island hosts many North American and European visitors. Data collection also coincided with the height of the growing season because gardens are typically planted with the first of the seasonal rains that begin in late May or early June.

Gender and tourism

As tourism has grown to become a critical sector of the global economy, so too has the recognition that gender differences are important in relation to work within a tourist economy (Kinnaird and Hall 1994; Sinclair 1997; Swain and Momsen 2002). A critical question that has been posed for gender research is, therefore, as follows: 'how does tourism articulate gender relations and how do gender relations inform and articulate the form of different tourism processes?' (Kinnaird and Hall 1994: 7). The case-study of the emerging tourist economy in Carriacou presented below is designed to explore this question.

Of particular interest is the work that has been done centring on gender and rural and farm tourism. The role of women in creating opportunities in these relatively new niches of the tourist economy is a subject that has begun to be explored. It is generally acknowledged that rural tourism will provide a new source of income and new jobs to areas where there are currently few employment opportunities in both the developed and the developing world (see Butler *et al.* 1998). With specific regard to rural women, these jobs tend to be in traditional roles, albeit expanded, rather than innovative in nature. For example, a woman may rent a room in her home to visitors and provide them with meals. In this case, the additional income generated from tourism results from an expansion of a woman's traditional role as housekeeper and cook within the existing family struc- ture and household workload. Work in the tourism sector is performed alongside many other tasks in the home. In such cases the income earned is rarely a means to economic independence for a woman but rather contributes to the total household budget.

To some extent, it could be suggested that participating in the tourist economy in rural or otherwise isolated areas gives women the opportunity to meet new people, have new experiences, and generally expand their horizons. This aspect of the interaction between tourist and host, however, can be uncomfortable or problematic, depending on the cultural constraints in the host society. The expansion of women's roles may be interpreted by some as trespassing on traditional cultures where strict rules govern the nature of interaction between women and men outside the household.

Momsen (1994) has outlined some of the connections and questions generated at the intersection of tourism, gender and development for the Caribbean. In general, women's roles in the region have been limited to filling the jobs for cook, maid or guesthouse keeper. There have been rela- tively few opportunities for women to enter the arena as entrepreneurs. For many parts of the rural Caribbean, particularly on the smaller islands where tourism is relatively recent or limited in scope, women are not in a socio-cultural position to exploit the opportunities of the (new) tourism.

Potential links for tourism and agriculture in the region

Regionally, local food production has, in some instances, made an important contribution to the balance of payments in tourist-dependent economies (Momsen 1998). Although much of the meat and alcohol served in hotels is imported, locally grown fruits, vegetables and herbs can more easily be substituted for imported items. As Momsen (1998) notes, however, the quality of the produce is the key to production for the tourist market (see also Conway, Chapter 11). Maintaining high-quality standards can often be difficult for small farmers because of inadequate packing, packaging, refrigeration and transport facilities.

With the growing emphasis on heritage and ecotourism in the Caribbean (see Found, Chapter 8 and Weaver, Chapter 10, respectively), there should be increasing opportunities for niche marketing of local produce that is grown without pesticides and in a way that encourages sustainability for the island environment. There is anecdotal evidence that socially sensitive and environmentally educated and interested tourists may be on the increase. These tourists are often more concerned with the politics of food production than the appearance of the food itself. However, this group is still quite small compared to the number of people travelling to large hotel destinations with high standards and expectations of sophisticated holiday dining. In addition, international tourism can be equated with periods of temporary increased consumption for travellers who are looking for a little extra of everything on their vacation.

Regardless, there has, in fact, been a trend in the Caribbean for small, locally owned, hotels and guesthouses to either raise their own produce or purchase it locally (Momsen 1998). Some progress has even been made in promoting direct arrangements between farmers and the large hotels, specifically on St Lucia. It has been reported that almost 80 per cent of hotels' vegetable needs on Grenada are being supplied locally (*Caribbean Basin Profile* 1997). Although the gender impacts of these changes in the market tend to be multifaceted (Momsen 1986), there is still a pattern whereby male farmers with large acreages are benefiting from the new linkages to the exclusion of women with smaller farms.

The growth of tourism on Carriacou

Carriacou is the largest and southernmost of the Grenadine islands, located between St Vincent and Grenada in the eastern Caribbean (Figure 15.1). Carriacou, and the nearby island of Petit Martinique, are administered by Grenada. Tourism in Carriacou was limited to a handful of visitors per year until the late 1980s. Until then, only two or three small hotels catered almost exclusively to regional business travellers. Since 1990, however, there has been a dramatic increase in tourism in the Grenadines, especially in Carriacou.

Much of the growth in tourism is attributed to the popularity of this area as sailing waters for both private yachts and 'bareboat charters'. Although exclusive, high-end accommodations for the wealthy have been available for some time on other Grenadine islands such as Mustique and Petit St Vincent, there has been an increase in more reasonably priced guest accommodations on Bequia, Union and Carriacou. As well, Internet advertising for small guesthouses and cottages has successfully penetrated a much wider market, an exercise that was less feasible in previous years.

Statistics surrounding tourism on Carriacou are difficult to ascertain as the numbers are generally aggregated up to the national level in St George's, Grenada. We do know that visitors to Carriacou increased from 1,590 people annually in 1988 to 7,238 people annually in 1997 (Table 15.1). Although the total number of annual visitors is small, the significance of the impact of 7,000 or 8,000 visitors on the island is best realised when one takes into account that the total population of Carriacou is approximately 5,000 people.

Tourist visits to Carriacou are concentrated in the 'high season' between Christmas and February and during the month of August. The high season coincides with cold weather in North America and Europe as well as with two important cultural festivals in Carriacou: 'Parang', which occurs around Christmas, and 'Carnival', which is celebrated in February. Tourism in the month of August generally centres around 'Regatta', which features the annual boat races that attract family members living abroad and people

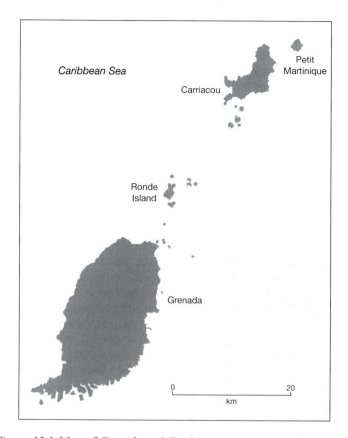

Figure 15.1 Map of Grenada and Carriacou

from nearby islands. Visitors to Regatta often combine attendance at that event with attendance at Carnival in Grenada, also celebrated in August to accommodate the transnational community. They return to celebrate these festivals and for marriages and other family-related gatherings.

Many visitors to Carriacou travel to the island for cultural events. In fact, and as a consequence, it can even be said that tourism is an extension of the transnational nature of Carriacou society. Like many small island economies, but perhaps even to a greater degree, Carriacouans rely on remittances from community members living and working abroad as well as the spending of returning migrants. The complex ties and system of reciprocity that has developed as part of the transnational network results in the presence of a lot of visitors during festival times (Mills 2002). Only the most preliminary marketing strategies for 'culture' or 'heritage' tourism are in place at present. It is easy to see how the annual community calendar might be exploited and commodified by agents who are encouraging the

Table 15.1 Visitors to Carriacou (1988–1997)

	1988	1989	1990	1991	1992	1993	1994	1995	1996	1997
Number of cruise-ship calls	234	218	343	401	431	382	420	446	393	323
Number of yacht calls	2,842	3,035	4,975	7,318	3,373	4,267	5,413	5,314	5,355	5,292
Hotel occupancy	42.41	43.3	49.32	61.01	58.03	67.33	70.13	67.17	61.11	62.1
Rooms available	1,019	1,085	1,105	1,118	1,114	1,428	1,428	1,652	1,669	1,775
Number of visitors	1,590	1,333	3,713	6,516	5,343	8,639	7,304	7,310	7,266	7,238

Source: Adapted from Grenada Board of Tourism *Annual Statistical Report* (1997).

development of tourism on Carriacou. However, Carriacou society has not yet been given over completely to what has been described in the tourism literature as 'staged authenticity' (MacCannell 1976) and traditional celebrations remain important to extended-family networks.

The nascent tourism industry in Carriacou is at a critical juncture in its relationship to the community and the economy. The concerns are both environmental and social. The increasing visits by private yachts and cruise ships have implications for both the reef and mangrove ecosystems in nearby waters. If boats are anchored in a careless manner a coral reef can be seriously diminished over time. Although the number of visitors arriving by boat has increased dramatically since the 1980s, there is as yet no regulation of anchorage by the Grenadian government or Carriacou officials. Similarly, if enough boats flush out their bilge and wastewater in Carriacou bays, mangrove ecosystems will continue to be negatively impacted. While tourism grows at a steady rate, there has been no response in the form of regulation from the Grenadian government regarding environmental impacts. Illegal dumping of solid waste from off island was contaminating government agricultural land in one location in 1998, but at that time the action went unchallenged by government. These circumstances are a good example of the problems inherent in the weakening of the state when global processes intersect local environment and economy. To some extent, this pattern can be observed throughout the less-developed (post-colonial) world.

A parallel situation exists with regard to cultural resources. Sand for a construction boom is being mined from the isolated beaches on the windward coast of the island, thus altering the immediate coastal environment and possibly destroying cultural artifacts. This activity goes unregulated. Archaeological resources are being lost or squandered when construction uncovers them because there is no institutional framework requiring mitigation, as evidenced in two cases involving construction of guesthouses in the early 1990s.

The riddle in the Carriacou example is, sadly, a familiar one in a period of growing international tourism. Formerly isolated communities are suddenly confronting the problems of coping with new economic, social and environmental challenges that tip whatever equilibrium may be operating in the community. If the state is weak, or poor, as is the case of the Grenadian government, the impacts go unanticipated and unregulated.

For Carriacou, it is yet unclear how far the effects of increasing tourism will be allowed to progress without the interference of the community as a whole. The question remains, however: what types of social regulations, if any, might develop in the absence of government regulations?

Tourism and agriculture on Carriacou

Almost every household on Carriacou has, at a minimum, a small garden that is traditionally planted in corn, pigeon peas and provision crops. The prominence of these gardens on the landscape is something that visitors to the island notice, in part because of the contrast with other islands where agricultural activity is often invisible to tourists. Despite the number of gardens, however, virtually all of the produce marketed on the island is grown in Grenada. Environmental constraints are, to some degree, responsible for this situation. Carriacou has no surface streams and the dry season that lasts several months has been known to extend into drought. The highest elevation on the island is just under 1,000 feet and this fact inhibits the land from benefiting from the orographic conditions that bring consistent rain to the higher elevations on larger islands such as Grenada and St Vincent. Consequently, year-round production on Carriacou must rely on irrigation.

Residents and visitors obtain produce from several sources:

1   fruits and vegetables grown in family gardens, usually quite close to the 'housespot' or in 'kitchen gardens';
2   produce purchased from the Government Marketing Board, which arranges to bring fresh fruit and vegetables from Grenada by boat twice a week and sets prices;
3   produce purchased from vendors or traders who set up stalls in the market square in the main town of Hillsborough, or from roadside vendors in some of the smaller villages; and
4   produce purchased by special arrangements with individual farmers on Carriacou.

The degree to which an individual is integrated into the Carriacou community or remains an 'outsider', as is the case of most tourists, determines their supply for fresh produce. For example, a housewife who has her own garden and a complex network of exchange relationships with other women in the community will purchase little of the produce for her household. Visitors and other outsiders must pay cash for produce and tend to buy from vendors first, largely because they are the most visible suppliers.

The largest single visitor demand for produce is from people arriving by yacht. However, informal interviews indicate that most people who stay on their private yacht have stocked up on nonperishable items on the larger islands before arriving at Carriacou. If they are staying for a short period of time they may buy from vendors in the main market. If they stay for an extended period they will buy produce from the Government Marketing Board.

There are two 'windjammers', or sailing cruise boats, that call weekly at Carriacou. In addition, there are two large motor yachts from Martinique and Guadeloupe that call regularly. All passengers come ashore in the main town of Hillsborough, where the main market is located, for at least part of the day. It was clear, however, that the vendors were generally unaware of the cruise boats' schedules. Their inventory is usually not, consequently, adjusted to the arrival of the boats.

Several hotel and guesthouse owners or managers were interviewed in order to determine how they were meeting their need for produce. Most buyers for guest accommodations on the island purchase primarily from the marketing board. (Only one individual insisted on special arrangements with individual vendors in the market square.) Vendors in the market square, individual purchase arrangements with local farmers, and retail 'shops' were secondary and tertiary sources. In two instances, the guesthouse owner was also a shop owner, so the arrangement was predictable. The hotel or guesthouse owner's kitchen garden supplied less than 25 per cent of the produce in three instances; otherwise, kitchen gardens were not cultivated.

Different strategies are employed when people cannot obtain what they want locally for their guests. In some cases, they simply do without. Other options are ordering produce from St Martin during the dry season or substituting with what is available. There are instances when accommodation providers telephone a large supermarket in Grenada and have produce sent to Carriacou. One individual even indicated that they buy 'smuggled' produce from St Vincent.

One individual male farmer on Carriacou had taken the initiative to respond to the opportunities presented by the increasing tourist market. This person's occupation was farming and his main source of income was from his garden produce. He had developed marketing arrangements with the windjammer cruise boats and three of the guesthouses. Eventually, he began to adjust the timing of his plantings to coincide with times of observed market shortages for particular crops. He was experimenting with new varieties and had acquired various types of lettuce and herb seeds from abroad and added them to his kitchen garden in an effort to meet the tastes of visitors. He noted a significant increase in his sales as a result of the recent increase in tourism. The land he cultivated was fragmented and held under a variety of tenure arrangements, none of which he regarded as an impediment to production.

Gender roles: hucksters and housewives

To summarise, most of the produce purchased on Carriacou comes from Grenada and most of that is sold through the Government Marketing Board. Although women are active as gardeners in almost all households, they do not market the surplus from their gardens. While there is a growing demand for local produce as a result of increasing tourism on Carriacou, the fact that no Carriacou housewives have seized the opportunities created by tourism is rooted in gender roles and socio-economic standing within the close-knit Carriacou community. Many women do, however, engage in complex exchanges of both produce and prepared food with other households. These exchanges function, in part, as a way of assuring their position within their community as well as meeting the personal obligations that women take on as part of their social network.

The vendors of produce in the market square are of a different socio-economic class than the Carriacou housewife/gardener. They are almost exclusively women who are selling produce cultivated in Grenada, and many of these vendors are actually from Grenada and therefore outside the sphere of Carriacou community. They travel back and forth between the two islands to trade. Some Carriacou women who sell produce on the roadside or in the market square in Hillsborough have special arrangements with the captains of vessels that call regularly between Grenada and Carriacou. The captains of some of the smaller vessels transport Grenadian produce (often from the town of Grenville on Grenada's north coast) to Carriacou for resale.

Observers have commented on the role of women traders in the Caribbean who are variously referred to as vendors, hucksters or higglers (Lagro 1990; Le Franc 1989). Women assume this role in various ways throughout the region. All operate in the 'informal sector' and trade in various items, primarily agricultural produce. This group has extended itself out of the Caribbean and includes women who trade between the metro-polises of North America and Europe and the islands, working as informal commercial importers. Le Franc has noted that more research is needed to illuminate the relationship between petty trading and economic and social mobility. Her Jamaican study suggests that the low economic status and marginalised position of higglers is due to the dual economic role into which they are forced; the line between enterprise and the family unit is blurred in a way that is debilitating to the former (Le Franc 1989).

Similarly, a United Nations report (Lagro 1990) concludes that women hucksters of Dominica, who trade agricultural produce between that island and St Martin, are entrepreneurs in all respects. However, they operate in an extremely difficult social and economic environment (Lagro 1990). They do not have access to the structure of the formal economy or its facilities. The economic solutions they are pursuing are a response to the structural limitations in which they function.

Those women tending market stalls in Carriacou, as well as those who set up informal stands weekly along the roadside in the smaller villages of the island, fall into this same category of petty traders. They are buying food in Grenada for sale in Carriacou. The prices they charge are almost always higher than those set at the marketing board. The women selling the produce appear to be among the poorer and more marginalised in the econ-omy and are, indeed, from outside the Carriacou community, in most cases.

Most households on Carriacou cultivate at least a small corn and pigeon pea garden, often intercropped with ground provisions, squash, pumpkins, and okra. Kitchen gardens with lettuce, tomato and herbs are common. Over half the households in Carriacou are headed by women. Tourism and construction of new housing, both tourist related and tied to return migra-tion, are at an unprecedented high. Given these circumstances, the question remains as to why more women are not involved in marketing of their surplus produce.

The answer lies at the intersection of the 'traditional' organisation of the island society, gender roles, and the fact that tourism, although increasing rapidly, has not reached a critical threshold. Although Carriacou culture is dynamic and constantly integrating and selecting influences from abroad, there are still some locally based social norms which shape activities in the community.

With regard to agriculture, the corn and pigeon pea garden is still important to Carriacouan identity. There is a ritual significance, albeit a declining one, to the corn harvest on the island. Similarly, 'maroons' (that is, collective work groups organised to accomplish a community or group project) are still held. Although these activities have no doubt evolved since they were first documented in the literature in the early 1970s (Hill 1977), it is important to note that agricultural production still does not follow 'rational economics'. There is as yet a great deal of exchange of goods, particularly food, in the villages. This fact complicates the seemingly straightforward idea of housewives wanting to market their surplus. The sharing and mutual aid within Carriacou society is no doubt part of life that will come under pressure if tourism and media influences from abroad increase. However, for the time being, many of these support networks and mutual aid mechanisms are still in place.

As an extension of this sensibility, there is a stigma attached to the women who are selling in the market or on the roadsides. It is easy to see how this perception might become a barrier to any 'respectable' housewife marketing her surplus. It would be unseemly and degrading to be found making the rounds to the hotels and guesthouses with produce.

The situation is quite different for men who engage in agricultural work on the island. A few Carriacou men identify themselves as farmers and spend several hours a day cultivating a garden. A Carriacou woman will not identify herself as a farmer even if her garden is quite large. The gardening is considered an extension of her household duties. The corn and pigeon pea gardens often surround the house itself, so that the home sits in the middle of the garden and a fence around the periphery keeps out free-ranging animals. This spatial arrangement reinforces the traditional perception that food production is a part of the domestic sphere and an extension of a housewife's responsibilities. The blurring of the domestic and entrepreneurial sphere (Le Franc 1989) would result if a housewife chose to market her surpluses. Housewives' responsibilities in the home do not allow them the freedom of movement and the time required to pursue business interests, such as the marketing of produce.

Conclusions

Carriacou society is at a crossroads as it engages a globalising world through increased international tourism and Internet travel advertising. There is a degree of social upheaval as well as economic opportunity resulting from the end of its isolation and its inclusion in a tourist circuit. In response to this increased exposure, people are making decisions in response to change at the community and the individual level.

It is clear from this example of unrealised opportunities in the agricultural sector that responses to change are influenced and informed by gender roles and socio-economic standing in the community. The petty trader, or huckster, remains marginalised within the Carriacou social framework. Women who make their living selling agricultural products are at the bottom of the economic and social scale. The majority of these women come from off the island and, to a large extent, are 'outsiders' in Carriacou's tiny community. This fact has implications for any woman entrepreneur, resident on Carriacou, who might want to seize the opportunities presented by new tourist markets. In other words, it would not be surprising if a woman from outside the island community recognised and pursued the opportunity to supply special produce to tourist markets, but it would be surprising if the impetus came from a woman well positioned within the Carriacou community.

Returning to a broader regional context, it can be said that these attitudes and circumstances are specific to Carriacou because of the island's small size and the nature of the close-knit community. There are currently no large hotels or resorts on the island. For now it seems that the critical number of tourists necessary to stimulate any major changes in the agricultural sector has not materialised. Change, however, is in the air. Carriacou is being marketed on the Internet as an alternative, small island destination. The number of tourist accommodations has grown since 1998. Continuing and increasing opportunities for the collision of local and global ideas and culture are part of the immediate future for the community. As it stands, social constraints keep women, who are cultivating much of the food grown on the island, from pursuing marketing schemes for the tourist market. It is preferable to continue to import food rather than expand local marketing options given these social constraints.

As a Caribbean example, social structure on Carriacou is representative of the overall picture for the region. Caribbean society displays its own unique characteristics in the larger context of gender and international tourism. The decision to expand the household sphere into the tourist economy through rural and farm tourism as seen in, for example, many European countries with vibrant and picturesque agricultural landscapes, is in stark contrast to the regional examples from the rural Caribbean.

First, the inescapable legacy of plantation agriculture and slavery casts its shadow over any entrepreneurial activities related to working the land. The perception remains that the road to success lies off the island and away from the land, through education and employment in the cities of the North. High migration rates and the Afro-Caribbean family structure mean that some women are alone raising their children and managing their households. In these cases the decision to engage the tourist economy is not a shared decision among family members because there is no husband or father present. It is a decision a woman may make out of economic necessity and often carries with it the stigma that comes from a lack of support from a man or others who are working abroad.

In the Carriacou example presented here, there is clear evidence of the way in which tourism articulates gender relationships. By providing new economic opportunities and options, tourism highlights the structure and dynamics of the community. The way in which individuals do or do not respond to these opportunities reflects the deeply embedded standards and norms of the society. This example from Carriacou highlights the obstacles to implementing economic change in the rural Caribbean where tradition is still firmly entrenched.

References

Belisle, F.J. (1980) 'Hotel food supply and local food production in Jamaica: a study in tourism geography', unpublished Ph.D. dissertation, University of Georgia.

Brown, H. (1974) 'The impact of tourist industries on the agricultural sector: the competition for resources and the market for food provided by tourism', in *Proceedings of the Ninth West Indian Agricultural Economics Conference*, St Augustine, Trinidad: University of the West Indies.

Butler, R., Hall, C.M. and Jenkins, J. (eds) (1998) *Tourism and Recreation in Rural Areas*, Chichester: John Wiley & Sons.

Caribbean Basin Profile (1997), Caribbean Publishing Company Limited and Caribbean Latin America Action.

Caribbean Tourism Organization (CTO) (2002) *Caribbean Tourism Statistical Report*, 2000–2001 edn, St Michael, Barbados: Caribbean Tourism Organization.

Gomes, A.J. (1993) 'Integrating tourism and agricultural development', in D.J. Gayle and J.N. Goodrich (eds) *Tourism Marketing and Management in the Caribbean*, London: Routledge.

Grenada Board of Tourism (1997) *Annual Statistical Report*, St. George's.

Hill, D. (1977) 'The impact of migration on the metropolitan and folk society of Carriacou, Grenada', *Anthropological Papers of the American Museum of Natural History* 54, Part 2.

Kinnaird, V. and Hall, D. (eds) (1994) *Tourism: A Gender Analysis,* New York: Wiley.

Lagro, M. (1990) 'The Hucksters of Dominica', Port of Spain: United Nations Economic Commission for Latin America and the Caribbean.

Le Franc, E. (1989) 'Petty trading and labour mobility: higglers in the Kingston metropolitan area', in K. Hart (ed.) *Women and the Sexual Division of Labour in the Caribbean*, Kingston: The Consortium Graduate School of Social Sciences.

MacCannell, D. (1976) *The Tourist: A New Theory of the Leisure Class*, New York: Schocken.

Mills, B. (2002) 'Family land in Carriacou, Grenada and its meaning in the transnational community: heritage, identity, rooted mobility', unpublished dissertation, University of California, Davis.

Momsen, J.H. (1986) *Linkages between Tourism and Agriculture: Problems for the Smaller Caribbean Economies,* Seminar Paper, No. 45, Department of Geography, University of Newcastle upon Tyne.

—— (1994) 'Tourism, gender, and development in the Caribbean', in V. Kinnaird and D. Hall (eds) *Tourism: A Gender Analysis,* New York: Wiley.

—— (1998) 'Caribbean tourism and agriculture: new linkages in the global era?', in T. Klak (ed.) *Globalisation and Neoliberalism: The Caribbean Context*, Lanham, MD: Rowman & Littlefield.

Sinclair, M.T. (1997) *Gender, Work, and Tourism*, London: Routledge.

Swain, M.B. and Momsen, J.H. (eds) (2002) *Gender/Tourism/Fun(?)*, New York: Cognizant Communications Corporation.

Part III

Future prospects

16 Post-colonial markets

New geographical spaces for tourism

Janet Henshall Momsen

Introduction

At the beginning of the third millennium, in an almost biblical way, the riders of the apocalypse descended on the travel industry in the form of death by plane crash (on 11 September 2001), war (with Afghanistan and Iraq) and now pestilence (SARS) (*Economist* 2003). Tourism is always a fickle industry but the impact of political crises seems more marked for American tourists than in other Caribbean tourist markets. This has had a long-term effect with the dominance of United States tourists declining rapidly since the Gulf War. The Caribbean has looked to new markets in Europe and South America to reduce the region's dependence on the colossus to the north, resulting in change in the product offered and increased inter-island variation in types of tourism. This chapter examines the growing impact of market specialisation in the Caribbean, within a framework of expanding regional integration of the tourism industry through organisations such as the Caribbean Tourism Organization (CTO) and the Caribbean Hotel Association.

In light of the discussions by McElroy (Chapter 3) and Duval and Wilkinson (Chapter 4), there is little question that the Caribbean is perhaps the one region in the world that is most dependent on tourism (Holder 1998). It has long been seen as the American backyard in which tourists from the US dominate, although this pattern has changed in the last decade. Various exogenous factors such as special air fares, package holidays, changing patterns of leisure and exchange rates have affected the relative attractiveness of the region. The result is that European visitors have become more common. Spatial segmentation of the market, once based on residual colonial ties or Cold War blockades, has grown more complex. The argument put forward in this chapter is that the changes in market patterns are not only a result of deliberate selling of new destinations by 'First World' travel agents, but are also brought about by both proactive and reactive responses in the tourism industry of the destinations. Change in the future will ultimately be a test of such responses.

The terrorist attacks that damaged the Pentagon in Washington, DC and reduced the World Trade Centre in New York to rubble in September 2001 had a marked effect on tourism, especially in areas such as the Caribbean which are most dependent on American tourists. The economic downturn had already led to a decline in tourist arrivals from the United States, but the events of 11 September 2001, and the resulting reduction in airline flights, led to cancellations of as many as 80 per cent of bookings for Caribbean holidays in September and October 2001. Double-digit declines in visitor arrivals were registered for the last four months of 2001, resulting in an average reduction of 18.8 per cent over the level seen for the same four months in 2000 (CTO 2002a).

The Bahamas was the hardest hit with a decline of one-third in the September to December period (CTO 2002a). This is a clear reflection of its heavy dependence on the United States market. Curaçao, on the other hand, where approximately two-thirds of tourists are Dutch, was the only country to report an increase in visitor arrivals in every month in 2001 (CTO 2002a). The immediate impact of 11 September 2001 was seen in closures of hotels and restaurants, with concomitant reductions in employment. Caribbean heads of state held an emergency meeting in the Bahamas in October 2001 to coordinate a regional response, as did tourism ministers in April 2003 in Jamaica, underlining the importance of tourism to the region (Gonzalez 2001, 2003).

In 2002 al-Qaeda terrorism specifically targeted tourist areas. Attacks on long-haul, tropical tourist sites in Bali, Mombasa and sub-tropical Tunisia reduced the global competition among traditional 'sun, sand and sea' destinations. These attacks were aimed at hitting other western, non-American tourist groups, specifically Australians, Israelis and Europeans. The Caribbean was seen as a safer non-Muslim area. However, a radical Muslim group in Trinidad and Tobago has been accused of links with al-Qaeda and of planning to attack US and British interests with biological weapons, if war was launched with Iraq. Consequently, the British Foreign and Commonwealth Office warned Britons against travelling to Trinidad and Tobago, as did Australia (James 2003). As a result, P&O lines, popular with British tourists, ordered four cruise ships not to call at Trinidad and Tobago (James 2003). However, the United States did not issue a travel advisory.

The year 2001 was the first in twenty years to witness negative growth in global tourism according to the World Tourism Organization, although the change was only a fall of 0.6 per cent after growth of 6.9 per cent in 2000. Despite the continuation of terrorist attacks, the World Tourism Organization reported that in 2002 international tourist arrivals surpassed 700 million for the first time, reaching almost 715 million. This was a rise of 3.1 per cent over 2001. However, tourists stayed closer to home, made shorter visits and spent less (Associated Press 2003). In 2003, growth in tourism among United States travellers had been expected, as those who stayed home, or took domestic holidays in 2001 and 2002, began once again to think more globally when planning vacations (Volgenau 2003). Yet airline bankruptcies and cutbacks have had an especially marked impact on destinations heavily dependent on air travel, such as small Caribbean islands, so accessibility, despite possible renewed demand, becomes a more critical issue.

The new epidemic, severe acute respiratory syndrome (SARS), is, as of April 2003, having a detrimental effect on tourism in Asia (Barboza 2003). Similarly, infections picked up on cruises have deterred elderly passengers. Overwhelmingly, however, the conflict in Iraq has put a further halt to international travel. As Micky Arison, chief executive of Carnival Cruise lines, said as American bombs began dropping on Baghdad: 'When people are buying duct tape, they aren't buying vacations' (Reuters 2003). After the brief Gulf War, it was eight months before travel began to return to normal. The growth of electronic booking, however, has made it easier for people to travel at short notice, so recovery may be faster after the recent conflict in Iraq. It would be logical for United States travel to Europe and anywhere within a thousand miles of the Persian Gulf to be most strongly affected in both 1991 and 2003. This could redirect American tourists to vacations in the western hemisphere, and thus benefit the Caribbean.

The effects of geographical proximity, however, appear to have less of an impact on US leisure travel than on that of Europeans, with only two (Aruba and Bermuda) of the top twelve destinations for American tourists in 2002 being in the Caribbean while the others were in Europe or Hong Kong, Australia and New Zealand (Volgenau 2003). Jamaica, where US tourists predominate, saw a 40 per cent fall in its future bookings after the start of the Iraq War and British tour groups have cancelled one-third of their bookings in the Dominican Republic (Gonzalez 2003). The smaller islands in the eastern Caribbean are especially vulnerable to such a decline, yet Caribbean tourism officials expressed relief that there had not been the kind of widespread cancellations that followed the 1991 Gulf War, or the paralysis of air travel after the 2001 terrorist attacks (Gonzalez 2003).

The Caribbean brand name

Such externalities are not recent, to be sure, and the image held of the Caribbean has strong historic roots (Sheller, Chapter 2). The word 'Caribbean' conjures up Kodachrome images of azure seas, green palm trees silhouetted against blue skies and unblemished white-sand beaches waiting for Robinson Crusoe's footprint. This is a picture unchanged since the first European tourist, Christopher Columbus, wrote home from the Bahamas of vegetation lush like that of Andalusia in April, of large flocks of parrots, of sweetly singing birds and plentiful, exotic, heavily laden and aromatic fruit trees. From Cuba, he wrote to his patron King Ferdinand in 1492, 'Sire, these countries far surpass all the rest of the world in beauty' (Watts 1987: 1). Thus the region's first publicist sold the image of an Edenic unspoiled paradise to attract investment and visitors half a millennium ago.

Officials in the Caribbean Tourism Organization (CTO) define the distinctive brand name they are selling only in terms of an agreement: for a country to be a member of the CTO it must have a coastline on the Caribbean.[1] Thus the CTO emphasises the co-importance of both land and sea in contemporary Caribbean tourism.[2] Reflecting this, and as Wood (Chapter 9) outlines, cruise business continued to be vibrant in the region in 2001 after strong historical growth in the past two decades. The combination of the introduction of larger vessels and more berths allocated to the region resulted in an over 15 per cent increase in cruise passenger arrivals in the first eight months of 2001, followed by a less than one per cent increase in the last four months of the year (CTO 2002a). The overwhelming majority of cruise passengers in the Caribbean are from the United States, so it is perhaps not surprising that the largest decline of 31 per cent in 2001 was recorded in the French territory of Martinique, where American tourists have to face non-English-speaking taxi drivers and shopkeepers (CTO 2002a).

Caribbean tourism markets in the 1990s

At the beginning of the 1990s, the West Indian Commission noted that 'Tourism is the dominant economic sector in many CARICOM countries; it is of growing importance in all' (West Indian Commission 1992: 191). In the world market, however, the Caribbean has been losing ground to East Asia/Pacific and the Middle East. These regions, which compete with the Caribbean as warm-weather destinations, recorded double digit compound annual growth rates over the period 1991–5: 14 per cent for the Middle East, 11 per cent for East Asia/Pacific and only 5 per cent for the Caribbean (Res and Co. 1999). Between 1998 and 1999, Singapore experienced a growth in tourists by 12 per cent, Orlando (Florida) by 10 per cent, the Maldives by 9 per cent, Southern Africa by 6 per cent and Taiwan by 5 per cent. Comparatively, the Caribbean region reported an increase of only 3.8 per cent (CTO 2001).

In 1999, the Caribbean was the fourth most important destination for US tourists (after Mexico, Canada and Europe), capturing 15 per cent of all outbound tourists (CTO 2001). Globally, the Caribbean as a whole is usually the sixth most important destination in terms of tourism receipts, after the United States, Italy, France, Spain and the UK (Jayawardena 2002). Not unlike the US, the Caribbean's share of Canadian total person trips abroad was 19 per cent in 1999. The Commonwealth Caribbean's share of British travellers overseas, excluding western Europe, was, however, only 4.3 per cent after the United States, Canada and North Africa, although another 4 per cent went to South America, which included several Caribbean destinations (CTO 2001). Thus, there is room for speculation that the recent outbreak of SARS in Asia, terrorist attacks in North Africa and Indonesia, and the wars in Afghanistan and Iraq, combined with accompanying anti-American feelings in many countries and anti-'old-Europe' attitudes among Americans, may actually benefit Caribbean tourism over the next few years.

The US market

In 1980, American tourists made up 58.7 per cent of Caribbean tourists, rising to 62.3 per cent in 1985, when European tourists were only 9.2 per cent of the total. By 1990, American tourists had fallen to 53.6 per cent while European tourists had increased to 16.1 per cent. This trend continued through the decade, with American tourists falling to 50.4 per cent in 1995. Since then they have become, for the first time in the last forty years, less than half of Caribbean tourists, forming 49.8 per cent in 2000 (see also Table 16.1). If the Mexican resorts of Cancun and Cozumel are excluded, then the figures for 1999 would be 45.4 per cent rather than 49.4 per cent American tourists and 27.7 per cent European rather than 26.8 per cent (CTO 2001). Over the last two decades, Canadian tourists have made up between 6 and 7 per cent of Caribbean tourists. Caribbean tourists have continued to constitute about 8–10 per cent. Visitors from elsewhere in the world, mainly South America, increased during the 1980s from 13.9 per cent in 1980 to 20.3 per cent in 1987, falling again in the following decade from 14.7 per cent in 1995 to 11.3 per cent in 2000.

Despite these changes in relative percentages, absolute numbers of tourists in all major markets have increased. Overall, tourist arrivals in the Caribbean increased from 12.8 million in 1990 to 20.3 million in 2000. The annual rate of increase was 4.7 per cent over the decade as compared to 4.3 per cent for tourists worldwide (CTO 2002a). These changes vary for individual territories throughout the Caribbean. The most rapid increases have been in European tourists while the lowest increases are seen in United States tourists during the 1990s (see also Table 16.1).

The dominance of tourists from particular markets influences the type of tourism offered (Table 16.2). Those countries with high concentrations of Americans tend to specialise in all-inclusive hotels and easy access while it is generally thought that European tourists, on the whole, are more interested in the local environment and in meeting local people. In 2001,

Table 16.1 Percentage growth in Caribbean main markets, 1990–2000

	US		Canada		Europe		Caribbean	
	Total	*Annual*	*Total*	*Annual*	*Total*	*Annual*	*Total*	*Annual*
1990–2000	37.6	3.2	46	3.9	138.1	9.1	59.4	4.3
1995–2000	18.6	3.5	32.3	5.8	38.9	6.8	18.6	3.5

Source: CTO (2002a).

Bermuda, Cancun, Cayman Islands, Puerto Rico, the Turks and Caicos and the United States Virgin Islands had at least three-quarters of their tourists coming from the United States, while Jamaica had 71 per cent (Table 16.2). These are clearly all northern tier areas, where proximity to the United States reduces the cost and time taken to travel. In 2000, 30 per cent of all United States tourists to the Caribbean went to the US possessions of Puerto Rico and the United States Virgin Islands and 37 per cent visited the English-speaking Commonwealth Caribbean. Reliance on the US market is less than 5 per cent in non-English-speaking Cuba, Martinique, Guadeloupe and Suriname, where European tourists are dominant. Canadian visitors are a small presence in most Caribbean countries but they are most important to Cuba, where they made up 24.9 per cent of visitors in 2000, and in the Dominican Republic where 19.9 per cent of visitors were from Canada.

The European market

European visitors are the largest group of visitors in Cuba and Barbados. In Guadeloupe and Martinique, French-speaking tourists predominate. In 2001, almost nine out of ten stayover tourists to Martinique came from France, with only 4.1 per cent from other European countries (primarily Belgium/Luxembourg, Germany and Switzerland) (ARDTM 2002). Language affinity is also important in the Dutch-speaking countries with Suriname getting 84 per cent of its tourist arrivals from Holland in 1999, Curaçao 28 per cent in 2000 and Bonaire 26 per cent in 2000 (CTO 2002a). However, Aruba, also a former Dutch colony, has targeted US visitors by

Table 16.2 Market dependence in various Caribbean territories (2000)

Market	Countries with strong dominance of tourists from a particular market
United States	Bahamas (82%), Cayman Islands (80%), US Virgin Islands (78%), Puerto Rico (78%), Cancun (78%), Bermuda (77%), Turks and Caicos (74%), Jamaica (71%)
Canada	Cuba (17%), Jamaica (9%), Dominican Republic (8%), Cancun (7%), Bahamas (7%)
Europe	Guadeloupe (87%), Suriname (86%), Martinique (84%), Cuba (54%), Barbados (48%)
Caribbean	Dominica (57%), Montserrat (51%), St Kitts and Nevis (39%), St Vincent and the Grenadines (32%)
South America[a]	Curaçao (24%), Aruba (22%)

Source: CTO (2002a).

Note
a Figures are for 1999 (CTO 2001).

arranging for US Customs clearance on the island so that, in 2000, 63.5 per cent of tourist arrivals came from the US and only 4.2 per cent from Holland (CTO 2002a). Saba, one of the Dutch Windward islands in the north of the region, had 29 per cent Dutch visitors, but 41 per cent US in 2000. Similarly, nearby St Martin/Sint Maarten, which has historical ties with both France and Holland, had 28 per cent European visitors and 43 per cent from the United States in 2002 (CTO 2002a). Overall, European visitors continue to gravitate towards destinations with which they have historical and linguistic ties:

- 60 per cent of visitors from the United Kingdom went to the Commonwealth Caribbean;
- 62 per cent of Dutch visitors went to the Dutch West Indies;
- 69 per cent of French tourists went to the French overseas departments of Martinique and Guadeloupe; and
- 93 per cent of Spanish-speaking visitors went to Cuba and the Dominican Republic (CTO 2002a).

Shifting markets

As highlighted above, language familiarity is a major attraction, while the actual location of their destination within the Caribbean makes very little difference to the flight time or cost for long-haul European visitors. For American tourists, distance is more important, as an hour's reduction in flight time can be a considerable saving, especially since these visitors usually have much shorter vacations than Europeans. Thus, tourists from the United States tend to concentrate on northern tier islands unless a southern tier country has reduced travel time, as in Aruba, by having US Customs clearance. Having an airport that allows Americans to fly direct to their destination also helps, as in the case of Sint Maarten and Jamaica. In this way, the Caribbean is evolving new geographies of tourism based on dominant market share.

Changes over time, as shown in Table 16.3, reinforce the continuing dominance of tourists from the United States in many northern tier countries. One exception is Cuba, where it is difficult for Americans to get permission to visit. The relative importance of tourists from the US has declined since 1980 in the Dominican Republic and Sint Maarten, where European tourists now dominate. In other northern tier countries, the proportion of American tourists has been stable or increased, as in Belize, Haiti and the Turks and Caicos Islands (Table 16.3). In the small islands of the eastern Caribbean there has been an increase in the proportion of US tourists over the last decade in the northern islands of Anguilla and St Kitts and Nevis and in St Lucia, where a new airport made it possible for large jets to fly there direct from the United States and Europe. In Antigua, Barbados and Martinique, American tourists have been largely replaced

Table 16.3 Percentage in market share of arrivals, by key markets, 1980 and 2000

Sector	From USA		From Canada		From Europe		From other countries	
	1980	2000	1980	2000	1980	2000	1980	2000
Northern Tier Islands								
Bahamas	74.8	81.9	11.0	5.7	9.7	8.0	4.5	4.5
Bermuda	86.5	77.3	7.1	9.3	5.1	10.7	1.3	2.6
British Virgin Islands	65.0	61.6	4.7	3.0	7.6	13.7	22.7	21.6
Cayman Islands	74.8	72.1	6.1	5.2	4.7	8.0	14.4	14.7
Cuba	5.4	4.3	21.8	17.3	10.9	53.5	61.9	17.7
Dominican Republic	69.4	19.5	3.0	8.4	8.0	43.2	19.6	28.9
Haiti	42.5	66.1	11.0	10.5	20.9	7.8	25.6	15.6
Jamaica	60.6	70.6	17.9	8.1	16.1	15.0	5.4	5.1
Puerto Rico	72.0	83.1	n.a.	1.1	n.a.	2.5	28	13.4
Saba	42.5 [b]	41	3.2 [b]	3.7	40.5 [b]	38.0	6.8 [b]	17.3
St Eustatius	15.9[a]	22.3[c]	0.0[a]	1.9 [c]	14.2[a]	46.8[c]	69.9[a]	29.0[c]
Sint Maarten	71.8[a]	44.7[c]	1.5[a]	6.7[c]	10.6[a]	28.3[c]	16.1[a]	20.4[c]
Turks and Caicos Islands	52.6[a]	73.9	6.3[a]	10.3	11.8[a]	7.7	29.2[a]	8.2
US Virgin Islands	88.1[a]	81.5	1.1[a]	0.7	1.7[a]	2.1	9.1[a]	15.8
Eastern Caribbean								
Anguilla	26.7	56.6	2.4	3.5	5.9	21.6	65.0	18.4
Antigua and Barbuda	39.7	28.5	10.2	6.8	23.5	43.5	26.5	21.2
Barbados	23.2	20.6	23.0	11.6	25.6	47.8	28.2	20.6
Dominica	15.9	21.7	5.6	3.1	23.9	16.0	54.6	59.2
Grenada	23.0	25.3	6.8	3.8	23.0	36.0	47.2	34.9
Guadeloupe	13.6	14.8	3.9	1.7	80.1	79.5	2.4	4.0
Martinique	24.7	1.0	6.2	1.0	59.0	84.2	10.1	12.8
Montserrat	38.3	15.1	11.1	3.3	11.4	25.7	39.2	55.9
St Kitts and Nevis	32.7	41.3[c]	5.4	7.0[c]	8.1	18.8[c]	53.8	33.0[c]
St Lucia	17.7	32.1[c]	16.3	5.1[c]	38.8	37.9[c]	27.2	25.1[c]
St Vincent/Grenadines	28.5	28.1[c]	11.0	6.6[c]	19.4	29.6[c]	41.1	35.7[c]
Southern Tier Islands								
Aruba	58.6	63.5	2.9	2.9	3.5	6.5	35.2	11.6
Bonaire	32.4	49.6	3.4	2.1	11.3	35.7	52.9	12.6
Curaçao	13.3	15.3	0.9	1.4	9.1	31.9	76.7	51.4
Trinidad and Tobago	27.7	33.3	14.1	1.2	14.7	20.7	43.5	34.0

Sources: Calculated from CTRDC (1982, 1985, 1988, 1989) and CTO (1992, 1997, 2001, 2002a).

Notes
a 1984 data.   b 1987 data.   c 1999 data.

by European visitors. For Martinique, the recent elimination of the island from American Airlines schedules has made it difficult to reach from the United States. Since the volcanic eruptions of the mid-1990s in Montserrat, most visitors have been from within the region. In southern tier countries, Dutch visitors are the main group going to Suriname but the Dutch territories of Aruba and Bonaire have increased their proportion of American tourists. Elsewhere, regional visitors remain a major segment of the market, especially in countries with land boundaries (e.g. Dominican Republic), in multi-island states such as Grenada, St Vincent and the Grenadines, Trinidad and Tobago, and St Kitts/Nevis, or where new and improved ferry connections have been developed (e.g. Montserrat, Dominica, St Lucia and Curaçao).

Length of stay

In addition to absolute and relative numbers of tourists from different markets, length of stay also varies. Linking length of stay with tourist numbers gives a much better overall understanding of tourist density and penetration ratios (see McElroy, Chapter 3). Breaking this down by market improves the understanding of the relative importance of different groups. On the whole, European tourists tend to stay longer as they have more paid holidays, travel further and the airfare is often a bigger proportion of their vacation costs than for Americans. The average length of stay for American tourists is about a week, compared to about two weeks for European visitors. While American tourists made up 49.8 per cent of visitors in 2000, European tourists accounted for 25.9 per cent. As a result, the difference in length of stay makes European tourists now the major market for the region.

There has been some change over time in length of stay, although the differences by source of tourists have remained fairly stable. In 1981, visitors stayed about a week on average in most parts of the region. As today, Puerto Rico had the shortest length of stay at about three days, followed by Sint Maarten at four days. The islands with the longest length of stay were Jamaica, Barbados, Cuba and St Vincent and the Grenadines with between nine and ten days on average (CTRDC 1988). Average length of stay in the Dominican Republic doubled from one to two weeks between 1981 and 1984, reflecting the beginning of package holidays from Europe (CTRDC 1988). In the mid-eighties most visitors stayed slightly longer than they do today. A likely explanation for this is because it took longer to get to the Caribbean and was relatively more expensive. The biggest change is seen in the French overseas departments (Départements d'Outre Mer (DOM)) of Martinique and Guadeloupe, where average length of stay was less than a week, except for Canadian tourists, in 1987. The increase in length of stay among Europeans, which occurred during the early 1990s, also reflects the growth of mass tourism and package holidays with cheap flights between Paris and the French islands. Differential flight costs make it cheaper to fly to Martinique from the west coast of the US via Paris rather than on a more geographically direct route. In Barbados, where European tourists are the majority, Canadian tourists tend to have a pattern of length of stay more like that of Europeans than like Americans (Barbados 1996). Dominica, which specialises in ecotourism, and St Vincent and the Grenadines, which attracts yachting visitors, have all increased their length of stay for American tourists (CTRDC 1989; CTO 1992, 2002a).

Reactive change: the impact of 11 September 2001

Following the attacks of September 2001, Caribbean tourism fell precipitously (Table 16.4). One cruise line went out of business but many ships were repositioned from the Mediterranean to the Caribbean (Glenton and James 2001; McDowell 2001). Bookings for cruises were 25 to 35 per cent below the previous year and occupancy levels fell although not by much as cruise lines were refusing to offer refunds for cancellations (McDowell 2001; CTO 2002b). However, most cruise ships move amongst the northern tier of islands and so the southern Caribbean saw few cruise ships. All-inclusive hotels fared better than hotels offering European Plan and also benefited from the diversion of business form Europe, the Mid-East, the Indian Ocean and North America. Employment per tourist is higher in all-inclusive hotels than in other types of accommodation so this was economically beneficial although 16,000 workers were laid off in Puerto Rico (CTO 2002b). Even countries where American tourists are a very small minority as in Cuba and Martinique complained of post-9/11 downturns (*Economist* 2001; Trucco 2001). The director of the Martinique Promotion Bureau in New York was reported to have said: 'Europeans aren't as willing to travel as before. This is a global situation' (Trucco 2001). Local airlines also suffered, with LIAT losing half a million US dollars in the first three weeks after 11 September and Air Jamaica losing US$11 million (CTO 2002b).

Many governments tried to assist their tourist industries. In Jamaica, Puerto Rico, Antigua and Barbuda and the Dominican Republic taxes were

Table 16.4 Tourist arrivals (thousands), 2001–2002

Season	All markets		Change (%) by major market			Cruise passengers	
	Arrivals	Change (%)	USA	Canada	Europe	Arrivals	Change (%)
2001							
Winter	7,875.5	+5.9	+10.9	−1.9	−5.8	7,267.5	+13.3
Summer	11,880.9	−7.7	−9.9	+7.1	−2.6	8,711.4	+7.5
January–August	14,835.8	+3.3	+5.1	−2.0	−2.4	n.a.	+15.2
September–December	4,924.1	−17.2	−23.2	−11.3	−6.6	n.a.	n.a.
Whole year	19,759.9	−2.7	−2.1	−4.5	−3.7	15,984.3	+10.1
2002							
January–February	n.a.	−13.1	−11.5	−19.2	−13.3	n.a.	−4.1

Source: CTO (2002b).

waived (CTO 2002b). Barbados provided a substantial aid package to the accommodation sector and introduced a new marketing plan (CTO 2002b). Widespread discounting in the hotel sector led to reduced earnings (Ferriss 2001; Gonzalez 2001; Trucco 2001). But the rich were little affected, continuing to fly in on Concorde to their winter homes in Barbados and elsewhere.

Following previous global disasters, American tourists have been more fickle than European tourists, but since 2001 both Europeans and Americans have reduced travel (Table 16.4). This may lead to a rebalancing of the national markets for Caribbean tourism. Visitors from the United States may see this once more as their backyard, and so be more willing to travel there than further afield. In order to compensate for lost long-stay European visitors, as soon as President Bush declared the Iraq War over, the Caribbean issued a special insert in the *New York Times* (Fine 2003) emphasising no-hassle direct flights on Air Jamaica and American Airlines, special deals for all-inclusive resorts such as Club Med, Hiltons and Sandals, and discounts on holidays of four to seven days.

Conclusion

While there has been an increase in European travellers to the Caribbean, this trend may soon come to an end. This is being reinforced by the European Union's Package Travel Directive which imposes on tour operators increased liability. Unfavourable publicity following food poisoning in the Dominican Republic led to increased liability and a subsequent wave of litigation, making European tour operators unwilling to send groups there. European tour operators are increasingly specifying standards which they expect their suppliers in the Caribbean to meet and these may be coupled with contractual obligations or recovery from the supplier of successful compensation claims against the tour operator. To deal with this new situation the Caribbean is developing higher standards for safety and hygiene. In fact, this is the focus of the Quality Tourism for the Caribbean project set up in June 2002 and involving Barbados, Jamaica, Trinidad and Tobago, the Bahamas and nine members of the Organisation of East Caribbean States. It is a joint venture with the Caribbean Alliance for Sustainable Tourism (CAST) and the Caribbean Epidemiology Centre (CAREC). The programme aims to strengthen the overall quality and competitiveness of the tourism industry in the Caribbean through the establishment and promotion of quality standards and systems designated to ensure healthy, safe and environmentally conscious products and services.

Clearly the Commonwealth Caribbean, for whom European tourists provide a major segment, is determined to continue to attract such tourists. However, French tourists may be seen as a separate market segment. In winter 2002–3, they were flying in increasing numbers to their possessions in the Caribbean, as a replacement for Mediterranean and North African holidays. In the DOM, they can escape Anglo criticism of their President's opposition to the war in Iraq and be in a safe and familiar but warm beach environment. However, French investors in the Caribbean tourism industry are threatening to pull out because of the inefficiencies of the islands.

The combination of discounts and improved offerings, including upgraded hotels, has to overcome declining airlift. American Airlines, the main carrier in the region, teeters on the edge of bankruptcy, and the recent decision to pull the Concorde from service will mean that it will not fly to Barbados beginning in the winter season of 2003–4. However, privately owned regional airlines are extending their reach, especially Butch Stewart's Air Jamaica and Allen Stanford's Caribbean Star. Once again the French connection differs. Air France flies two jumbo jets daily from Paris to the French Caribbean and they have been packed in 2003 as have the daily flights by three other French airlines.

Currently, tourism provides 2.5 million jobs in the Caribbean, about one-quarter of all employment. In 2000 tourism generated US$35.3 billion in economic activity and was responsible for 25 per cent of the region's GDP. Since then terrorism, war and pestilence have reduced the growth of tourism in the Caribbean. But the region is fighting back in both United States and European markets, by preserving the paradise its brand name suggests and so aiming to keep tourists coming. The recognition of two separate main markets: a European one with long stay for both mass package tourists and high-end elite visitors; and an American market looking for short no-hassle vacations in all-inclusives involving minimum travel time, has produced a transnational regionalisation of the Caribbean. In some cases, especially in the French- and Spanish-speaking countries, post-colonial links have been strengthened, yet elsewhere widespread use of English and improved facilities have proved attractive to a range of markets. Adjustment to specialised tourist markets is producing a new kind of spatiality with a new geography mapped onto flows of different groups of tourists.

Notes

1   Bermuda is a member of the CTO although it lies in the North Atlantic rather than the Caribbean.
2   The CTO has 33 members including mainland countries of Suriname, Guyana, Venezuela and Mexico. Tourism statistics for Venezuela are not included in the CTO *Annual Statistical Report* and data for Mexico are limited to figures for Cancun and Cozumel. The Central American Caribbean coastal areas are not included in CTO data.

References

Agence régionale de développement touristique de la Martinique (ARDTM) (2002) *Bilan 2001 du tourisme à la Martinique: statistiques*, Schoelcher, Martinique: ARDTM.

Associated Press (2003) 'World tourism rises', *New York Times*, 28 January: W1.

Barbados (1996) *Digest of Tourism Statistics*, Bridgetown, Barbados: Statistical Service.

Barboza, D. (2003) 'Fears of war and illness hurt tourism in Asia', *New York Times*, 28 March: W1.

Caribbean Tourism Research and Development Centre (CTRDC) (1982) *Caribbean Tourism Statistical Report, 1981*, Christ Church, Barbados: CTRDC.

—— (1985) *Caribbean Tourism Statistical Report, 1984*, Christ Church, Barbados: CTRDC.

—— (1988) *Caribbean Tourism Statistical Report, 1987*, Christ Church, Barbados: CTRDC.

—— (1989) *Caribbean Tourism Statistical Report, 1988*, Christ Church, Barbados: CTRDC, Barbados.

Caribbean Tourism Organization (CTO) (1992) *Caribbean Tourism Statistical Report, 1991*, St Michael, Barbados: Caribbean Tourism Organization.

—— (1997) *Caribbean Tourism Statistical Report, 1996*, St Michael, Barbados: Caribbean Tourism Organization.

—— (2001) *Caribbean Tourism Statistical Report, 1999–2000*, St Michael, Barbados: Caribbean Tourism Organization.

—— (2002a) *Latest Statistics 2001*. Online: <www.one-caribbean.org> (accessed 28 October 2002).

—— (2002b) *37th Report of CTO to the Board of Directors*, New York, 30–31 May 2002.

Economist (2001) 'Cuba's economy: blaming the victim', *The Economist*, 27 October: 37.

—— (2003) 'Business travaillers – crisis in corporate travel', *The Economist*, 5 April: 56.

Ferriss, S. (2001) 'Mexican tourism industry hurting as bookings drop, resorts start cutting prices and staff', *Sacramento Bee*, 23 November: 4.

Fine, B. (2003) 'Find your place in the Caribbean sun', Advertising supplement to *New York Times*, 15 April.

Glenton, B. and James, J. (2001) 'Conflict forces cruise lines to change course', *Financial Times*, 13/14 October: xv.

Gonzalez, D. (2001) 'More bad economic news for the faltering Caribbean: attacks in US are keeping tourists away', *New York Times*, 7 October: A6.

—— (2003) 'War and a slowdown empty Caribbean beaches', *New York Times*, 8 April: A6.

Holder, J. (1998) 'Consulting our enlightened self-interest', in *Proceedings of the first Caribbean Hotel and Tourism Investment Conference*, Bridgetown: CTO.

James, C. (2003) 'Trinidad terrorist threats drive off cruises', *Financial Times*, 28 January: 2.

Jayawardena, C. (2002) 'Future challenges for tourism in the Caribbean', *Social and Economic Studies* 51(1): 1–24.

McDowell, E. (2001) 'Ships go trolling for passengers', *New York Times*, 14 October: TR11.

Res and Co. (1999) 'Jamaica, master plan for sustainable tourism development: diagnostic and strategic options', Draft Report, Office of the Prime Minister Tourism, Kingston, Jamaica.

Reuters (2003) 'Cruise shares up on Carnival news', *Financial Times*, 22/23 March: 10.

Trucco, T. (2001) 'The islands feel an autumn chill: travel apprehension has hit the Caribbean hard, and a rebound in the prime winter season will be crucial', *New York Times*, 28 October: TR11.

Volgenau, Gerry (2003) 'Travel experts predict a strong 2003 unless new Gulf War intervenes', *Sacramento Bee*, 26 January: M7.

Watts, D. (1987) *The West Indies: Patterns of Development, Culture and Environmental Change since 1492*, Cambridge: Cambridge University Press.

West Indian Commission (1992) *Time for Action: Overview of the Report of the West Indian Commission*, Barbados: West Indian Commission.

17 Future prospects for tourism in the Caribbean

David Timothy Duval

Introduction

In accepting the systemic nature of tourism (Mill and Morrison 1998), an important consideration arising from the preceding chapters is that any assessment of tourism in the Caribbean requires consideration of existing global, regional and local conditions of social, political and economic realities. Without question, the nature of tourism in the Caribbean is complex and intricate in its arrangements, policies and strategies. Quite often, the solutions are therefore just as complex as the problems. In fact, Parker (2000) speaks of 'multi-dimensional and interdependent problem domains' with respect to destination growth, and the management of tourism in island environments has received considerable attention in the literature (e.g. Henderson 2000; Ryan 2001; Wilkinson 1987).

Not only are the conditions that influence tourism almost always in a state of flux, the extent to which they permeate regions, nations and communities also fluctuates. Consequently, the economic and social context within which tourism is situated shifts considerably, and governments may find themselves, often through no fault of their own, addressing 'yesterday's problems'. Nonetheless, given the extent to which tourism is manifested in the Caribbean, and because many island states rely heavily upon it for economic growth, there exist substantive issues that can indeed be addressed.

In the Caribbean context, perhaps the most pressing issue is the extent to which political and economic arrangements, centred within global contexts, maximise economic returns from tourism (McElroy, Chapter 3). The degree of government involvement is, as Wilkinson outlines in Chapter 5, a question of active versus passive involvement. As Poon (1988a) argues, much of the future of tourism in the Caribbean will be based on innovation and the region's ability to adapt to changes in key generating markets. Elsewhere, Poon (1988b) argues that the decision ultimately needs to be made over the extent to which tourism becomes fatalistic or flexible. By extension, Mather and Todd (1993) argue that future trends in Caribbean tourism will be based around quantity versus quality, relating specifically to:

- whether a country should aim to attract large numbers creating a high aggregate level of expenditure, or fewer numbers of relatively high-spending tourists;
- whether both types of trade can happily coexist in one destination;
- whether a destination can successfully promote two types of tourism;
- whether relatively small countries can develop different types of visitor attractions.

Related to this are concerns over seasonality and product development. The northern hemisphere's traditional holiday period of May to August acts as a key season for the Caribbean. Consequently, alternative (or 'non-mainstream': Widfeldt 1996) product development such as heritage tourism (Found, Chapter 8) or ecotourism (Weaver, Chapter 10) is at least an attempt at normalising such seasonality, and as an added bonus it addresses the concerns raised over the apparently unsustainable nature of mass tourism. It begs the question, however, whether a mass, seasonal concentration of tourists is any less environmentally threatening than steady, year-long alternative tourists visiting sites that are physically and/or culturally sensitive. Whether this is empirically sound reasoning or not remains to be seen in the Caribbean, although the consensus seems to be that the negative impacts of tourism in general seem to be amplified in island environments (Widfeldt 1996).

Balancing quality versus quantity in Caribbean tourism inevitably leads to the question of sustainability. Much research has been devoted to the environmentally sustainable nature of tourism in island environments (e.g. Fotiou *et al.* 2002; Davies 1996) and in the Caribbean in particular (e.g. Wilson 1996). In their discussion of small-island policies directed at up-market visitors, as an attempt to enhance the sustainable nature of tourism overall, Ioannides and Holcomb (2003: 42–3) argue that the infrastructure required to cater to this segment (e.g. 'air conditioning, heated swimming pools, luxurious spas and frequent linen changes') is often less sustainable than that which is required by and provided to mass tourists (see also Olsen 1997). Instead, Ioannides and Holcomb (2003: 45) argue that 'It is far more important for destinations to adopt a planning/policy framework that treats tourism not in isolation but as an integral component of the entire development process.' One might also question the uniformity of sustainable approaches to tourism development in the broader cultural context, especially given the variability and dynamic nature of community (and cultural) response(s) (Dogan 1989). To this end, the assertion by Milne and Ewing (Chapter 12) for sustainable forms of tourism that empower communities in order to ultimately reflect community values is critical.

What does the future hold for Caribbean tourism? It is a complex question that is, to some extent, unanswerable, but certain indicators can be used to point to critical issues on the horizon. Over twenty years ago,

Holder (1980) argued that the future success of Caribbean tourism will hinge around issues such as political stability, monitoring of community attitudes and change, strong investment (both local and foreign), regional marketing and adequate training. For the most part, these conditions, while nonetheless generic in their utility, still apply today. Perhaps what is missing is how the image of the Caribbean is constructed. Historically, as discussed by Sheller (Chapter 2), the Caribbean's image as a 'paradise' with 'natural' and 'cultivated' scenery can be tied to 'ways of viewing Caribbean people'. The local social realm, therefore, would seem to be of critical importance. This chapter attempts to draw some conclusions regarding future prospects for Caribbean tourism. Ahead of addressing such prospects, what follows is a discussion and assessment of past predictions.

Past predictions: the benefit of hindsight

In 1993, John Bell, then the Executive Vice President of the Caribbean Hotel Association, offered a series of predictions for the performance of tourism in the Caribbean in the year 2000 (Bell 1993: 234–5). Bell's predictions point to how tourism, at that time, was viewed in the context of regional development. Bell's (1993: 234) initial prediction was that the region will continue to function as 'the world's premier warm weather destination, with perhaps twice the number of accommodation units as are available today'. For the most part, tourism growth throughout the 1990s has been steady. While the region has seen annual tourism growth rates of approximately 4 to 5 per cent in the 1990s (CTO 2002), other regions, especially warm-weather destinations such as the Mediterranean and Mexico (e.g. Clancy 2001), have also witnessed substantial growth in arrivals and government-directed development projects. Growth in accommodations in the Caribbean has also been significant. Between 1980 and 2000, the number of rooms in those Caribbean island states considered in this book (i.e. excluding Cancun, Cozumel, Belize, Suriname and Guyana) increased from nearly 82,000 to almost 217,000, representing an increase of 165 per cent (CTO 2002).

A somewhat related prediction by Bell focused on what he referred to as a 'shake down' of the Caribbean hotel industry, with five 'basic' segments characterising accommodations in the region: luxury mega-resorts, full-service hotels with recognisable brand-name identification, all-inclusive resorts, exclusive boutique hotels and family inns (Bell 1993: 234). All-inclusive resorts continue to flourish in the region (Issa and Jayawardena 2003). As Coles (Chapter 14) illustrates, many resorts operate within a Fordist mode of production, thus broadly offering what appears on the surface to be benign and homogeneous experiences. In reality, however, multiple sub-products are on offer, but as Coles points out these choices are 'regimented within institutionalised networks of supply and spatial consumption patterns'. Bell also predicted that the number of cruise ships in the Caribbean would double by 2000. Without question, cruise tourism has grown substantially. In 1990 the region hosted 7,750,000 passengers, while in 2000 that figure almost doubled to 14,518,000 (CTO 2002). As Wood (Chapter 9) has outlined, however, issues of control (ownership, operations, management, itineraries) are paramount for cruise tourism in the Caribbean.

A related prediction by Bell involved the pace of development. Bell (1993: 234) envisioned a tourism sector where certain countries with 'carefully integrated, professionally managed, tourism plans, coupled to a realistic regulatory and tax climate' would prosper, while those countries with somewhat 'fragmented' public and private economies and where tourism subsidises other economic sectors would be 'moribund'. The accuracy of this prediction is difficult to ascertain. While tourism is certainly the mainstay of many economies in the region, much of this is the direct result of conscious efforts by local governments. In other words, Bell's prediction almost presupposes that there is almost a conscious, uniform desire to move away from tourism as an economic mainstay in some situations. Of course, this is not always the case. Many island states' efforts at boosting tourism earnings does not necessarily reflect or imply spiralling trends in other economic sectors, but rather demonstrates a directed policy aimed at capturing the benefits of one form of development (see Wilkinson, Chapter 5). The decline of other economic sectors, then, may not have influenced increased attention devoted to tourism development *per se*, but rather function as a consequence of that attention (Pattullo 1996).

The uncertainty over adequate airlift was the focus of Bell's (1993) prediction that a regional airline will have emerged by the year 2000 to service the entire Lesser Antilles. To some extent, this prediction has been realised, but such regional coverage has been manifested as a veritable patchwork of smaller airlines forming both regional and external alliances. In fact, several alliances have recently been formed in order to cut costs and streamline operations throughout the region:

- The CaribSky alliance was formed in early January 2001 by LIAT, WINAIR, Air Caraïbes, Tyden Air and Caraïbes Air Transport. In late 2002, however, WINAIR backed out of the alliance.
- WINAIR and US Airways began code-sharing flights in January 2003, allowing the latter airline to expand its destination coverage to include key destinations served by WINAIR (Anguilla, Saba, St Eustatius, St Kitts, Nevis and St Barthelemy) (M2 Communications 2002).
- Caribbean Star airlines, based out of Antigua, and WINAIR, based out of St Martin, announced an alliance in mid-February 2003 in an attempt to stimulate travel through the region (BBC 2003).
- BWIA and LIAT signed an alliance in January 2002 in an effort to reduce costs and embark upon joint marketing efforts.

- GoCaribbean, a marketing initiative matched with broadening route access from North America, was launched in June 2002 and incorporates Caribbean Star Airlines, Nevis Express, WINAIR and US Airways. The programme was designed to facilitate inclusion of the US Airways' Dividend Miles frequent traveller programme, US Airways Vacations and code sharing to some destinations (US Airways 2002).

Future prospects and management assessments

In the spirit of Bell's assessment of the future of Caribbean tourism, some broad considerations of the future of tourism in the region can be offered. While some focus on individual island states, others more regional in scope can also be considered.

Competition, market share and product development

There is little question that competition from other destinations, regardless of product range, will continue to have a strong impact on tourism in the Caribbean. Strategic marketing by Hawai'i in the west coast regions of the United States, for example, continues unabated. Likewise, an increased presence in the North American tourism market by Mexico could have significant implications for the Caribbean. In the case of Mexico, targeting Europeans could be quite beneficial, but perhaps to the detriment of the Caribbean. Strong intra-regional competition for international visitors should not be discounted. As Tewarie (2002: 442) argues, Cuba is intentionally positioned as a low-price destination in the Caribbean:

> This strategy has contributed to develop Cuba's reputation as a destination appealing to the lower end of the market. As a result, there are those in the region who are worried about Cuba as a competitor. . . . What is clear is that Cuba is embarking on a major thrust in tourism . . . Cuba may now be seeking to build a tourism platform purely on the basis of price, in the hope that it can take a much more strategic approach later. It is instructive, for instance, that on average a tourist spends about $70 in Cuba, among the lowest amounts in the region.

As Haywood and Jayawardena (Chapter 13) point out, the region has recently addressed global competition with its 'Life needs the Caribbean' campaign, consisting of two television spots and magazine ads aimed at residents of the United States earning more than US$70,000 per annum. The emphasis of this campaign is on the region as a whole and not individual island states. The question remains as to the effectiveness of a regional approach to tourism promotion in the Caribbean. While the Caribbean Tourism Organization (CTO) plays a vital role in both the promotion of the Caribbean as a destination and the level of integrated tourism development throughout, criticisms over its direction and mandate have recently been raised. In October 2002, the Minister of Tourism for St Kitts and Nevis argued that the CTO is neglecting small eastern Caribbean states in favour of larger countries such as Cancun, Jamaica and the Dominican Republic (Caribbean Media Corporation 2002). The Minister called for a separate body, focused primarily on smaller eastern Caribbean states, to be established. Whether such a proposal is implemented remains to be seen. In essence, such criticisms reveal an underlying consequence of a geographically fragmented region exhibiting distinct political and policy-framed tourism promotions at local levels. The attention of markets for which worldwide destinations are vying does not come cheap. The CTO must continue to act as the avenue through which promotion and image-management decisions are made. While regionally focused campaigns often cannot adequately address the broad range of products on offer, the reality is that many island states do not have the resources available to mount their own individual campaigns. By extension, local accommodation providers are often faced with substantial construction and operating costs, and thus little capital is left over for marketing (*Salt Lake Tribune* 2000).

As discussed by Momsen in Chapter 16, the problem of shifting markets (where it is even apparent that the European market appears to be one of the fastest growing market segments for the region (CTO 2003)) begs further consideration of the nature of the tourism product itself. For example, several up-market accommodation units have recently been constructed throughout the region, but the question remains as to whether these are designed to supplant the previous up-market visitors originally targeted by all-inclusive resorts. In Nevis, for example, the Villa Paradiso offers a gated community of luxury villas featuring four-bedroom units with marble bathrooms, maid service and 24-hour security. Other 'alternative' products (perhaps more indicative of a maturing destination as opposed to non-mass forms of development) include spas and beauty baths, but these are not being introduced in more 'traditional' tourism destinations such as Antigua or Barbados. Instead, small islands such as Mustique, St Barts (a dependency of Guadeloupe), Anguilla and the British Virgin Islands are attempting to secure access to individuals of high net worth in both North America and Europe. The intent, it seems, is to play above the head of what would normally be classified as the competition. Further, festivals and events, such as the St Lucia Jazz Festival (widely regarded as one of the most worthwhile showcases of Jazz music in the world), will continue to draw tourists to the region. Annual Carnivals are also occasions which draw numerous expatriate visitors to the region.

While the targeted emphasis, in both promotions and development initiatives, on non-mass-tourism products (especially ecotourism, as Weaver discusses in Chapter 10), has been somewhat successful region wide, the real impact of target marketing upon the existing mass tourism infrastructure remains to be seen. If, for example, the Caribbean's image as an idyllic destination for sun, sand and sea continues to erode as a consequence of strident marketing by other 'sun 'n' fun' destinations with strong (and cheap) airlift capacities, the economic sustainability of tourism in the region will be called into question, especially in those particular island states which rely heavily on high volumes of visitors. Some resorts have already been forced to drastically cut prices in an effort to entice travellers in the wake of the terrorist attacks in the United States in 2001. For example, in December 2002 several were slashing prices by as much as 35 per cent in a bid to entice more visitors from the United States (*Los Angeles Times* 2002; see also Lee 2002). Granted, such a move can be attributed to the lack of confidence by United States travellers to travel in an immediate post-11 September environment, it remains to be seen whether the effects of such catastrophic events could be duplicated with mere shifts in consumer preference. Other markets can certainly be targeted. The British Virgin Islands, for example, recently opened a tourist office in London, following a strong redirection of advertising towards Europe. Further, it is quite possible that Cuba will become a strong contender in visitor arrivals from the United States if socialism is replaced. This could cause significant harm to the overall share of United States visitors to other islands in the region, particularly Jamaica and the Dominican Republic.

As the tourism product in the Caribbean is certainly diversified, the degree to which the entire Caribbean can be stereotyped as a 'sun 'n' fun' destination should be seriously questioned. Numerous products abound, including, for example, health-care and spa tourism (Goodrich 1993) and indigenous tourism (Duval 1998; Slinger 2000). In fact, tourists visiting all-inclusive resorts even have the option of short excursions that present historic and natural histories (Coles, Chapter 14). As Weaver (Chapter 10) points out, the development of ecotourism and ACE tourism is often intentional due to the lack of natural resources associated with 3S tourism. In light of the above, it can be concluded that product mix attained at regional level in the Caribbean is reasonably diversified and, to some extent, complementary.

Access and airlift

A further consideration of future prospects, and one which is tangentially related to the issue of new and expanding marketing efforts, centres upon the nature of transport in the Caribbean. As Timothy (Chapter 7) points out, while the region is served by numerous international airlines (e.g. Air Canada, British Airways, Air France, United Airlines, US Airways, American Airlines), the actual transport of passengers largely remains an externality. The recent trouble surrounding United Airlines and Air Canada highlights the fragile nature of routes and passenger mobility from the

United States and Canada. Perhaps even more perilous are the recent financial struggles at American Airlines, largely because the airline controls upwards of 70 per cent of the market to the Caribbean from the United States (Smith 2002). While LIAT and BWIA continue to operate in the region, the extent to which adequate airlift exists seems to be more dependent on both government involvement (financially, logistically or both) and the degree to which operations can be integrated into existing global alliance networks. In a press release from October 2002, the CTO's Aviation Committee suggested that existing carriers need help from the 'marketing machinery' of external airlines. It was also suggested that existing carriers meet regularly and work towards ways in which costs can be reduced. Examples include group purchases of airlines and harmonising airport operations and maintenance.

As a consequence, it is likely that a formal emphasis on curtailing substantial competition from numerous, small airlines and subsequently shaping a regional transport plan of one or two strong RPT (Regular Passenger Transport) providers will transpire within the next five years. The need for regional control over airlift is certainly evident. At a recent conference Bahamian Premier Perry Christie suggested that 'there can be no rejuvenation of Caribbean tourism without the simultaneous reinvention of Caribbean air transportation' (Smith 2002). Some argue against the outright mergers and instead favour more cooperation, although some elements of cooperation have already been established. BWIA and LIAT, for example, have even been code-sharing flights since July of 2002.

Panacea or delusion?

The degree to which small islands will either continue, or come, to rely on tourism is questionable, although there is every reason to believe that tourism will remain as the most important form of economic activity in many of these small countries (Royle 2001). While it is evident that domestic production of specialty foodstuffs for tourists and regional consumption has prospered (Conway, Chapter 11), tourism continues to be seen as the economic saviour for depressed and vulnerable economies in the region. The remaining question, then, is whether the region is overdependent on tourism (Jayawardena and Ramajeesingh 2003), despite some hints at positive linkages being established with other local economic sectors such as agriculture. High degrees of foreign exchange leakage still occur, and resource constraints within smaller islands means significant importation of goods and services to build and serve the very infrastructure that tourists pay to use (Jayawardena and Ramajeesingh 2003).

The question surrounding the suitability of tourism as a form of development in the Caribbean looms large. Throughout the region, social problems, such as squatting, prostitution, illegal vending, harassment and crime could force potential tourists to other destinations (see Duval and Wilkinson, Chapter 4). To some extent, such shifts have already started. Crime in St Croix, for example, seems to have diverted some tourists to other smaller islands, especially Americans (Lee 2002). Local attitudes towards tourism vary. Some residents (and politicians) see tourism as inherently destructive, while others will see opportunity (see Mills, Chapter 15). Community-based involvement in tourism development is one technique that could help destinations manage such issues (Boxill 2000; Brown 2000; see also Milne and Ewing, Chapter 12). Environmental problems will also likely dominate discussion of sustainable approaches to tourism development, primarily in the context of marine impacts as the majority of development occurs in coastal zones (Allen 1992; Baldwin 2000; Burac 1996).

Conclusion

At the 25th annual CTO conference held in Grand Bahama in 2002, the Secretary General of the CTO, Jean Holder, noted that the future direction of tourism in the Caribbean is in the hands of external forces:

> A combination of poor global economic performance and a continuing sense of international insecurity have conspired to create an environment completely inimical to travel and tourism and a high price has been paid in loss of profits, loss of jobs, loss of revenue, loss of service, inter alia, by every business and every sub-sector of the industry.
>
> (*Guyana Chronicle Online* 2002)

Holder also announced that a Strategic Plan developed by the CTO identifies short-term recovery actions from recent slumps in visitor arrivals and the various challenges facing various island states in the region. In particular, aspects of sustainability will be addressed, as will air access, marketing, human resource development and cruise tourism. The CTO's plan also seeks to:

- increase annual tourist expenditure by at least 5 per cent each year over the next ten years;
- increase stayover arrivals by at least 1 per cent above the world average growth;
- increase the conversion of cruise tourists to stayover tourists;
- enhance linkages between tourism and other sectors of the economy; and
- increase the level and range of employment opportunities and tourism education and training.

(*Guyana Chronicle Online* 2002)

As Holder (1996) has previously outlined, the success of tourism in the Caribbean will, in the future, depend on a variety of factors, including product quality (with special emphasis on environmental resources), profitability, effective regional promotion, the provision of competitive air access, the provision of a secure environment in terms of tourist safety and host acceptance, the strengthening of linkages between tourism and other sectors, and combined regional efforts to address competition.

Without question, tourism in the Caribbean will face significant challenges in the future. Poon (2001) suggests that the most pertinent will be the degree of competitiveness that the region is able to exhibit in the face of global competition, sustainability, marketing and promotional activities, air access, the maximising the economic impact of tourist expenditures. Poon (2001) (see also Timothy in Chapter 7) rightfully acknowledges that political and procedural barriers to regional movement will need to be as liberal as possible. Such considerations, however, require a broad framework from which they can be addressed. This is especially important in wider regions in which various micro-destinations are politically and geographically disconnected. As Haywood and Jayawardena (Chapter 13) point out, the management of tourism business in the region must be focused on establishing trust and credibility. Similarly, Jordan's call (Chapter 6) for more transparent and streamlined interrelationships in twin-island states can also be applied to a region-wide context. While many of the challenges raised by Poon (2001) can, to some degree, be controlled, some external forces cannot. As Momsen illustrates in Chapter 16, the recent terrorist attacks in the United States serve as a reminder of the fragility of Caribbean tourism.

In many respects, tourism in the Caribbean joins other economic activities as almost an exercise in peripherality, but it is also important to recognise that the wider peripheral nature of tourism within the region (Weaver 1998) can be applied within some destinations themselves (Weaver 1993). As such, perhaps the debate over whether tourism in the Caribbean will function as either an inevitability or a predetermined, consciously directed economic sector can be ended. The role of tourism is certainly important enough to warrant considerable future attention by local governments (McDavid and Ramajeesingh 2003), but perhaps the new debate that will frame the next ten years is the extent to which the Caribbean will be reliant on tourism for regional, and certainly local, economic well-being (Poon 2001). More importantly, the scope of tourism development, or even the degree to which impacts are managed, is critical in any forecasting of whether or not tourism can continue to fit in with other economic alternatives (e.g. Holder 1980). The question, alluded to by Wilkinson in Chapter 5, is whether island states in the region will opt for direct control over their own tourism sectors or seek to manage external forces where possible. The next ten years of tourism in the region will no doubt be turbulent, and its successes will need to be measured at a variety of levels, including community, state and region.

References

Allen, W.H. (1992) 'Increased dangers to Caribbean marine ecosystems: cruise ship anchors and intensified tourism threaten reefs', *BioScience* 42: 330–5.

Baldwin, J. (2000) 'Tourism development, wetland degradation and beach erosion in Antigua, West Indies', *Tourism Geographies* 2: 193–218.

BBC (2003) 'Caribbean Star and WINAIR announce strategic alliance', BBC Monitoring Americas, 21 February.

Bell, J.H. (1993) 'Caribbean tourism in the year 2000', in D.J. Gayle and J.N. Goodrich (eds) *Tourism Marketing and Management in the Caribbean*, London: Routledge.

Boxill, I. (2000) 'Overcoming social problems in the Jamaican tourism industry', in J. Maerk and I. Boxill (eds) *Tourism in the Caribbean*, San Rafael: Plaza y Valdés.

Brown, K.L.A. (2000) 'Physical and socio-economic impacts of tourist recreational activities in Montego Bay, Ocho Rios and Port Antonio', in J. Maerk and I. Boxill (eds) *Tourism in the Caribbean*, San Rafael: Plaza y Valdés.

Burac, M. (1996) 'Tourism and environment in Guadeloupe and Martinique', in L. Briguglio, R. Butler, D. Harrison and W.L. Filho (eds) *Sustainable Tourism in Islands and Small States: Case Studies*, New York: Pinter.

Caribbean Media Corporation (CMC) (2002) 'Complaint that Caribbean tourism body neglects smaller member states', CMC news agency, 30 October.

Caribbean Tourism Organization (CTO) (2002) *Caribbean Tourism Statistical Report*, 2000–2001 edn, St Michael, Barbados: CTO.

—— (2003) *Caribbean Tourism Statistical Report: 2001–2002 Edition*, St Michael, Barbados: CTO.

Clancy, M. (2001) *Exporting Paradise: Tourism and Development in Mexico*, Amsterdam: Elsevier.

Davies, B. (1996) 'Island states and the problems of mass tourism', in L. Briguglio, B. Archer, J. Jafari and G. Wall (eds) *Sustainable Tourism in Islands and Small States: Issues and Policies*, New York: Pinter.

Dogan, H.Z. (1989) 'Forms of adjustment: sociocultural impacts of tourism', *Annals of Tourism Research* 16: 216–36.

Duval, D.T. (1998) 'Alternative tourism on St Vincent,' *Caribbean Geography* 9: 44–57.

Fotiou, S., Buhalis, D. and Vereczi, G. (2002) 'Sustainable development of ecotourism in small islands developing states (SIDS) and other small islands', *Tourism and Hospitality Research* 4: 79–88.

Goodrich, J. (1993) 'Health-care tourism in the Caribbean', in D.J. Gayle and J.N. Goodrich (eds) *Tourism Marketing and Management in the Caribbean*, London: Routledge.

Guyana Chronicle Online (2002) www.guyanachronicle.com/news.html (accessed 13 November).

Henderson, J.C. (2000) 'Managing tourism in small islands: the case of Pulau Ubin, Singapore', *Journal of Sustainable Tourism* 8: 250–62.

Holder, J. (1980) 'Buying time with tourism in the Caribbean', *International Journal of Tourism Management* 1: 76–83.

—— (1996) 'Maintaining competitiveness in a New World Order: regional solutions to Caribbean tourism sustainability problems', in L.C. Harrison and W. Husbands (eds) *Practicing Responsible Tourism: International Case Studies in Tourism Planning, Policy, and Development*, New York: John Wiley & Sons.

Ioannides, D. and Holcomb, B. (2003) 'Misguided policy initiatives in small-island destinations: why do up-market tourism policies fail?', *Tourism Geographies* 5: 39–48.

Issa, J.J. and Jayawardena, C. (2003) 'The "all-inclusive" concept in the Caribbean', *International Journal of Contemporary Hospitality Management* 15: 167–71.

Jayawardena, C. and Ramajeesingh, D. (2003) 'Performance of tourism analysis: a Caribbean perspective', *International Journal of Contemporary Hospitality Management* 15: 176–9.

Lee, G. (2002) 'Added flights, hotel deals make Caribbean vacations attractive', *The Record*, Northern Jersey Media Group, Inc., 10 November.

Los Angeles Times (2002) 'Prices dropping on Caribbean trips', 8 December.

McDavid, H. and Ramajeesingh, D. (2003) 'The state and tourism: a Caribbean perspective', *International Journal of Contemporary Hospitality Management* 15: 180–3.

Mather, S. and Todd, G. (1993) *Tourism in the Caribbean*, Special Report No. 455, London: The Economist Intelligence Unit.

Mill, R.C. and Morrison, A.M. (1998) *The Tourism System*, 3rd edn, Dubuque: Kendall/Hunt Publishing.

M2 Communications (2002) 'US Airways and Windward Island Airways begin code share relationship', M2 Presswire, 24 December.

Olsen, B. (1997) 'Environmentally sustainable development and tourism: lessons from Negril, Jamaica', *Human Organization* 56(3): 285–93.

Parker, S. (2000) 'Collaboration on tourism policy making: environmental and commercial sustainability on Bonaire, NA', in B. Bramwell and B. Lane (eds) *Tourism Collaboration and Partnerships: Politics, Practice and Sustainability*, Clevedon: Channel View Publications.

Pattullo, P. (1996) *Last Resorts: The Cost of Tourism in the Caribbean*, London: Cassell.

Poon, A. (1988a) 'Innovation and the future of Caribbean tourism', *Tourism Management* 9: 213–20.

—— (1988b), 'Flexible specialization and small size – the case of Caribbean tourism', DRC Discussion Paper No. 57, Science Policy Research Unit, Falmer: University of Sussex.

—— (2001) 'The Caribbean', in A. Lockwood and S. Medlik (eds) *Tourism and Hospitality in the 21st Century*, Oxford: Butterworth–Heinemann.

Royle, S.A. (2001) *A Geography of Islands: Small Island Insularity*, London: Routledge.

Ryan, C. (2001) 'Tourism in the South Pacific – a case of marginalities', *Tourism Recreation Research* 26: 43–9.

Salt Lake Tribune (2000) 'There's trouble in Caribbean paradise', 6 May: B7.

Slinger, V. (2000) 'Ecotourism in the last indigenous Caribbean community', *Annals of Tourism Research* 27(2): 520–3.

Smith, M. (2002) 'Caribbean mulls merging airlines', Associated Press Online, 12 November.

Tewarie, B. (2002) 'The development of a sustainable tourism sector in the Caribbean', in Y. Apostolopoulos and D.J. Gayle (eds) *Island Tourism and Sustainable Development: Caribbean, Pacific and Mediterranean Examples*, Westport: Praeger.

US Airways (2002) 'US Airways launches "GoCaribbean" network with three regional carriers'. Online: <http://www.usairways.com/about/press/nw_02_0603.htm>.

Weaver, D. (1993) 'Model of urban tourism for small Caribbean islands', *Geographical Review* 83: 134–40.

—— (1998) 'Peripheries of the periphery: tourism in Tobago and Barbuda', *Annals of Tourism Research* 25: 292–313.

Widfeldt, A. (1996) 'Alternative development strategies and tourism in Caribbean microstates', in L. Briguglio, R. Butler, D. Harrison and W.L. Filho (eds) *Sustainable Tourism in Islands and Small States: Case Studies*, New York: Pinter.

Wilkinson, P.F. (1987) 'Tourism in small island nations: a fragile dependence', *Leisure Studies* 6: 127–46.

Wilson, D. (1996) 'Glimpses of Caribbean tourism and the question of sustainability in Barbados and St Lucia', in L. Briguglio, R. Butler, D. Harrison and W.L. Filho (eds) *Sustainable Tourism in Islands and Small States: Case Studies*, New York: Pinter.

Index